PALGRAVE STUDIES IN THE HISTORY OF
SCIENCE AND TECHNOLOGY

Series Editors

James Rodger Fleming (Colby College) and Roger D. Launius (National Air and Space Museum)

This series presents original, high-quality, and accessible works at the cutting edge of scholarship within the history of science and technology. Books in the series aim to disseminate new knowledge and new perspectives about the history of science and technology, enhance and extend education, foster public understanding, and enrich cultural life. Collectively, these books will break down conventional lines of demarcation by incorporating historical perspectives into issues of current and ongoing concern, offering international and global perspectives on a variety of issues, and bridging the gap between historians and practicing scientists. In this way they advance scholarly conversation within and across traditional disciplines but also to help define new areas of intellectual endeavor.

Published by Palgrave Macmillan:

Continental Defense in the Eisenhower Era: Nuclear Antiaircraft Arms and the Cold War
By Christopher J. Bright

Confronting the Climate: British Airs and the Making of Environmental Medicine
By Vladimir Jankovic'

Globalizing Polar Science: Reconsidering the International Polar and Geophysical Years
Edited by Roger D. Launius, James Rodger Fleming, and David H. DeVorkin

Eugenics and the Nature-Nurture Debate in the Twentieth Century
By Aaron Gillette

John F. Kennedy and the Race to the Moon
By John M. Logsdon

A Vision of Modern Science: John Tyndall and the Role of the Scientist in Victorian Culture
By Ursula DeYoung

Searching for Sasquatch: Crackpots, Eggheads, and Cryptozoology
By Brian Regal

Inventing the American Astronaut
By Matthew H. Hersch

The Nuclear Age in Popular Media: A Transnational History
Edited by Dick van Lente

Exploring the Solar System: The History and Science of Planetary Exploration
Edited by Roger D. Launius

The Sociable Sciences: Darwin and His Contemporaries in Chile
By Patience A. Schell

The Surveillance Imperative

Geosciences during the
Cold War and Beyond

Edited by
Simone Turchetti and Peder Roberts

First published in 2014 by
PALGRAVE MACMILLAN®
in the United States—a division of St. Martin's Press LLC,
175 Fifth Avenue, New York, NY 10010.

Where this book is distributed in the UK, Europe and the rest of the world,
this is by Palgrave Macmillan, a division of Macmillan Publishers Limited,
registered in England, company number 785998, of Houndmills,
Basingstoke, Hampshire RG21 6XS.

Palgrave Macmillan is the global academic imprint of the above companies
and has companies and representatives throughout the world.

Palgrave® and Macmillan® are registered trademarks in the United States,
the United Kingdom, Europe and other countries.

ISBN: 978–1–137–43872–0

Library of Congress Cataloging-in-Publication Data is available from the
Library of Congress.

A catalogue record of the book is available from the British Library.

Design by Newgen Knowledge Works (P) Ltd., Chennai, India.

First edition: September 2014

10 9 8 7 6 5 4 3 2 1

Contents

Section III   Seeing the Sea—From Above and Below

Section IV   Surveillance Technologies

Section V   From Surveillance to Environmental Monitoring

Illustrations

Figures

Table

Acknowledgments

Research for this edited collection was made possible by the project "The Earth Under Surveillance" funded by the European Research Council (ERC) with the grant n. 241009 (http://teus.unistra.fr/). The project's principal investigator, Simone Turchetti, and his collaborators (Matthew Adamson, Soraya Boudia, Lino Camprubí, Roberto Cantoni, Sebastian Grevsmühl, Néstor Herran, Peder Roberts, Sam Robinson, and Leucha Veneer) would like to thank the ERC for the support received in the last five years. The plans to complete an edited volume were finalized during the workshop "Cold War/Blue Planet," organized in Manchester, on June 27–29, 2012 (http://www.chstm.manchester.ac.uk/newsandevents/conferences/coldwarblueplanet/). On that occasion not only some of the original research material for the collection was presented for the first time, but other scholars who took part in the meeting (Roger Launius and Robert Poole) decided to contribute to the completion of this volume. We would like to thank them, and all the other participants in the workshop, for contributing to our research efforts. We would also like to thank the anonymous reviewer who provided precious advice on how to improve the chapters in this collection and the staff at Palgrave Macmillan for helping them throughout the publication process.

In addition, Sebastian Grevsmühl would like to thank his colleagues at the Centre Koyré, and Simone Turchetti and Peder Roberts for comments and questions. He also acknowledges the CNRS for support in completing parts of his research. Soraya Boudia would like to express special thanks to Peder Roberts and Simone Turchetti for their contribution to her article, both in substance and form.

Néstor Herran acknowledges the support of the 7th Framework Program of the European Commission (PIEF-GA-2009-235012). He also thanks Peder Roberts and Simone Turchetti for comments on the paper; Soraya Boudia for suggestions and help with the UNSCEAR archival materials; and the members of the "Eurorisk" research group (Sandrine Revet especially) for valuable remarks in the early stage of the manuscript.

Robert Poole would like to thank Simone Turchetti and the organizers of the "Cold War, Blue Planet" conference for helping to bring his paper to life and keep it in order, and to pay tribute to the long-standing academic hospitality of the Centre for the History of Science, Technology and Medicine at the University of Manchester (CHSTM), particularly the history of the physical sciences reading group.

Peder Roberts acknowledges that research for his chapter was made possible by a travel grant from the American Institute of Physics. He also wishes to thank Simone Turchetti, Julia Lajus, and above all Wilford F. Weeks for their comments on the manuscript, Janet Martin-Nielsen for requesting and obtaining materials from Libraries and Archives Canada, and Sam Robinson for gathering materials from the National Archives of the United Kingdom.

Sam Robinson would like to thank personnel at the National Oceanographic Centre of Southampton and at the National Archives of the United Kingdom at Kew for their assistance and Lino Camprubí for helping to retrieve Spanish archival documentation.

Simone Turchetti acknowledges the CHSTM and the Max Planck Institute for the History of Science (MPIWG) for support in completing the edited volume. He would also like to thank all his collaborators in the TEUS project for their hard work, Peter Marshall, Alan Douglas, and David Davies for exchanges of views on forensic seismology, Silvia Filosa and Graziano Ferrari for access to archival papers of the Italian Institute of Geophysics, and Valentina della Gala for precious assistance in retrieving Italian periodicals on atomic energy.

Contributors

Matthew Adamson, Director of Academic and Student Affairs at the Budapest campus of McDaniel College, has researched the history of nuclear technology and institution-building in France since 1945. He holds a Ph.D. in history and philosophy of science from the University of Indiana. As part of the TEUS project, he has examined uranium prospecting and its relationship to geopolitics and the growth of the geophysical sciences in the Cold War era.

Soraya Boudia is professor of sociology at the University of Paris-Est. She previously held the positions of director of the Curie Museum in Paris and professor of history of science at the University of Strasbourg. Her main area of research is the history of radioactivity and its applications. She currently studies the global expertise and regulation of nuclear risks and environmental hazards. She is also preparing a book on the history of risk and risk society.

Lino Camprubí, obtained his Ph.D. in history from the University of California, Los Angeles, in 2011. His dissertation explored the links between science, technology, and the transformation of the Spanish landscape in the first years of Franco's dictatorship. He joined the TEUS project to expand his dissertation's perspective onto a transnational narrative, aiming to place geophysical research in the broad Cold War context. He will be a research scholar at the Max Planck Institute for the History of Science from September 2014.

Roberto Cantoni, who recently received a Ph.D. award at the Centre for the History of Science, Technology and Medicine (University of Manchester) is currently researching the history of geophysical technologies for oil prospecting, on the history of oil diplomacies and on security issues related thereto. He holds Masters degrees from the Université Denis Diderot (Paris VII) and the Scuola Internazionale Superiore di Studi Avanzati di Trieste (SISSA). He collaborates with the webzine *OggiScienza* edited by the SISSA and has worked as CERN Bulletin Editor at the CERN (Geneva).

Sebastian Vincent Grevsmühl is a postdoctoral research fellow at the Pierre et Marie Curie University (UPMC) of Paris and works in the TEUS project researching the history of satellites for geophysical exploration. He holds a Ph.D. from the Ecole des Hautes Etudes en Sciences Sociales (EHESS/Centre Alexandre Koyré).

Néstor Herran is associate professor at the UPMC, Paris, France. He obtained his Ph.D. from the University Autònoma of Barcelona. His main area of interest is the history of physical sciences in the twentieth century, with particular interest on the history of radioactivity and nuclear technology. His current research deals with the history of environmental monitoring of radioactivity.

Roger D. Launius is currently associate director of Collection and Curatorial Affairs at the Smithsonian Institution's National Air and Space Museum in Washington, DC. Between 1990 and 2002 he served as chief historian of the National Aeronautics and Space Administration. A graduate of Graceland College in Lamoni, Iowa, he received his Ph.D. from Louisiana State University, Baton Rouge, in 1982 and worked as a civilian historian with the United States Air Force until 1990.

Robert Poole is currently Guild Research Fellow at the University of Central Lancashire, an associate of the Centre for the History of Science, Technology and Medicine at the University of Manchester, Poole was formerly a reader in history at the University of Cumbria and Leverhulme senior research fellow at the University of Manchester and an associate member of 'The Future in the Stars' research programme, Friedrich-Meinecke Institut, Freie Universität, Berlin. He is the author of *Earthrise: How Man First Saw the Earth* (Yale, 2008). He is currently working on the history of the scientific belief in extra-terrestrial intelligence and the film *2001: A Space Odyssey*.

Peder Roberts is a researcher at the Division of History of Science, Technology, and Environment at KTH Royal Institute of Technology, Stockholm, Sweden. He holds a BA (Hons) and MA from the University of New South Wales, and Ph.D. in history from Stanford University. He has previously worked at the University of Strasbourg as part of the TEUS project. His current research interests include the history and geopolitics of science during the Cold War, particularly involving the oceans, along with the past and future of the polar regions.

Sam Robinson, CHSTM, University of Manchester, Manchester, UK, is currently working on completing a Ph.D. on the history of military oceanography in Western Europe during the Cold War, focused on the growth, development, and ultimate collapse of the committees that came to define the relationship between oceanographers and their military patrons in Britain from 1955 to 1975. He completed his MA at the University of Aberdeen.

Simone Turchetti is lecturer at the Centre for the History of Science, Technology and Medicine at the University of Manchester and the principal investigator of the five-year project "The Earth Under Surveillance" funded by the European Research Council. He is interested in the history of twentieth century science and technology with special emphasis on the geophysical sciences. He has previously worked as researcher and teaching fellow at the Universities of Bristol and Leeds covering a variety of issues that relate to Cold War science from atomic espionage to Antarctic research.

Leucha Veneer took her Ph.D., which focused on the applications of geology in early nineteenth-century Britain, at the University of Leeds. Leucha has worked in the context of the TEUS project on the geophysics of oil in the Cold War, following earlier research on the relationship between geology and mining and the history of the earth sciences more generally.

Introduction

Knowing the Enemy, Knowing the Earth

Simone Turchetti and Peder Roberts

Surveillance is a subject on many lips. Thanks to Edward Snowden's revelations, commentators around the world have questioned if anything remains undetected by the surveillance networks set up by the world's most powerful nations. Documentation leaked by the former Central Intelligence Agency (CIA) contractor has revealed electronic ears and eyes spreading across the planet, enabling the rapid transfer of massive amounts of data to an army of intelligence operators, aided by some of the fastest computing machines on earth and their capacious hard drives. While emblematic examples such as German chancellor Angela Merkel's tapped Nokia handset evoke the gadget-oriented espionage of an early 007 movie, the sheer scale and sweep of the operations have caused the greatest concern for most members of the public. Not only has it become apparent how much private information transferred through mobile phones, e-mails, Web portals, and social networking websites can be tapped into by security agencies, but we now also know that intelligence operators do not always discriminate between enemies and allies in tapping operations—something that has come to light in the most embarrassing circumstances for the Obama administration.[1]

While the Snowden case has thus put in plain sight the truly global reach of surveillance operations and networks, the historical provenance of this powerful system of global monitoring continues to be virtually unknown. Watching over enemies (political and otherwise) has been an essential feature in the exercise of power since time immemorial, and knowledge of the earth and its resources has long been useful to statecraft: consider the strategic value possessed by the Map Room of Britain's Royal Geographical Society as late as World War II. But the transformation that took place during the Cold War involved putting the *entire earth* under surveillance, altering the scope, the nature, and above all the extent of scientific interrogation of the planet and its environs.

Both superpowers, especially the US administration, conceived the capacity to monitor the earth within a framework of control through strategic influence, without the need for explicit sovereignty over colonial spaces. This led to the establishment of infrastructures that routed signals from overseas outposts to central homeland units devoted to their analysis. Human communications were—and still are—a miniscule part in this traffic, which includes data from the oceans, the surface and interior of the earth, and the sky (and more recently celestial space) above it. So from the beginning of the Cold War onward, gathering new information on enemies or potential enemies has been intimately linked to gathering information about the earth.

Despite a recent flourish of studies on the relationship between states and the scientific research they patronized during the Cold War, especially in the United States,[2] surveillance is too often regarded as a discrete activity linked to concrete state aims rather than a more general imperative to understand and control both the earth and its inhabitants. Surveillance networks owe their existence, or at least their sophistication and extent, to the dramatic expansion of funding to the geosciences after 1945. Their contribution was decisive not only in making it possible to analyze the activities of potential enemies through traces upon the earth's environment, but also to understand that environment as an end in itself. In this light we might fruitfully think of environmental surveillance as a means to detect signals, packets of data that could be unpacked to reveal intelligence with value in multiple contexts.

The intimate connection between science and surveillance was neatly captured through *Sputnik*, the first artificial satellite to enter the earth's orbit and a powerful symbol of the central role of science and technology in Cold War strategies.[3] Artificial satellites had long been discussed as part of the International Geophysical Year (IGY, 1957–58), an event that simultaneously demonstrated the power of the geosciences to understand the whole earth (and its environs) and showcased competition as well as cooperation between the superpowers.[4] *Sputnik* not only provided a platform for observing the earth, but also created a new category of objects that themselves required surveillance as potential threats, in terms of both data collection and military strikes. This in turn stimulated the field of upper atmospheric research and the development of tracking technologies, in addition to sparking significant political debate and strategic deliberation.

Yet the *Sputnik* launch was only one aspect of a pervasive concern for understanding the earth, its ocean, and its atmosphere within the context of state security. How could the Cold War West establish an effective detection system for enemy missiles, having already invested massively in early warning systems for conventional aircraft? Could satellites detect sensitive military systems on the earth's surface and even in its oceans? Could the extent of sea ice be reliably forecast in order to supply Arctic bases? How could foreign nuclear tests be reliably located and identified? As this volume demonstrates, addressing these questions led chiefs of government and their scientific advisers to envisage modern forms of global surveillance and helped to establish the geosciences in Cold War strategic planning. Knowledge about the circulation of jet streams and ocean currents assisted in the improvement of antiaircraft defense and antisubmarine warfare measures. A major injection of funds into the study of earthquakes was premised explicitly upon the need to monitor underground nuclear tests.

Studies of the atmosphere, the oceans, and the inner earth thus coupled the desire of scientists to acquire new knowledge of the earth's features with the need to better know the enemy. As this knowledge had the potential to transform more traditional methods of surveillance, detection, and intelligence-gathering, the enormous influx of state funding for the geosciences during the 1950s and 1960s helped researchers to accumulate vast data sets and derive important new insights that furthered research agendas within specific disciplines, while also providing benefits—either directly or indirectly—for states. The chapters in this volume examine the rise of the geosciences during the second half of the twentieth century through the lens of this "surveillance imperative." Using surveillance as the central analytic concept and shifting the focus beyond American borders, the set of chapters that follow explain how a constellation of disciplines, namely the geosciences, benefitted from this search for novel means to monitor the enemy instigated by the confrontation between superpowers. Disciplines that eventually became imbued with "green" values—especially through environmental monitoring—flourished within a geopolitical context in which watching over enemy states and alliances was at least as important as the "assault on the unknown."[5]

The Surveillance Imperative

Surveillance has long been an important concept in historical and sociological analyses of science, technology, and society, from the philosopher and historian Michel Foucault's early study of penitentiaries to the historian and activist Mike Davis's more recent portrayal of CCTV-controlled Los Angeles.[6] The field of surveillance studies now has many of the trappings of disciplinary success, including university centers and departments, an international research network, and a burgeoning literature. The great majority of these works are concerned with the relationship between individuals and the states, armed forces, and corporations that desire to know, predict, and, perhaps most worryingly, control their actions.[7]

Yet in the earth and environmental sciences, surveillance typically does not connote the same sense of malevolent intent. A quick scan (we nearly wrote a survey) of recent literature reveals reference to the surveillance of coral reef fauna, the monitoring of marine conditions to assist navigation and quickly detect pollution, and the reconnaissance of territories potentially infested by disease-vector mosquitoes in the context of biomonitoring operations.[8] Observing a person, a citizen, or a politician carries a set of legal and moral concerns that do not exist for an iceberg or for the composition of the earth's mantle, despite the fact that each can produce information with relevance to statecraft. The uneasy relationship between technologies designed to ascertain facts about individuals and organizations of governance seems to fade when the targets of surveillance are objects rather than subjects, phenomena to be ordered through science rather than citizens within a polity. This distinction hinges upon the separation of the natural and the human, a distinction grounded in the possession of political agency, but which implicitly supposes that surveillance of objects is unproblematic because the consequences of that action are limited to the target. The interdependence of the human and the natural, and between the observer and the observed, is a reminder that putting a thing rather than a person under surveillance does not render the action politically neutral.[9]

The bifurcation between surveillance as the stereotypical Orwellian challenge to free society and as a set of seemingly innocent scientific practices that have to do with the gathering of environmental knowledge draws a moral distinction that obscures common origins. Since the Cold War, intelligence ambitions have been embedded in novel methods of scientific inquiry targeting the earth and its features rather than using human agents alone. The coupling of intelligence and scientific goals has also created the opportunity to pioneer new forms of environmental monitoring, as some of the chapters in this volume show.

Surveillance also remains an underutilized concept in analyses of the Cold War from both international relations and history of science perspectives. In 2001, international relations scholar Robert Jervis correctly identified the "security dilemma" as a key tenet of Cold War geopolitics, suggesting that "as each state seeks to be able to protect itself, it is likely to gain the ability to menace others."[10] Jervis has understood the ways in which security challenges were met mainly in terms of the expansion of military and nuclear capability rather than through the growth of information-gathering structures, but these too exemplify his point: competition for supremacy spread far beyond the confines of missile silos and armaments depots. Cold War policymakers and science planners devoted enormous resources to developing early warning and monitoring systems, and to developing the careers of scientists within disciplines from seismology[11] to physical oceanography[12]—many of whom relished the leverage they obtained over state patrons through the perceived strategic relevance of their own disciplines and the specter of other states leaping ahead within them due to greater resources.

The IGY was the preeminent example of a wider phenomenon. Surveillance of the planet through the lens of the geosciences involved prospecting foreign territories to determine the availability of strategic natural resources; reconnaissance overflights to chart military facilities and the geomorphology of potential combat sites; surveys and satellite programs to gather atmospheric, terrestrial, and oceanographic data; seismic observatories and atmospheric monitoring stations to detect nuclear tests; deep-ocean studies to facilitate submarine detection; and much more. Investment in such projects increased the amount (and diversity) of new knowledge about the earth as a whole, establishing an infrastructure for ongoing research and on occasion pushing rivals to counter with their own initiatives.

Several authors have related espionage and reconnaissance to the development of the Cold War earth sciences from different scholarly perspectives and these perspectives certainly have informed studies in the history of contemporary science and technology.[13] David van Keuren was correct to draw attention to the relationship between "science in black" (the world of classified knowledge production) and its open cousin, "science in white," citing the dual value of the abortive 600-foot diameter "big dish" in West Virginia for both radioastronomy and intelligence-gathering.[14] John Krige has highlighted the importance of international scientific meetings during the 1950s to assessing the state of science behind the Iron Curtain, not least in the field of atomic energy.[15] Our claim in this volume is that while specific incidents such as these cast valuable spotlights, the coexistence of scientific and intelligence ambitions should be regarded as ubiquitous rather than episodic. State support for the earth sciences recognized the value of the earth itself to Cold War strategy—that the quest to obtain information for state advantage involved interrogating the planet in addition to spying on those who inhabited its surface.

As the earth was placed under surveillance through the gaze of the geosciences, state strategy provided both context and motivation.[16] Competition between states, as much as collaboration between them, provided the fuel for research. Alan Needell has shown that during the 1950s, the American physicist and science administrator Lloyd Berkner vigorously advocated for improved US surveillance of its military rivals while acting as a key organizer of international geophysical research.[17] Berkner insisted that his country respond to the Soviet nuclear threat through aggressive stances based on intelligence gathering, monitoring, and reconnaissance, materialized for instance in new radar-based interception systems such as the Distant Early Warning (DEW) Line (see Figure 6.2, p. 130).[18] Such concerns also shaped Berkner's vision for the IGY: international scientific endeavors of such a magnitude offered a wealth of new data on foreign environments and scientific activities.

The chapters in this volume demonstrate that the growth of the geosciences in Western Europe, in addition to North America, was to a considerable extent driven by a security dilemma in ways beyond those described by Jervis. State demands for increased vigilance were manifested through a range of research programs from covert national missions to open international collaborations. A new perspective may thus be gained upon the history of the earth and environmental sciences during the Cold War—and beyond—by using surveillance as a conceptual linchpin, and by foregrounding the international and transnational dimensions of the geosciences during these years.

Beyond the National

The term "transnational" emphasizes flows across borders, connecting simultaneous, often coproduced developments in different national contexts.[19] Natural resource extraction frequently involved state or multinational actors operating in territory far from corporate headquarters, feeding markets around the world. Events such as the IGY were international in the sense that individuals and groups acted on behalf of state sponsors within the overall frame of a larger collaboration, but also transnational in the sense that such events helped furthering research in certain areas of the earth, notably Antarctica, as targets for investigation uniting different national groups, regardless of sovereignty claims. Secrecy nevertheless occupied a central role in practices across the spectrum of the geosciences. As Michael A. Dennis has argued, secrecy shapes research environments across academic and industrial domains, structuring the process as well as the dissemination.[20] A transnational perspective captures the cross-border nature of the surveillance imperative and its role in sparking activity in different states: developments within one national context were often directly related to developments in another, for reasons of political as much as intellectual rivalry.[21] While the superpower face-off was the most prominent example, we emphasize that such rivalries involving European states—and their former imperial territories—could also be powerful drivers for the Cold War geosciences. Despite this fact, existing work on North American–Western European scientific relations during this period has focused only briefly on the geosciences, and far less has been written on Western Europe and its crumbling empires than about the superpowers.[22]

A transnational approach also enables us to revisit the role of science as an instrument of political power through the tail end of the long history of European imperialism, bringing the years of decolonization into the same historiographic conversation as the heyday of science as a handmaiden of empire.[23] The function of science as an instrument of European imperial authority, like the structures of early twentieth-century international science with their focus on national delegations, reflected a "realist" view that recognized the nation and its interests as the fundamental frame for political action.[24] After 1945 it quickly became clear that the pre-1939 status quo, with its inscription of European nationality upon global political geography, no longer corresponded to the emerging superpower-dominated world order. US policymakers, diplomats, and scientists alike regarded geophysical knowledge as a powerful tool to gain knowledge of foreign environments in order to facilitate their control, without necessarily involving territorial annexation.

The growth of the geosciences and the pursuit of global surveillance not only overlapped, but were also mutually reinforcing. The establishment of scientific outposts in far-flung lands, the collection of data from satellites, and the intensification of research on the high seas led to the collection of geophysical data on a truly global basis. In some cases, such as the monitoring of foreign nuclear tests, establishing friendly relations with foreign governments made it possible to covertly foster monitoring projects. Of course, the covert ambitions of earth data collection programs were not distinctive of the US intelligence community alone: most notably, Soviet and British intelligence had similar ambitions. Yet no other state could afford to promote these programs to the same extent. By contrast, despite the continued assertion by many European politicians (especially Charles de Gaulle)[25] that the nation-state remained the natural unit of political authority, the years after 1945 saw a decline of formal European imperialism. Intra–Western European integration through political-economic organizations such as the European Coal and Steel Community and Euratom was also manifested through scientific bodies such as CERN (the European Organization for Nuclear Research).[26] Successive US administrations encouraged these integration projects and sought to align them with the United States's own national interests, as integration in broader alliances strengthened its role as the Machiavellian Prince of the Cold War world.[27] John Krige has famously argued that during the Cold War, the funding of new research programs in Europe helped American patrons to forge cultural synergies across the Atlantic and spread American values, thus setting the conditions for alliances that embedded political and military goals within a common cultural and economic stratum.[28]

The goal of fostering European integration did not preclude the United States from acquiring classified data from these allies with direct relevance to military or economic goals, either in the 1950s and 1960s or indeed the present—as the Snowden revelations demonstrate. Interactions with European partners were often informed by such knowledge. We learn in this volume that a number of undercover US agents were dispatched to European territories to monitor deposits of strategically important natural resources such as oil and uranium, and to gain information on the intentions of corporations and governments. These activities blurred the distinction between scientific experts, diplomats, and intelligence agents as their roles became contiguous and, at times, overlapping. Such data helped the US government evaluate requests for assistance or collaboration from European partners in fields where the United States often competed with European states for access to resources and profits. US science administrators quarantined more sensitive scientific data (for instance, on nuclear weapons and Soviet nuclear tests), making them accessible only to a few allies. Data from the geosciences thus functioned as valuable commodities in terms of building American political and economic strength, and in turn validated the importance of the geosciences more generally.

It is worth dwelling further on the dual role of data from the geosciences as sources of privileged information and avenues for collaboration across national borders. Monolithic interpretations focused on superpower decision making miss the often-messy relations within (and occasionally across) Cold War geopolitical alliances.[29] As recent revelations of American spying upon the leaders of friendly states reminds us, surveillance of the political world is almost as pervasive as surveillance of the natural world. New security challenges could be catered for and collaborative deals offered that might provide mutual benefit—though the stronger party inevitably set the terms for exchange.[30] Secretive transfers of environmental knowledge stirred tensions in Europe, as described by Roberto Cantoni and Leucha Veneerin in this volume (Chapter 2), exacerbated by the ongoing process of decolonization and its attendant challenges to established political and economic systems. While bonds with research communities in the United States enabled European scientists to access additional support (intellectual and material), erstwhile imperial powers maintained or even sought to expand their scientific presence in many former colonies.

As newly decolonized states joined bodies such as the United Nations Educational, Cultural and Scientific Organization (UNESCO) and the International Council of Scientific Unions (ICSU), the quest to know and control regions from the poles to Africa and Asia continued apace.[31] The bonds of empire continued to be relevant, as in the case of geological prospecting in Africa, but claims to national authority were complicated both by the global ambitions of the superpowers and rivalries between European states. Italy undermined France's control in North Africa by letting Italian firms offer scientific information on oil deposits to Algerian rebels, for instance, while cooperative uranium prospecting in Europe and Africa alternated between uniting and dividing the atomic research organizations of France, Italy, Spain, and the United States.[32]

Equally, the existence of internationally structured scientific events was often (perhaps invariably) consistent with the military-strategic goals of states in addition to the research agendas of scientists. Propagandized as an enlightened event that bucked the confrontational atmosphere of the early Cold War, the IGY nonetheless straddled the military/civilian domains by instigating studies coupling science with intelligence work.[33] When Soviet research groups began transmitting reports to international IGY organizations from 1955, the United States IGY Committee promptly forwarded this information to State Department and the CIA's Office of Scientific Intelligence. Intelligence reports on Soviet advances in oceanography, meteorology, and rocketry were subsequently shared with other states, enhancing policymaking but also helping scientists in the Cold War West to demand increased expenditures premised on the need to compete.[34] Nor did the accumulation of vast, openly accessible data sets at the official World Data Centers mean the data within held equal value to all states. As Jacob

Hamblin has pointed out in the case of oceanography, the United States recognized that because global data sets were so much more valuable than the partial sets it could acquire through its own resources, international data sharing was to its military advantage—even if this meant providing data to rivals.[35] Global data held greatest value to states with ambitions to global power.

The most notable event of the IGY—the *Sputnik* launch—renewed fears among scientists and politicians alike in the Western bloc that the Soviet Union had achieved supremacy in key fields of science and technology. The crisis that *Sputnik* precipitated focused American minds on the importance of science and ensured new funding opportunities for the geosciences—not least through the North Atlantic Treaty Organization (NATO), which could mobilize European scientific cooperation in addition to boosting capacity in strategically useful fields. A good example was, again, oceanography. NATO entrusted European experts with conducting surveys in the Atlantic and the Mediterranean, with a particular focus on areas with direct relevance to antisubmarine warfare strategies, such as the Straits of Gibraltar and the Faroe-Shetland Channel.[36] Cooperation on the acquisition of strategically relevant scientific data took place despite continued disputes between American and British naval officials about how those spaces should be patrolled.

By considering surveillance in both scientific competition and collaboration, and examining in greater detail the flow of scientific knowledge across and beyond national borders, the chapters in this collection thus go beyond the narrative of escalation defined by nuclear deterrence or the "two scorpions in a bottle" scenario.[37] While few could deny that the Cold War was a conflict between two distinct blocs, the historical examples discussed in this volume complicate the picture in interesting ways, providing new perspectives on the strategic value of the earth sciences within the ever-changing historical landscape of the Cold War conflict—and into the present.

From Science in Khaki to Science in Green?

The surveillance imperative contributed to a new image of the earth as a series of systems (and even, some would argue, as a single system). As Robert Poole shows in this volume (chapter 10), by the 1970s space missions had returned a wealth of data, including photographic images that revolutionized our previous understanding of the earth and resonated with an emerging environmental consciousness.[38] Along with important new research in fields such as atmospheric chemistry (such as the Keeling curve), this consciousness contributed to a reassessment of research priorities in the earth sciences, which increasingly came to be associated with the green of modern environmentalism rather than the khaki of military science. Ronald E. Doel has demonstrated that the growth of the geosciences in the United States after 1945 environmental sciences was strongly linked to military strategy, knowledge of the earth's surface an essential prerequisite for controlling it.[39] Manifested also far beyond the borders of the United States, this trend emphasized the power of a global scientific vision.

Paradoxically, the surveillance imperative that thrived in the context of superpower competition helped create an image of the earth as a fragile, complex entity and to highlight the power of human agency to harm the planet. The goals of earth scientists were aligned with those of states during the Cold War—especially in its first two decades—while also providing a conceptual thread to the present, where knowledge of the earth and its systems has become central to debates about climate change. As Naomi Oreskes has put it, military funding contributed to "a period of unprecedented scientific productivity" in the earth sciences that must be located within the context of the times: "military concerns were naturalized, and the extrinsically motivated became the intrinsically interesting."[40] The word "state" might be used with equal effect instead of the word "military." The wide range of relationships that contributed to that burst of productivity in some cases persisted through the later part of the century. In many cases, this took the form of continued active support, but in others new research agendas could be developed upon infrastructure made possible by lavish Cold War funding. Notably, Paul Edwards's comprehensive study of the history of climate science reveals the close relationship between technologies of surveillance and the theories that allowed scientists to make use of them for constructing climate models throughout the twentieth century.[41]

This question of infrastructure invites a reexamination of a hoary question: whether military funding, particularly in the post-1945 United States, distorted science from its "true" trajectory or "generously supplemented pre-existing trajectories."[42] The former position, advocated most notably by Paul Forman and Stuart Leslie, requires proof of deviation from a "natural" research path.[43] The case is at best difficult to prove (though Forman marshaled compelling evidence in the context of quantum electronics research) and at worst nearly impossible, given the reliance upon proving divergence from an inherently hypothetical path. Like Kai-Henrik Barth, Ronald E. Doel, Naomi Oreskes, and (we strongly suspect) the majority of scholars working today, we lean toward a more nuanced position that preserves agency for scientists while emphasizing the importance of patronage in shaping the environments within which research questions are chosen and investigation conducted.[44]

As the term "distortion" implies, arguments about the extent of the military's role in shaping the research it sponsors are inevitably also loaded with claims to moral and intellectual superiority, a question of "who was using who?" accompanied by a whiff of skepticism about how much the science was thus by definition compromised. Links between basic research and specific applications are notoriously difficult to predict, and the post-1945 earth sciences offer particularly strong examples of military funding being used on research that produced immense advances in fundamental scientific understanding, most notably the theory of plate tectonics.[45] Military funding was a topic of contention among scientists from the outset of the Cold War,[46] but the difficulty of determining how research deviated from a hypothetical "natural" trajectory leads us to prefer questions about the trajectory that we *do* know of—the geophysical sciences becoming associated with a form of environmental surveillance that today is widely considered as a force for good.

Whether or not the military distorted the earth sciences, some of its key players sought to harness the tools and training they gained thanks to generous postwar funding to explicitly environmental problems. Partly this was serendipity; as Sebastian Grevsmühl (chapter 8) demonstrates in this volume, satellites designed to address defense research problems could aid in the assessment of meteorological and ice conditions, providing valuable data on warming in the polar regions. The deliberate effort of Cold War planners to train earth scientists produced new generations of scholars who directed the surveillance imperative toward environmental monitoring, most notably of climate change. An independent British scientist, James Lovelock, exchanged views with CIA and MI6 officials on how to find people by covertly labeling them with chemical compounds and then using a device to detect its presence. On his way to the United States, where he was to report on his surveillance gizmo, Lovelock met with NASA's Dian Hitchcock and their collaboration would break new ground in the understanding of the earth as a system also chiming with the environmental discourse (as Poole shows in chapter 10).[47] Individual careers in Europe as well as the United States can reveal such transitions clearly. To take but one example, the Norwegian physical oceanographer Ola M. Johannessen began his studies under Håkon Mosby—a key figure in NATO's oceanographic community—and spent time at the NATO Supreme Allied Command Atlantic Anti-Submarine Warfare Centre at La Spezia in Italy, before leading a number of large-scale environmental monitoring projects and founding the Nansen Environmental and Remote Sensing Center in Bergen, Norway. Johannessen's career path is not particularly unique and indeed mirrors the institutional milieu in Bergen, where NATO money helped reinvigorate a world-renowned hub for geophysical research that suffered with Norway's relative poverty after 1945.[48]

These transitions from "khaki" to "green" do, however, present problematic issues. Naomi Oreskes and Erik Conway have recently documented how a small group of "Cold Warriors" hindered the acceptance of climate change research that they considered politically problematic.[49] Their work is in some ways a rejoinder to accusations by climate change deniers that research into global warming is an attempt to extract money from states under false pretenses—a position taken seriously by a disturbing number of political figures.[50] Yet so much of the infrastructure (material and intellectual) that underpins modern climate research grew out of the Cold War and the strategic decisions made by science administrators such as Lloyd Berkner and Frederick Brundrett and statesmen such as Dwight D. Eisenhower and Charles De Gaulle.[51] Meteorology and atmospheric research, holding the promise of accurate weather prediction but potentially also control of weather systems,[52] have come to underpin both our understanding of climate change processes and dreams of geoengineering projects to ameliorate those changes. The hubristic belief that global surveillance could lead to global-scale intervention had great appeal to military planners half a century ago, and a similar faith can be seen today.[53]

Finally, the use of artificial satellites to chart major environmental changes (deforestation, for instance) was the result of lobbying in the US Congress and elsewhere for the release of hitherto classified data.[54] But, as Roger Launius shows in this volume (chapter 7), the question of how far the US surveillance state could develop in the future thanks to spy satellites is yet to be answered. Since an increase in surveillance is often accompanied by the deployment of new weaponry, Launius argues that even space, the last frontier of surveillance, may not escape weaponization. The dual power of satellites to know the enemy and to know the earth is to a significant extent replicated in unpiloted aircraft (drones), which have come to symbolize the new face of warfare while also being touted as flexible and powerful surveillance platforms, for good or ill.[55]

Perhaps the more difficult question is how the political transitions toward the end of the Cold War connected to the emergence of new political priorities, including the monitoring of environmental changes. While full historical assessment awaits the release of further archival evidence, it seems clear that from the 1970s traditional Cold War urgencies embraced new environmental problems. Following President Richard Nixon's "environmental diplomacy," NATO supported a new program on the Challenges of Modern Society, which sought to offer solution to problems such as air and sea pollution.[56] The scientific shield that the defense alliance wielded seemed now to offer protection to the planet as much as the Cold War West, invoking a discourse of environmental security that remains prominent today. While Nixon's attempt was met with resistance, similar efforts led to the creation of the United Nations Environment Programme (UNEP), which has pushed for global environmental policymaking.[57] As Soraya Boudia shows in this volume (chapter 9), new systems of environmental monitoring adopted in the UNEP context drew on existing surveillance networks, replicating similar attitudes toward scientific and technological prowess and the importance of amassing an arsenal of environmental data. And in the final years of the Soviet Union, Mikhail Gorbachev embraced environmental monitoring as a tool of international cooperation in the Arctic, leading to the Arctic Environmental Protection Strategy and helping prepare the ground for contemporary initiatives in the region such as the Arctic Council.[58]

The surveillance imperative proved equally useful for detecting the enemy and for protecting the planet, either from the threats of military enemies—culminating in the mooted Strategic Defense Initiative (aka the "Star Wars" program)—or from modern industrial civilization more generally. The power of surveillance to know nature, and thus to facilitate its control must not be underestimated, or the technology regarded as unproblematic simply because the cause of environmental monitoring is regarded as enlightened. Just as powerful tools of surveillance such as CCTV cameras in modern cities have rightly been located within discourses of political control,[59] knowing the environment remains a critically underestimated source of power, moral as well as practical, for decisions on the future of the planet and its inhabitants, human and otherwise. The capacity of technologies to furnish information can serve to naturalize political decisions when the uses to which that information is put become reduced to inevitable outcomes of technological development. Critical and historically aware analyses of the origins of modern environmental monitoring technologies are essential to understanding why as well as how such technologies have been adopted, and to ground informed decisions on their usage: as Melvin Kranzberg famously put it, technology is neither good nor bad, but neither is it neutral.[60]

Seen from one perspective, the possibility of truly global environmental monitoring has enabled a problem caused by actors within specific geographic contexts—notably the traditional European empires, but also the Cold War superpowers and new industrial giants such as China—to become regarded as a global political responsibility. This is good inasmuch as it confirms the effect of local actions upon global stages, with potentially catastrophic consequences

at the planetary scale, but the history of asymmetric contributions to environmental damage—disproportionately the responsibilities of rich countries and
former imperial powers—risks becoming obscured.[61] Pointing to the severity of
the current crisis and its potential consequences has failed to produce significant
action in the rich world while providing arguments for restraining industrial
development elsewhere, and thus potentially entrenching injustice.

Moreover, the disjuncture between acceptance of data indicating climate
change and acceptance of the possibility of remedial action has fostered narratives of inevitability, rendering human agency secondary to environmental
change. Nowhere is this clearer than in the Arctic, where retreating sea ice and
glaciers are painted as necessarily leading to increased shipping and extractive
industry, as though the climate has assumed the power of political decision
making. Such perspectives reflect the considerable inertia of the hydrocarbon
economy but also the frame through which the neoliberal gaze views the global
environment. The threat of global environmental catastrophe can be mobilized
to justify global restraint, despite the fact that imperial powers exploited global
energy resources before, during, and after the Cold War and used them to grow
rich while polluting the planet. Asymmetries in power derived from history thus
can fade from view, especially when many of the most severe effects of global
climate change are likely to be felt by those least able to cope.

The converse problem is even more troubling. If the consequences of climate
change are deemed incompatible with the political and economic goals of the
rich world, data that might be incorporated into a narrative of inevitable progress becomes an obstacle to be challenged. Spells of cold weather in specific
locations still lead individuals to claim that local experience contradicts global
warming narratives: oddly enough, such claims tend to be made by those with
political views most hostile to global environmental regulation.[62] Others have
argued that organized climate change skepticism amounts to disinformation
campaigns based on ideology rather than facts—with similarities to tactics used
by the tobacco industry.[63] To label this the politicization of a neutral process is
simplistic: the political character of *all* research findings is latent, and moments
of conflict reveal rather than create this condition. Today perhaps more than
ever, understanding the relationship between the geosciences and the global
surveillance imperative is crucial to risk perception and thus to informed decision making.

Structure of the Volume

The book is divided into five sections, each focusing on a different aspect of the
surveillance imperative and the Cold War earth sciences. The chapters highlight
how new surveillance priorities informed the rise of specific disciplines and fields
of expertise while also molding new images of the earth.

The first section, Surveillance Strategies to Control Natural Resources,
considers how geophysical prospecting methods were enrolled in the shifting
geopolitical dynamics of post-1945 Europe. The concept of resource security possessed both domestic and foreign dimensions, posing challenges to
existing networks of colonial influence within the overarching shadow of the
nuclear-armed superpowers.[64] Oil was critical to domestic stability and national

economic prosperity in addition to military capacity. As Roberto Cantoni and Leucha Veneer (chapter 2) demonstrate, ensuring its discovery and delivery was a matter of state interest in both France and Britain. The search for uranium involved even more intensive state surveillance, a topic investigated by Matthew Adamson, Lino Camprubi, and Simone Turchetti (chapter 1). From the late 1940s the US government viewed attempts by other states to locate uranium reserves as a potential threat to its own security, and the United States Atomic Energy Commission sponsored intelligence work to seek and control sources in other countries. As the demand for strategic resources like oil and uranium grew, geoscientists developed new radioactivity-based methods for mapping the earth and its mineral contents.[65] This reinforced the image of the earth as a storehouse of resources, an enduring theme in European geographical and geopolitical thought that developed in concert with the view of the world as a space to be known and then controlled. Deploying ever more sophisticated techniques to interrogate the earth was a central component of state planning for both international conflict and domestic security.

The second section, Monitoring the Earth: Nuclear Weapons Programs, examines atmospheric and seismological surveillance of the geophysical traces of nuclear testing. Effective surveillance required both theoretical knowledge and an extensive network of monitoring stations. At the same time, questions that previously held primarily academic interest—from the early uses of radionuclides as tracers to the nature of the earth's interior and ways to transmit seismic waves—became fields in which state strategic interest made intelligence agents of scientists. As Néstor Herran (chapter 3) shows, the recognition that atmospheric radioactivity could have significant public health consequences prompted the creation of international networks devoted to its measurement. But radiological techniques also played a key role in gathering information on foreign nuclear weapons programs, leading to concerns at the highest level of state administration over what could or could not be divulged in scientific meetings without jeopardizing national security. The 1963 Partial Test Ban Treaty banned atmospheric explosions, establishing the importance of underground tests—and the seismic methods that could detect them, as shown by Simone Turchetti (chapter 4).[66] Openly stated goals to ban nuclear tests were thus coupled with more secret ambitions such as knowing, in the context of official meetings, the level of enemy expertise in seismic analysis. The planet came to be perceived as a signals transmission device, with the results of seismic analysis relevant to both intelligence-gathering and the advance of academic research agendas.[67]

The third section, Seeing the Sea—From Above and Below, examines how the surveillance imperative shaped oceanography and sea-ice research during the 1950s. In addition to boosting surveillance of the earth through geophysical research, geoscientists themselves became objects of interest due to their specific expertise, even as events such as the IGY reinforced the advantages to the superpowers of open data sharing. Sam Robinson (chapter 5) reconsiders the problem of relations between "special" allies by examining conflicts between British and US naval leaders over military strategy in the North Atlantic, and the role of oceanographic surveillance in underpinning such strategies. Open collaboration helped produce large-scale data sets, but it did not dictate either shared visions for how that data would be used or even how it should be acquired.

Peder Roberts (chapter 6) explores how charting and forecasting sea ice became a major concern for North American military planners after 1945, as the Arctic became an important potential military theater—and its waters became vital supply lanes for both superpowers. Observations of sea ice across the Northern Hemisphere, both historical and contemporary, became crucial foundations for accurate forecasting, and for the development of predictive techniques. The perceived strength of Soviet researchers in this field made knowledge of their techniques strategically important. Placing the Arctic under surveillance included picking the brains of those who studied it, and assessing the strength of enemy research capacity.

The fourth section, Surveillance Technologies, considers how new technologies were used to produce and establish new images of the earth from space. Roger Launius (chapter 7) reconstructs the history of satellite deployment and its underlying surveillance dimensions, arguing that surveillance was in fact the driver behind technological innovation in the satellite field. Sebastian Grevsmühl (chapter 8) contends that in addition to their initially envisaged uses for espionage and communications, satellites quickly evolved in unexpected ways to become resources for assessing environmental conditions and performing global environmental monitoring. The relationship between satellite imagery and conceptions of global systems is a particularly striking illustration of the surveillance imperative's connection with modern environmental consciousness. No longer just a medium for processing and interpreting otherwise obscure signals providing information about Cold War enemies, the earth became, through the interpretation offered by newly available photographic evidence, the fragile system that we are more familiar with.

The final section, From Surveillance to Environmental Monitoring, takes up the connection between surveillance and environmental consciousness with analyses of new global systems (of both monitoring and thought). Robert Poole (chapter 10) examines the impact of photographs of the earth from space in framing perceptions of the earth as a global system, from the IGY to *Apollo* and beyond. In addition to providing data with application for both civilian and military statecraft, images from space helped create a new mindset toward the earth as a discrete entity, the possibility of surveying it as a whole unit augmenting the fragility revealed by the "blue marble" *Apollo* images. These images fueled the ongoing shift from traditional surveillance monitoring practices adopted at global level to the creation of new systems for environmental monitoring, and to a new set of international organizations, a story picked up by Soraya Boudia (chapter 9). The Global Environmental Monitoring System (GEMS) has come to embrace a range of technological systems across a broad transnational framework, encompassing issues from water quality to biodiversity to atmospheric chemistry. With its roots in the drive to place the earth as a whole under surveillance, GEMS represents both the evolution and culmination of a process that has persisted from the Cold War into the present.

Notes

1. Ian Traynor, Philip Oltermann, and Paul Lewis, "Angela Merkel's Call to Obama: Are You Bugging My Mobile Phone?" *The Guardian*, October 24, 2013.

2. See, among others, Ronald E. Doel, "The Earth Sciences and Geophysics," in John Krige and Dominique Pestre (eds), *Companion to Science in the 20th Century* (London: Routledge, 2003), 391–417; Kai-Henrik Barth, "The Politics of Seismology: Nuclear Testing, Arms Control and the Transformation of a Discipline," *Social Studies of Science* 33:5 (2003): 743–781; Jacob Darwin Hamblin, *Oceanographers and the Cold War: Disciples of Marine Science* (Seattle: University of Washington Press, 2005); James Roger Fleming, *Fixing the Sky: The Checkered History of Weather and Climate Control* (New York: Columbia University Press, 2010); Doel and Naomi Oreskes, "The Physics and Chemistry of the Earth," in Mary Jo Nye (ed.), *The Cambridge History of Science*, Vol. 5 (Cambridge: Cambridge University Press, 2012), 538–557.

3. On *Sputnik*, see for instance Rip Bulkeley, *The Sputniks Crisis and Early United States Space Policy: A Critique of the Historiography of Space* (London: Macmillan, 1991); and John M. Logsdon, Roger Launius, and Robert W. Smith (eds), *Reconsidering Sputnik: Forty Years Since the Soviet Satellite* (London: Routledge, 2000).

4. On the IGY, see, for instance, Walter Sullivan, *Assault on the Unknown. The International Geophysical Year* (New York: McGraw-Hill, 1961).

5. This is the term coined by journalist Walter Sullivan to describe the IGY. See ibid.

6. Michel Foucault, *Discipline and Punish* (New York: Pantheon, 1977); Mike Davies, *City of Quartz: Excavating The Future in Los Angeles* (New York: Verso, 1992); and more recently Torin Monahan (ed.), *Surveillance and Security: Technological Politics and Power in Everyday Life* (Abingdon: Routledge, 2006).

7. A general survey is in Kirstie Ball, Kevin Haggerty, and David Lyon (eds), *Routledge Handbook of Surveillance Studies* (Abingdon: Routledge, 2012). A list of relevant centers devoted to surveillance studies is in www.surveillance-studies. net (accessed January 20, 2013). For a thoughtful reflection on the state of the field, see also Kirstie Ball and Kevin Haggerty, "Editorial: Doing Surveillance Studies," *Surveillance & Society* 3 (2005): 129–138 (available at: www.surveillance-and-society.org/Articles3(2)/editorial.pdf, accessed January 20, 2013).

8. M. Dirnwoeber, R. Machan, and J. Herler, "Coral Reef Surveillance: Infrared-Sensitive Video Surveillance Technology as a New Tool for Diurnal and Nocturnal Long-Term Field Observations," *Remote Sensing* 4:11 (2012): 3346–3362; D. G. M. Miller, N. M. Slicer, and Q. Hanich, "Monitoring, Control and Surveillance of Protected Areas and Specially Managed Areas in the Marine Domain," *Marine Policy* 39 (2013): 64–71. On biomonitoring, see for instance: National Research Council, *Human Biomonitoring for Environmental Chemicals* (Washington DC: NRC, 2006).

9. For an analysis of the interdependence between observer and observed in the case of the first photographs of the earth from space, see Robert Poole, *Earthrise: How Man First Saw the Earth* (New Haven, CT: Yale University Press, 2010).

10. Robert Jervis, "Was the Cold War a Security Dilemma," *Journal of Cold War Studies* 3 (2001): 36–60. See also Charles L. Glaser, "The Security Dilemma Revisited," *World Politics* 50 (1997): 171–201; and Alan Collins (ed.), *Contemporary Security Studies* (Oxford: Oxford University Press, 2007), 18.

11. See for instance Turchetti, this volume; Barth, "Politics of Seismology."

12. See, among many others, Ronald E. Doel, Tanya J. Levin, and Mason K. Marker, "Extending Modern Cartography to the Ocean Depths: Military Patronage, Cold War Priorities, and the Heezen–Tharp Mapping Project 1952-1959," *Journal of Historical Geography* 32 (2006): 605–626; Hamblin, *Oceanographers and the Cold War*; Ronald Rainger, "Constructing a Landscape for Postwar

Science: Roger Revelle, the Scripps Institution and the University of California, San Diego," *Minerva* 39:3 (2001): 327–352; Gary Weir, *An Ocean in Common: American Naval Officers, Scientists and the Ocean Environment* (College Station: Texas A&M University Press, 2001).

13. See, for instance, important contributions to science studies such as David H. DeVorkin, *Science with a Vengeance: How the Military Created the US Space Sciences after World War II* (London: Springer, 1993). Other contributions from intelligence studies include: Jeffrey Richelson, *Spying on the Bomb: American Nuclear Intelligence from Nazi Germany to Iran and North Korea* (New York: W.W. Norton & Co., 2006); Charles A. Ziegler and David Jacobson, *Spying Without Spies: Origins of America's Nuclear Surveillance System* (Westport, CT: Praeger, 1995); Michael S. Goodman, *Spying on the Nuclear Bear: Anglo-American Intelligence and the Soviet Bomb* (Stanford: Stanford University Press, 2007). On the transition from human to scientific intelligence see also: Kristie Macrakis, "Technophilic Hubris and Espionage Styles During the Cold War," *Isis* 101 (2010): 378–385.

14. The key reference here is Ronald E. Doel, "Scientists as Policymakers, Advisors, and Intelligence Agents: Linking Contemporary Diplomatic History with the History of Contemporary Science," in Thomas Söderqvist, ed, *The Historiography of Contemporary Science* (Amsterdam: Harwood, 1997), 215–244. See also David Van Keuren, "Cold War Science in Black and White: US Intelligence Gathering and Its Scientific Cover at the Naval Research Laboratory, 1948–1962," *Social Studies of Science*, 31 (2001): 207–229.

15. John Krige, "Atoms for Peace, Scientific Internationalism, and Scientific Intelligence," in Krige and Kai-Henrik Barth (eds), *Global Power Knowledge: Science and Technology in International Affairs: Osiris* 21 (2006): 161–181.

16. As the wording makes obvious, this point draws upon Naomi Oreskes, "A Context of Motivation: US Navy Oceanographic Research and the Discovery of Sea-Floor Hydrothermal Vents," *Social Studies of Science* 33 (2003): 697–742.

17. See on this Alan Needell, *Science, Cold War and the American State: Lloyd V. Berkner and the Balance of Professional Ideals* (Amsterdam: Harwood/NASA, 2000).

18. On the history of the DEW Line, see, for instance, P.W. Lackenbauer, Matthew Farish, and Jennifer Arthur-Lackenbauer, *The Distant Early Warning (DEW) Line: A Bibliography and Document Resource List* (Calgary: Arctic Institute of North America, 2005).

19. Key references in the emerging and still contested field of transnational history include (among many others) Akira Iniye, *Global Community: The Role of International Organizations in the Making of the Contemporary World* (Berkeley: University of California Press, 2002); Ian Tyrrell, "American Exceptionalism in an Age of International History," *The American Historical Review* 96 (1991): 1031–1055; Christopher A. Bayly, Sven Beckert, Matthew Connelly, Isabel Hofmeyr, Wendy Kozol, and Patricia Seed, "AHR Conversation: On Transnational History," *American Historical Review* 111 (2006): 1441–1464.

20. Michael A. Dennis, "Secrecy and Science Revisited: From Politics to Historical Practice and Back", in Ronald E. Doel and Thomas Söderqvist, eds. *The Historiography of Contemporary Science, Tecnology and Medicine: Writing Recent Science* (London: Routledge, 2006), 172–184.

21. See the essays in the edited collection Néstor Herran, Soraya Boudia, and Simone Turchetti (eds), *Transnational History of Science: Special Issue of the British Journal for the History of Science* 45 (2012).

22. John Krige, *American Hegemony and the Postwar Reconstruction of Science in Europe* (Cambridge, MA: MIT Press, 2006). There are, of course, works detailing

interactions between the US and Western European states, especially in space research. See, for instance, Michelangelo De Maria and Lucia Orlando (eds), *Italy in Space: In Search of a Strategy, 1957–1975* (Paris: Beauchesne, 2008). See also John Krige and Arturo Russo, *A History of the European Space Agency*, 2 Vols. (Noordwijk: ESA, 2000) and Lorenza Sebesta, "The Good, the Bad and the Ugly: U.S.-European Relations and the Decision to Build a European Launch Vehicle," in Andrew J. Butrica (ed.), *Beyond the Ionosphere: The Development of Satellite Communications* (Washington DC: NASA, 1997).

23. On science and empire, see amongst many others Peder Anker, *Imperial Ecology: Environmental Order and the British Empire, 1895–1945* (Cambridge, MA: Harvard University Press, 2001); Jorge Canizares-Esguerra, *Nature, Empire, and Nation: Explorations of the History of Science in the Iberian World* (Stanford: Stanford University Press, 2006); Richard Drayton, *Nature's Government: Science, Imperial Britain, and the "Improvement" of the World* (New Haven, CT: Yale University Press, 2000); Michael Osborne, "Science and the French Empire," *Isis* 96 (2005): 80–87; and John Stafford, *Scientist of Empire: Sir Roderick Murchison, Scientific Exploration and Victorian Imperialism* (Cambridge: Cambridge University Press, 1989).

24. Historians refer to this conception as a "Westphalian paradigm," tracing it back to the 1648 Peace of Westphalia. For an overview of the Westphalian system, see, for instance, Paul D'Anieri, *International Politics: Power and Purpose in Global Affairs*, 2nd ed. (Boston: Wadsworth, 2011), 28–30. We follow Mario Telò in preferring the term "Westphalian paradigm" to "Westphalian system," reflecting its continued conceptual applicability to the post-1945 world. Telò, *International Relations: A European Perspective* (Farnham: Ashgate, 2009), 2–3.

25. See, for instance, Sebastian Reyn, *Atlantis Lost: The American Experience with Charles de Gaulle, 1958–1969* (Amsterdam: Amsterdam University Press, 2010).

26. On the history of CERN, see A. Hermann et al., *History of CERN Volume 1*.

27. On CERN, see Krige and Dominique Pestre, "Some Thoughts on the History of CERN in the 50s and 60s," in P. Galison and B. Hevly (eds.), *Big Science: The Growth of Large Scale Research* (Stanford: Stanford University Press, 1992), 78–99.

28. John Krige, *American Hegemony*. Ronald E. Doel has argued for greater attention to intelligence-related matters in the treatment of US–Europe relations in Doel, "Does Scientific Intelligence Matter?" *Centaurus* 52 (2010): 311–322.

29. See, for instance, John Lewis Gaddis, *We Now Know. Rethinking Cold War History* (Oxford: Oxford University Press, 1997).

30. John Krige, "Hybrid Knowledge: the Transnational Co-production of the Gas Centrifuge for Uranium Enrichment in the 1960s," *BJHS* 45 (2012): 340.

31. On ICSU see Frank Greenaway, *Science International: History of the International Council of Scientific Unions* (Cambridge: Cambridge University Press, 1996). On the coexistence of territorial ambitions and international scientific cooperation, see, for instance, the case of Antarctica discussed in S. Turchetti, S. Naylor, K. Dean, and M. Siegert, "On Thick Ice: Scientific Internationalism and Antarctic Affairs, 1957–1980," *History and Technology*, 24 (2008): 351–376.

32. See Adamson, Camprubí, and Turchetti and Roberto Cantoni, "Oily Deals. Exploration, Diplomacy and Security in Cold War France and Italy," PhD diss. submitted at the University of Manchester, 2014, chap. 4.

33. On this, see Needell, *Science, Cold War and the American State*, 146.

34. The most important references are Needell, *Science, Cold War and the American State*; and Ronald E. Doel and Allan A. Needell, "Science, Scientists and the CIA: Balancing International Ideals, National Needs, and Professional Opportunities," *Intelligence and National Security* 12 (1997): 59–81.

35. Hamblin, *Oceanographers and the Cold War*. See also Elena Aronova, Karen Baker, and Naomi Oreskes, "Big Science and Big Data in Biology: From the International Geophysical Year through the International Biological Program to the Long Term Ecological Research Program, 1957 to Present," *Historical Studies in the Natural Sciences* 40:2 (2012): 183–224.

36. Simone Turchetti, "Sword, Shield and Buoys: A History of the NATO Sub-Committee on Oceanographic Research, 1959–1973," *Centaurus* 54 (2012): 205–231.

37. This famous image was first used by J. Robert Oppenheimer to describe the emerging nuclear stand-off between the United States and the Soviet Union in 1953. Oppenheimer, "Atomic Weapons and American Policy," *Foreign Affairs* (July 1953): 529.

38. See Robert Poole, "What Was Whole about the Whole Earth?" and Poole, *Earthrise*.

39. Doel, "Constituting the Postwar Earth Sciences The Military's Influence on the Environmental Sciences in the USA after 1945," *Social Studies of Science* 33,5 (2003): 635–666.

40. Oreskes, "A Context of Motivation," 699, 730.

41. Paul Edwards, *A Vast Machine: Computer Models, Climate Data, and the Politics of Global Warming* (Cambridge, MA: MIT Press, 2010).

42. David Hounshell, "The Cold War, RAND, and the Generation of Knowledge, 1946–62," *Historical Studies in the Physical and Biological Sciences* 27 (1997): 239.

43. Paul Forman, "Behind Quantum Electronics: National Security as Basis for Physical Research in the US, 1940–1960," *Historical Studies in the Physical Sciences* 18 (1985): 149–229; Stuart W. Leslie, *The Cold War and American Science: The Military-Industrial-Academic Complex at MIT and Stanford* (New York: Columbia University Press, 1993). For a critique, see Daniel J. Kevles, "Cold War and Hot Physics: Science, Security, and the American State, 1945–1956," *Historical Studies in the Physical Sciences* 20 (1990): 239–264.

44. Barth, "Politics of Seismology."

45. See Naomi Oreskes and Homer Le Grand (eds.), *Plate Tectonics: An Insider's History of the Modern Theory of the Earth* (Boulder: Westview, 2002).

46. See, for instance, Jessica Wang, *American Science in an Age of Anxiety: Scientists, Anticommunism and the Cold War* (Chapel Hill: UNC Press, 1999).

47. James Lovelock, *Homage to Gaia* (Oxford: Oxford University Press, 2000), 169 and 173.

48. Gunnar Ellingsen, "Instrumentutvikling med NATO-bistand," in Edgar Hovland (ed.), *I vinden: Geofysisk Institutt 90 år* (Bergen: Fagbokforlaget, 2007), 112–115.

49. Naomi Oreskes and Erik M. Conway, *Merchants of Doubt: How a Handful of Scientists Obscured the Truth on Issues From Tobacco Smoke to Global Warming* (New York: Bloomsbury, 2010).

50. The most notable is Senator James Inhofe (R-OK), the title of whose recent book accurately captures the essence of his views. Inhofe, *The Greatest Hoax: How the Global Warming Conspiracy Threatens Your Future* (Washington, DC: WND Books, 2012).

51. Edwards, *Vast Machine*.

52. Kristine C. Harper, "Climate Control: United States Weather Modification in the Cold War and Beyond," *Endeavour* 32,1 (2008): 20–26; Fleming, *Fixing the Sky*.

53. On geoengineering see Fleming, *Fixing the Sky*, and also Clive Hamilton, *Earthmasters: The Dawn of the Age of Climate Engineering* (New Haven, CT: Yale University Press, 2013).

54. Thanks especially to US Congressman George E. Brown Jr. (D-CA). See Gary Chapman, "Rep. Brown Left a Legacy for Science," *Los Angeles Times*, August 2, 1999.

55. See, among many others, Rachel L. Finn and David Wright, "Unmanned Aircraft Systems: Surveillance, Ethics and Privacy in Civil Applications," *Computer Law & Security Review* 28, 2 (2012): 184–194; and Kathleen Bartzen Culver, "From Battlefield to Newsroom: Ethical Implications of Drone Technology in Journalism," *Journal of Mass Media Ethics: Exploring Media Morality* 29,1 (2014): 52–56.

56. On this, see Jacob Hamblin, "Environmentalism for the Atlantic Alliance," *Environmental History* 15,1 (2010): 54–75.

57. This includes key decisions on the banning of CFCs through the Montreal Protocol and the establishment of the Intergovernmental Panel on Climate Change. See Shardul Agrawala, "Context and Early Origins of the Intergovernmental Panel for Climate Change," *Climatic Change* 39 (1998): 605–620.

58. See, for instance, Carina Keskitalo, *Negotiating the Arctic: The Construction of an International Region* (New York: Routledge, 2004).

59. On this, see, especially chapter 4 "Fortress L.A.," in Davies, *City of Quartz*.

60. Melvin Kranzberg, "Technology and History: 'Kranzberg's Laws,'" *Technology and Culture* 27,3 (1986): 544–560

61. On risk perception, see for instance Wiebe Bijker, Roland Bal, and Ruud Hendriks, *The Paradox of Scientific Authority: The Role of Scientific Advice in Democracies* (Cambridge, MA: MIT Press, 2009). Although the Kyoto Protocol was designed as a two tier system distinguishing between developed and developing countries in enforcing carbon emission cuts, it did not recognize damages to underdeveloped countries, something recently addressed as the 2012 Doha Climate Change conference. On this, see: Fiona Harvey, "Doha Climate Change Deal Clears Way for 'Damage Aid' to Poor Nations," *The Observer*, December 8, 2012.

62. For a good recent example, see Boris Johnson (Mayor of London), "It's Snowing, and It Really Feels Like the Start of a Mini Ice Age," *Daily Telegraph* (January 20, 2013).

63. James Hoggan and Richard Littlemore, *Climate Cover-Up: The Crusade to Deny Global Warming* (Vancouver: Greystone Books, 2009).

64. For an overview on energy security, see for instance: Benjamin K. Sovacool and Marylin A. Brown, "Competing Dimensions of Energy Security: An International Perspective," *Annual Review of Environment and Resources* 35 (2010): 77–108; and Robert J. Lieber, "Energy, Economics and Security in Alliance Perspective," *International Security* 4 (1980): 139–163.

65. On novel geophysical techniques, Louis A. Allaud and Maurice H. Martin, *Schlumberger: The History of a Technique* (New York: Wiley & Sons, 1978) and Henry Faul, *Nuclear Geology* (New York: Wiley & Sons, 1954).

66. See Turchetti in this volume. For an overview on these negotiations and the rise of seismology, see also Barth, "Politics of Seismology."

67. On signals intelligence, see, for instance, Richard Aldrich, *GCHQ: Britain's Most Secret Intelligence Agency* (London: HarperCollins, 2011).

Section I

Surveillance Strategies to Control Natural Resources

Chapter 1

From the Ground Up: Uranium Surveillance and Atomic Energy in Western Europe

Matthew Adamson, Lino Cambrubí, and Simone Turchetti

Uranium was one of the key strategic materials of the Cold War, when its overall production worldwide increased exponentially.[1] During this period, the US administration and other Western governments interested in the exploitation of atomic energy invested enormous resources in charting deposits worldwide, setting up surveying operations, mining and refining uranium ores, and putting rival prospecting efforts under surveillance. This gave uranium exploration a peculiar importance. Uranium prospecting required innovative geological work and often occurred as part of intelligence missions assessing the availability of minerals.[2] It became a product of knowledge hybridization and a technopolitical asset.[3] The exigencies of uranium reconnaissance also helped prospecting specialists to promote their research and create new disciplines such as nuclear geology and radiogeology. In short, prospecting uranium became an emblematic Cold War technoscientific endeavor.

By analyzing untapped archival materials from several repositories and illuminating new aspects of the international circulation of uranium and other atomic matériel and knowledge in Italy, France, and Spain, this chapter fills an important gap.[4] Literature on atomic energy projects has neglected the role played by strategic minerals in these programs. Only Jonathan Helmreich's monograph on its wartime and postwar diplomacy and Gabrielle Hecht's recent volume on its African mining complex explicitly focus on uranium.[5] Our chapter falls between these works both chronologically and thematically. It reveals that during the Cold War the control of uranium production brought together scientific experts and intelligence agents and their interaction informed the status of uranium as a resource and commodity. Available only on the black market at the end of World War II, uranium soon became more readily available. Using the knowledge that both prospecting experts and intelligence agents offered, the US administration attempted to forestall the expansion of its global market. The result was "artificial scarcity," that is, new circumstances typified by a surplus in uranium coupled with constraints making it difficult to acquire.[6]

This chapter also shows that the missions of scientists and intelligence agents had repercussions for international relations, since uranium functioned as a political device in the construction of American hegemony in Western Europe.[7] We take a transnational approach in order to document these ramifications; rather than draw comparisons between national cases, we look at how national programs were shaped by flows of scientific knowledge, expertise, and restricted information internationally.[8] Most Western European governments perceived the development of an independent national nuclear program as a priority. US intelligence work and offers of collaboration in the atomic energy field countered the ensuing independent efforts and disrupted rival projects and partnerships. American propositions and covert activities helped to reconfigure the Italian nuclear project, but did not prevent the establishment of alternative schemes in Spain and France. Hegemony, as we shall see, worked differently in different places.

The Quest for Independent Uranium Supplies

After the atomic bomb made its appearance in global politics, Western European leaders asked themselves whether the future sovereignty of their countries depended on getting access to atomic technology. The resultant ambitions were from the outset marked by different goals and research trajectories. France aspired to experiment with nuclear energy without limitations, in theory excluding nothing, not even nuclear weapons. Italy, on the other hand, was forbidden from doing so by the Paris Peace Treaty signed on February 10, 1947. Spain was not in this way restricted, but its impoverishment and isolation left it little hope of "going nuclear." Despite these differences, all three countries embarked on efforts to acquire uranium. They began to manoeuver through alliances and divisions at a moment when uranium minerals were thought to be rare, readily available in only a few rich deposits.[9]

The French government took action rapidly, by the end of 1945 suspending the granting of uranium mining permits to private firms and creating a state agency for atomic energy, the Commissariat à l'Energie Atomique (CEA). The CEA's charter defined its mission in the realms of science, energy production, and national defense, and, combined with associated decrees, gave it a final say in staking claims on uranium deposits. The French supply, however, amounted to two modest caches of uranium recovered at the end of the war.[10] Approaches to the Belgians to purchase uranium were unsuccessful and a deal with a Portuguese firm for uranium from Mozambique yielded little due to the vendor's unreliability and the ore's poor quality.[11]

Prospectors' findings in La Crouzille (1948), Grury (1950), and promising indications in the Vendée and in the Upper Rhine shales (1951) suggested that more, if limited deposits were waiting in France (see Figure 1.1). Meanwhile, findings in the French colonies were disappointing. Up to 1951 less than half a ton of uranium was mined in Madagascar despite earlier promising reports.[12] In that year, the state secretary for atomic energy ordered a review of uranium production. By then the CEA had collected just 46 tons of uranium (half from La Crouzille). The director of its mining division, Marcel Roubault, promised 50 more, barely enough for an industrial program.[13]

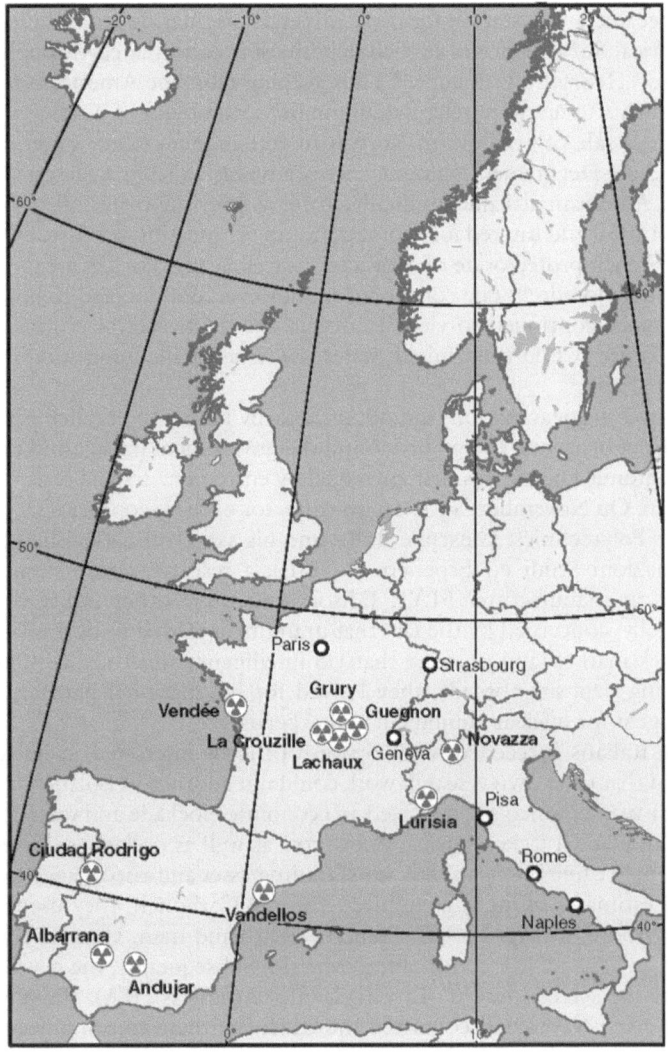

Figure 1.1   Uranium deposits in Western Europe and other nuclear sites of interest. Map by Hans van der Maarel, Red Geographics

The Americans were not detached observers of the CEA effort, and the communist affiliation of CEA high commissioner Frédéric Joliot-Curie made them especially wary. In 1947 the US Atomic Energy Commission (AEC) established its Division of Raw Materials, and its response to foreign prospecting work was to dispatch overseas experts (at times undercover) to reconnoiter territories, gain information, and negotiate deals with local companies. The division took over from the Manhattan District foreign intelligence section, which underestimated the potential of deposits in mainland France and overrated those of its colonies. AEC technicians now began to collect samples and the American embassy in

Paris became a way station for these minerals originating from Western Europe and Africa. Surveillance of French deposits was coordinated through the CIA's "wizard" Henry Lowenhaupt.[14] Thus keeping tabs, the Americans learnt about the French attempt to purchase uranium in Mozambique. And they succeeded in blocking a sale of uranium to Norway in exchange for heavy water, as reported by the State Department's atomic energy adviser, Gordon Arneson.[15]

The American attitude gradually changed after the dismissal of Joliot-Curie in April 1950 and catered for a joint uranium prospecting venture, SOMAREM, in the French protectorate of Morocco, as well as visits of CEA experts to AEC mines in Colorado.[16] The collaboration, however, did not lead to broader synergies. France continued to view its atomic energy project as crucial to its Cold War posture, while the United States saw it as challenging its dominance in nuclear affairs.[17]

Other European nations did not necessarily follow the French path. In Italy, the subject of atomic energy had found the government passive and the exploitation of atomic energy was first envisaged by entrepreneurs and researchers based in Milan. On November 26, 1946, the director of the Geophysics Department at Milan's Polytechnics, Giuseppe Bolla, and his assistants established the Centro Informazioni Studi ed Esperienze (CISE), a research consortium sponsored by the car manufacturer FIAT, Edison, and three other utility companies.[18] Eventually, concerned by the fact that uranium appeared to be available only on the black market (and unaware that US intelligence stealthily assisted the AEC in seizing deposits abroad), they looked for international partners willing to exchange it for nuclear equipment and expertise.

The Italians succeeded when Spanish officials interested in nuclear energy realized that their own research work could not advance in isolation. In the early postwar years, Franco's Spain faced an economic blockade and was left out of the Marshall Plan. It lacked nuclear know-how as well as dollars to buy equipment. Uranium became the Spaniards' chief trading asset and encouraged by the (optimistic) estimates of mining engineer Antonio Carbonell, they used it as a bargaining tool.[19] In April 1948, Franco's right-hand man, General Luis Carrero Blanco, invited a CISE consultant to Madrid. Subsequently, the newly appointed head of the secret Junta de Investigaciones Atómicas (JIA), the physicist José Maria Otero Navascués, contacted the CISE chairman to bring negotiations to a successful conclusion.[20]

The final agreement granted CISE up to half a ton of uranium from the Albarrana mine (Córdoba, see Figure 1.1). In exchange, Italian experts went to Spain and from October 1948, geologist Luigi Trevisan (University of Pisa) and chemist Giorgio Marinelli (University of Rome) led prospecting work. Three Spanish researchers were also invited to the CISE to study radioactivity measurement techniques and instrumentation.[21] Despite all good prospects, however, the collaboration was undermined from the ground up. In May 1949, Otero travelled to Italy with less than promised, just 213 kilograms of uranyl nitrate. And by then JIA physicists invited to the Italian laboratory were already designing gammascopes more sensitive than their CISE equivalents.[22]

As the collaboration with the CISE ended, the Spaniards looked for more partners. In 1951, the Junta Energía Nuclear (JEN) replaced the JIA. It was entrusted with exclusive rights over Spain's uranium, and its leadership abandoned secrecy. Otero, now its managing director, paid visits to CEA's research centers hoping to collaborate with the French. Spanish nuclear engineers were invited to tour CEA facilities, but French administrators stopped short of a collaborative agreement, sensing that they had more to give than gain.[23] JEN's relationship with the Germans were more productive. Physicists Werner Heisenberg and Karl Wirtz visited JEN offices in Spain, pleased with the opportunity to return to nuclear work. They were instrumental in making it possible for Spanish researchers to work in German laboratories; the Spanish came away from the exchange with new information and equipment, especially on uranium processing.[24]

Unsurprisingly, these Spanish efforts put US officials on alert once again. Spain was (wrongly) thought to possess the fifth largest reserve of uranium as a 1949 *Herald Tribune* article openly divulged.[25] Aiming to learn more about Spanish reserves, secure as much uranium as possible, and counter attempts to stockpile it by other nations, AEC technicians traveled to Spain in great secrecy to gain access to JEN's uranium prospecting work. In keeping with that policy, the AEC provided Spanish laboratories with better gammascopes than those made in Italy, while the US National Bureau of Standards offered assistance to certify the grade of JEN's uranium samples.[26] In 1953 a prominent AEC official, John A. Hall, visited Spain to offer assistance in prospecting, and the AEC eventually authorized JEN scientists to visit its laboratories.[27]

Since Italy, in contrast to Spain and France, had yet to pass legislation ensuring a state monopoly on uranium deposits, the AEC could check Italy's nuclear ambitions by offering deals to private uranium mining concerns. US surveillance operations in Italy informed the AEC's approach. The Division of Raw Materials had learnt all it could about Italian uranium. A Resources and Strategic Minerals consultant appointed in the framework of the Marshall Plan had provided preliminary data. In 1950 Arneson dispatched the geologist Clarence Wendel to work as a "consultant" (meaning an undercover expert) in the US High Commissioner's Office of Political Affairs (Germany). Wendel was eventually called in to secretly patrol uranium-related businesses in Western Europe and three years later he surveyed Alpine uranium ores.[28] When Wendel learnt about the prospecting efforts of two Italian companies on the largest uranium deposit then known in Italy (in Lurisia, near Cuneo, see Figure 1.1), he promptly informed the division, triggering the AEC to make an offer to one of these firms and buy its output.[29]

The AEC's mounting interest in Italian uranium elicited a secret clash between Italian and US administrations. The Italian Ministry of Industry had established the Comitato Nazionale Ricerche Nucleari (CNRN); Naples' chemistry professor Francesco Giordani became its chairman, while the physicist Edoardo Amaldi and the geologist Felice Ippolito became members and a contract was signed with the CISE in May 1952.[30] Finding the CNRN unable to block AEC efforts to buy Italian uranium, Giordani turned to the CEA to gather information about the laws granting to the state the French uranium. He presented a similar piece of legislation in the Italian Parliament.[31] In early 1953, US diplomats informed the Italian prime minister that opposition to uranium export would challenge defense agreements between the two countries.[32] The result was a standoff. The AEC could not export the Italian uranium that it was promised, but the CNRN could not buy it either.

By this time, though, the AEC approach to uranium trading was changing considerably. Now the Americans aimed to grant Western European countries greater access to nuclear technologies, all the while attempting to make it difficult for these states to formulate independent schemes.

The Scramble for International Collaboration

Until 1953 the US administration's efforts primarily aimed at gathering intelligence on foreign uranium assets with the ultimate goal of limiting the potential of other countries' nuclear programs. But the growing availability of uranium worldwide made this strategy untenable.[33] Eisenhower's Atoms for Peace speech paved the way for asserting nuclear hegemony differently, through the offer of collaborative agreements that molded foreign programs in the ways the AEC advocated. Meanwhile, the Americans continued to keep foreign atomic projects and collaborations under close scrutiny.

The 1954 Atomic Energy Act changed course and allowed the trade of nuclear equipment abroad. In addition, at the 1955 Conference on the Peaceful Uses of Atomic Energy in Geneva, high-ranking US government and industry officials disclosed information on uranium reserves and nuclear techniques to scientists and administrators of other countries. Importantly, the US reactors they were now offering for sale only worked with *enriched* uranium and only the AEC (and its Soviet competitors) had the isotope separation technologies needed to manufacture it.[34] In the next five years, the AEC provided reactors to dozens of countries that consequently became dependent on the American enriched uranium technology. In this way, the United States dissuaded these nations from developing independent atomic projects.[35] Exporting the peaceful atom became the pillar of a new American strategy aimed at maintaining control of the uranium trade.

The US scheme produced a situation of artificial scarcity. The AEC continued to control the mineral's transnational flow, thus making it harder for competing agencies to acquire it independently, *and* at the same time offered its enriched version at cheaper rates to dissuade other countries from autonomously searching for it. Under these new circumstances, the AEC did not fear sharing information on its prospecting activities precisely because if collaborating countries could be dissuaded from going nuclear independent of the United States, then the natural uranium they extracted could be sold only to the AEC in exchange for other atomic matériel and knowledge. In 1956 the commission took European technicians on a tour of uraniferous regions in the United States when just a year earlier foreign geologists had complained about its secretive approach. The power and the vast resources the Americans sunk into uranium prospecting and mining were plain to see. Now directed by Lewis Strauss, the commission tasked John Hall with controlling foreign distribution through a newly established Office of International Affairs.[36]

Responses to the American initiative varied. The CEA countered US plans by invigorating its own search for uranium. French uranium should now have "a permanent market value," the new CEA high commissioner Francis Perrin argued. His administrative counterpart Pierre Guillaumat (now the real man in charge, due to changes in the CEA's constitution) regarded uranium provision as a matter of resource security. He was ideally suited to the task of supervising France's uranium acquisition on the world stage, having served as an intelligence officer in North Africa during the war and, after that, overseeing French oil interests worldwide.[37] Guillaumat made the like-minded, energetic mining engineer Jacques Mabile director of CEA's Mining Division. Mabile's geologists and prospectors spread to virtually every corner of the *Métropole* and made sweeping searches in the colonies. Their prospecting techniques multiplied. Car-borne prospecting now covered more ground and airborne prospection was tested in the Grury and the Vendée provinces. They discovered numerous low-grade deposits, the ores from which were by 1955 feeding a new chemical concentration plant at Gueugnon.[38] In fact, by the end of the 1950s, with the added production of private entrepreneurs, the CEA stockpile had grown to 559 tons.[39] At the same time, prospecting in Madagascar eventually paid off in the form of a substantial supply of uranothorianite. Meanwhile, a geological study of North Africa's Precambrian shields, featuring aerial surveys, led to the discovery of massive deposits in Gabon (1956–57) and Niger (early 1960s). Africa was to become France's principal source of uranium.

With its uranium supply assured, France, like Britain, could sidestep the American scheme to control circulation of uranium and build its own industrial-scale nuclear reactors fuelled with natural uranium.[40] Moreover, with the increased weight its uranium lent it, France could leverage for bilateral relations, operating with increasing effectiveness transnationally, while, at the same time, turning down collaboration with the AEC. The CEA sent mining engineers to inspect deposits and examine techniques in Sweden, West Germany, and Italy. The Italian prospecting program drew heavily on French methods and uranium extracted from Sweden's bituminous shales was sent to France for processing and then shipped back to fuel a Swedish plutonium breeding reactor. French uranium with no strings attached was sold to Denmark and Israel.[41]

This improved supply capability made the United States anxious. Its officers kept an eye on these transactions, asserting control over the international circulation of uranium and isolating the French. Hall was especially wary of a joint JEN-CEA project. The US administration cemented its dominant influence in Spanish nuclear affairs by increasing nuclear assistance to Franco's Spain in the broader context of the 1953 military agreements between the two countries. In the next three years the number of American prospectors in Spain doubled, helping it to find two rich pitchblende veins.[42] In 1957, a US–Spain bilateral agreement catered to the purchase of a General Electric reactor fuelled with enriched uranium. However, the Spanish nuclear program advanced rapidly at the expense of its autonomy. Otero disliked the dependence on American enriched uranium, but felt the need to compromise in order to get the program going.[43] Indeed, the growing AEC influence on Spanish atomic affairs neutralized previous collaborations. The terms of the Spanish–American bilateral agreement prevented the Spanish from producing spare parts with other countries' assistance, hindering joint research with the Germans. And just as Hall had hoped, the AEC presence in Spain kept the French at bay. In August 1956, Otero was compelled to inform a CEA administrator that the JEN would only collaborate with the AEC.[44]

The US administration watched with concern the influence that France had in Italy's fast-growing uranium prospecting program. During the 1950s, the search for uranium in Italy gained momentum in order to match the CISE request of no less than five tons of uranium to start a research reactor. Stockpiling the mineral was not easy though, especially since the Spanish uranium was no longer available and half of Lurisia's output was promised to the AEC.[45] To find additional uranium, the CNRN established a Divisione Geo-Mineraria. Ippolito directed its program, planning surveys and mapping uranium resources nationwide.[46] Informed by visits to the CEA's La Crouzille deposits, Ippolito and his co-workers formed *squadre volanti* (flying squads) equipped with GM counters (see Figure 1.2) and started to make maps charting iso-radioactivity curves in uranium-rich areas. Mabile and Ippolito's teams exchanged visits and the Italians saw the CEA's Craëlius borehole drilling machine in operation at Lachaux.

Thanks to French assistance and the CISE uranium prospecting project in Spain, the Italian nuclear geologists could diversify their efforts. The CNRN's division commissioned an aerial survey of the Lurisia area. The geologist Trevisan, who had explored the Albarrana uranium mines, sought to extract uranium and thorium from Tyrrhenian sands in the context of a project developed at the University of Pisa. Another group based at Rome's Institute of Chemistry (including Marinelli, the other returnee from Spain) researched the possibility of obtaining radioactive substances from volcanic lavas.[47]

These exchanges between Italian, Spanish, and French researchers led to the rise of uranium geology in Europe. French geologists were particularly keen

Figure 1.2   Radiometric prospecting in Italy

Source: F. Ippolito, "Dieci Anni di Ricerca Uranifera in Italia" *Notiziario CNRN* 9:7 (1963), 25.

to examine the Alpine uranium formations in Lurisia since they had hitherto focused on massifs. Ippolito agreed with his French colleagues that it would be valuable to determine whether the band of mineralized rock extending through the Maritime Alps continued into French territory. This sort of research question stimulated broader synergies. Prospecting began to more evidently shape disciplinary identities and redefine research subjects. Out of the uranium geologists' work emerged a new discipline, "nuclear geology," bringing together the geochemistry of uranium and thorium and the measurement of radioactivity in rocks in the name of determining composition and mineralization characteristics. A 1954 symposium organized in Strasbourg (France) catered for its expansion internationally; Italy's first meeting of nuclear geologists took place in 1955.[48]

While US authorities had nothing against the establishment of new disciplinary perspectives, concrete French assistance to the CNRN made them anxious. Ippolito asserted at the 1955 Geneva conference that up to 100 tons of uranium could be mined in Italy, enough to fuel an independent nuclear program.[49] Ippolito's statement certainly caught the attention of US informers. In fact, the US embassy in Italy had already relayed secret information on "several tens of kilograms" of uranium transferred from Lurisia to the CISE laboratory in Milan.[50] To prevent further transfers Arneson's successor at State Department, Gerald Smith, had already agreed with Hall on an assistance plan. And in March 1955 Hall met a CNRN delegation to negotiate supplies of heavy water and enriched uranium and arrange for the purchase of a US-made research reactor.[51]

Much like the Italians, Spanish officials now found themselves lured by American technology even as they fought to regain their project's lost autonomy. "We need to win the battle for Spanish fuel," wrote Otero in 1958, "and to this goal the JEN devotes all its energies."[52] He believed that Spain needed to remain open to bilateral agreements with other countries. A long exchange with the CEA's director of external affairs Bertrand Goldschmidt stirred new life into JEN–CEA relations.[53] The following year Otero and the JEN president, Hernández Vidal, visited Paris and eventually Hernández and Guillaumat drafted a bilateral treaty.[54] Otero now wondered if French assistance might allow the JEN to build a natural uranium reactor and free Spain from its dependence on American fuel (see Figure 1.3).

As a consequence the JEN did not downsize its uranium prospecting program and even tried to get private interests involved. Its prospectors continued to visit facilities in the United States, Sweden, Canada, Portugal, and France.[55] US and French experts helped convince the chief of JEN's Mining Division, Demetrio Santana, that his division ought to deploy new geophysical techniques.[56] These were provided by the CEA at a reduced cost in the hope of undercutting US influence on Spanish nuclear affairs. The Spanish geologists were now able to independently prepare maps for uranium-bearing areas in Andújar and Ciudad Rodrigo (see Figure 1.1).[57]

Much like the Spaniards, Italian officials sought to avoid technological dependence on the United States. The CNEN continued to look for ways to recover natural uranium and at the 20th International Geological Congress (Mexico City), Ippolito, Marinelli, and their co-workers reported on morphological features and transformations typifying the Alpine uranium veins. Their study anticipated systematic explorations undertaken from 1956 to 1958.[58] A report prepared for the

Figure 1.3   Guillaumat examining the mineral at the Spanish uranium factory, 1958
Source: Archivo General Administración (Medios de Comunicación).

Italian prime minister asked for a considerable investment in prospecting, noting that the division had expanded considerably.[59] Ippolito was about to be rewarded for this expansion as he succeeded Giordani as CNRN's head.

Meanwhile conversations between European experts on nuclear geology continued. The newly established European Atomic Energy Society (EAES) provided for exchanges between these experts outside "official" national atomic agencies' channels. Goldschmidt was instrumental in the JEN's entry into the forum and in May 1957 the Junta hosted the society's first meeting devoted, notably, to uranium geology. Delegates compared geological structures in different countries, thus promoting a better understanding of uranium-bearing formations.[60] Now CEA's André Lenoble and Jaque Geffroy extended the collaboration that had typified their relations with the Italians to other European countries. In the 1957 Madrid workshop they advanced their ideas, envisioning for the first time a *province uranifère européenne*, and justifying pan-European cooperation in the nuclear field.[61] Other papers presented at the Madrid workshop envisaged geological links with geopolitical resonance: Portugal shared mountain chains with Spain, as did Italy with France, and France with Spain. European integration could be built from the ground up. But in fact, as we shall now see, it was constructed from high above.

Geopolitical Oligopolies in an Age of Surplus, 1958–1964

By the end of the 1950s, worldwide prospecting efforts had succeeded beyond anyone's expectations. The availability of natural uranium had increased dramatically; world uranium reserves amounted to 1.5 million tons. Canada and

South Africa had expressed an interest in selling excess output, and the US government now decided against renewal of contracts with these countries as well as with Belgium, the other major world supplier of natural uranium.[62] World uranium production peaked in 1962 and then slumped for five years.[63] The status of global uranium as a commodity thus shifted once again, now pivoting on the distinction between its "natural" form and the enriched one.

As the global uranium supply grew, buyers found themselves with more leverage, which, again, preoccupied Hall and his colleagues at the AEC. Their European strategy was no longer focused on how to prevent individual nations from exploiting alternative collaborative frameworks, but instead concentrated on concocting new alliances whose agenda would be in line with US interests. Thus another framework for acquiring uranium and many sorts of nuclear technologies emerged: Euratom, which entered force on January 1, 1958. First suggested at the 1955 Messina Conference, Euratom advanced Western European unification in the way the US administration wanted.[64] A supply agency was established to assure access to uranium as well as to American-licensed technologies. The treaty catered for revising US bilateral agreements so as to allow private companies in Europe to purchase American nuclear reactors for energy production.[65] Euratom actually had ownership, not of natural uranium, but of fissile materials, enriched uranium and plutonium, but this "theoretical" ownership (Goldschmidt's words) did not extend to military establishments. Watertight controls from the International Atomic Energy Agency (IAEA), while present, did not extend to these French establishments, meaning that the French could continue their weapons program.

The path leading to the establishment of Euratom is telling: US interests dominated its final form. The French had proposed a European uranium enrichment plant—a proposal that greatly displeased the AEC, since such a plant would threaten its new monopoly. Intelligence on the plan convinced US officials to intervene. As the Secretary of State John Foster Dulles put it in 1955, Euratom had produced "unforeseen political and security problems" since enrichment not only put the US-enriched uranium monopoly at risk, but also gave the Europeans an opportunity to jump-start any military program.[66] The CIA was busy monitoring the French atomic weapons program and feared especially that French, Italians, and Germans—the chief three Euratom promoters—would unite in an effort to build atomic weapons. A later CIA report on the subject clarified that an enrichment facility would enable France to produce 350 kilograms a year of weapons grade uranium.[67] US diplomats worked hard to dissuade the French and their partners from embarking on the enrichment path. They succeeded: by the time Euratom's three wise men (Louis Armand, Giordani, and Franz Etzel) prepared Euratom's constitutive document, enrichment was no longer the priority. An American proposal to guarantee the provision of enriched uranium to Euratom member states doomed an independent, collective European enrichment capability.[68]

Under these circumstances, the French redoubled their efforts to locate and secure natural uranium and demonstrate to the world that "we are the incontestable leaders of the Euratom powers" (as a 1958 internal CEA report put it).[69] For the CEA leadership, "free" trading of natural uranium was paramount—witness French opposition at the 1956 IAEA negotiations to safeguards being extended to the raw material.[70] Thriving uranium prospection and mining activities formed the leading edge of French atomic foreign policy. A 1958 CEA report stressed that France must use its "mining advantage" to press upon countries its growing arsenal of atomic technologies and its European neighbors were encouraged to call upon its mining and prospecting expertise.[71] Following the exploitation of the Nigér deposit, Mabile was heard to say: "France is ready to take a seat among the great powers, something never before seen in the history of mineral substances."[72] (see Figure 1.4)

Nevertheless, the French felt their lack of enriched uranium keenly. US officials had scrutinized this French deficiency for years. The forlorn attempt to

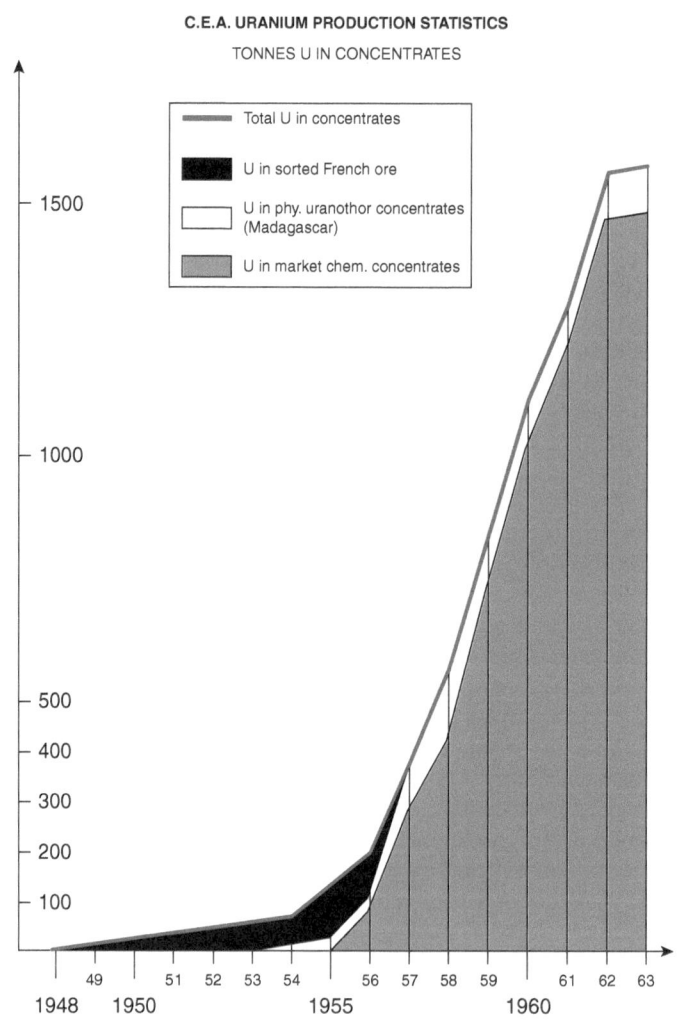

Figure 1.4   CEA production statistics

Source: CEA, *The Uranium Mining Industry. Commissariat a l'Energie Atomique. 1964–1965* (Lyon: CEA, 1966).

design a natural uranium-fueled nuclear submarine and French frustration with the underperformance of their third research reactor suggested how deleterious this lack could be. In fact, France purchased limited amounts of enriched uranium from the United States (and Britain), and at the end of the 1960s, completed at great expense a uranium enrichment plant at Pierrelatte, in large part for the benefit of their military program.[73]

Still, natural uranium brought France and Spain closer than ever. The JEN maintained working ties with the CEA's Mining Division into the 1960s and the collaboration gave the Spanish the opportunity to reduce their dependence on US atomic technologies by purchasing a French natural uranium reactor.[74] The French offer, first made in 1963, highlighted that natural uranium, in contrast with enriched uranium, assured an independent nuclear fuel supply and the possibility of making use of in-house reserves, as a Spanish report highlighted: "The discovery in France of significant uranium deposits is what has allowed our neighbor to build a national nuclear industry [our translation]"[75]

This is what Otero wanted. In 1958 he had produced detailed calculations to back an argument for a reactor fuelled with natural uranium (Figure 1.5).[76] That year the JEN opened its first uranium treatment plant and Otero was eager to use its output to fuel Spanish reactors. However, high costs doomed the project.[77] The 1963 agreement with the CEA took up where Otero's failed initiative left off, giving the JEN the opportunity to use its surplus of natural uranium to fuel Spain's third power plant, Vandellós I.[78] To the French, the sale represented the first export of their reactor technology, and a break of the US monopoly. They offered very good conditions to the Spaniards.[79] However, the French natural uranium reactors were not attractive to Spanish private companies, who found them expensive to operate. They made little sense from the mid-1970s onward, when enrichment technologies had become widespread. Even France would opt, 10 years later, for US reactors in order to supply its national electric power grid.

In this changing scenario of transnational flows, countries where neither bombs nor major uranium outputs were foreseeable had no choice but to "shop abroad." This was the case for Italy, where the need to carry forward prospecting activities faded after some unexpected twists. In 1960 a new law replaced the CNRN with a new National Committee on Nuclear Energy (CNEN). Ippolito, its new head, hoped to follow the French strategy of using natural uranium to propel an independent nuclear project. A new atomic energy law declared that natural uranium came under exclusive state ownership, enabling the state-owned Società Minerali Radioattivi Energia Nucleare (SOMIREN) to gain priority in uranium prospecting in Italy. Administered by oil tycoon Enrico Mattei, SOMIREN had appeared on the prospecting scene in 1955, collaborating with the CNRN in surveys in Sardinia and Calabria. By 1959 its well-practiced prospectors had found the largest uranium deposit in Italy, near Novazza (Bergamo). Its output was expected to be 1,000 tons.[80]

But unlike Spain, Italy had invested far too much in US technology to pursue in-house reactors fuelled with natural uranium. On the one hand, Mattei succeeded in purchasing one of these reactors in Britain, thus using some of the natural uranium SOMIREN had found. On the other hand, Italy's CNEN did not have the mining and prospecting complex of the French CEA. When in 1962 world uranium production peaked, the economic incentive to search

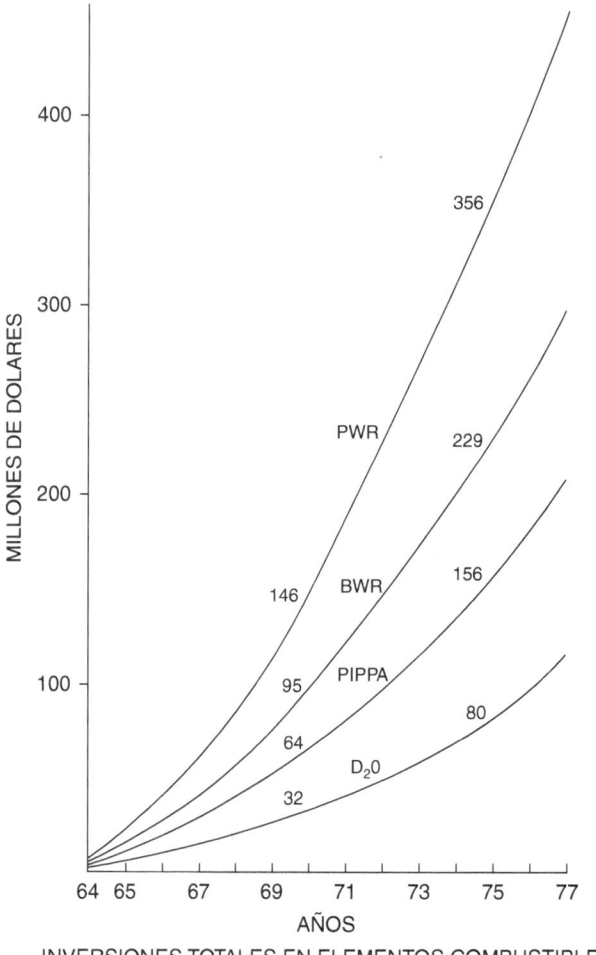

Figure 1.5 Otero's graphs comparing the mid-term costs of possible reactors, two fuelled with natural uranium (below) and two with enriched uranium

Source: Otero Navascués, "Necesidades españolas de elementos combustibles", *Energía nuclear* 5 (1958): 18–34, 25.

for and mine more uranium vanished, as did any hope of exploiting Italy's uranium reserves. Although by 1965 Italy could have produced up to 350 tons of natural uranium, it never did. In 1962 the Lurisia uranium mine shut down, the Novazza deposit was never excavated, and the plan to extract thorium from Tyrrhenian sands came to nothing.[81]

When in 1963 Ippolito documented the achievements of 10 years of uranium prospecting in Italy, he acknowledged first of all that the uranium market had changed considerably. But, faithful to his early strategy, he claimed that the search for uranium should continue anyway. Europe, he argued, "must look for its own deposits to stabilize its uranium market [our translation]."[82] Then, abruptly, accusations of mismanagement led to investigations of Ippolito's activities and eventually to his imprisonment for unlawful use of public funds in 1964.[83]

Following this affair, the CNEN was reorganized and the geo-mineralogy division dismantled. With uranium prospecting no longer at the top of the CNEN's agenda, a geo-mineralogy laboratory was established in the CNEN center La Casaccia (near Rome) and a nuclear geology laboratory at the University of Pisa. These research centers broadened the set of applications emerging from earlier nuclear geology work without prioritizing new uranium search methods.[84] Nuclear geology was now described as "a very new field" uniting geologists, petrographers, cosmic ray researchers, nuclear physicists, meteorologists, oceanographers, and volcanologists.[85] Similarly, in France and Spain, radiometric techniques developed from uranium prospection became accepted practice in other pursuits, leading to the rejuvenation of radioactive geology after its initial appearance in the 1930s. In Spain, expertise and equipment in radioactive measurements had advanced dramatically since the first maps of radioactive areas of the 1920s and 1930s.[86] In France, *radiogéologie*, while still associated with mapping of uranium minerals, also branched out into the radiometry of all classes of rocks, the study of radioactivity in the environment, and the measurement of geological time.[87]

Conclusion: Open and Secret Ambitions

By the end of the 1960s, their 20-year efforts to acquire uranium had decisively shaped the nuclear programs of France, Italy, and Spain. As we have shown, these efforts were based on finding common interests, building networks of trust, and establishing or impeding the circulation of materials, technologies, and experts. European countries established alliances and relations of mutual dependency in a context marked by the global transition from uranium scarcity to surplus and by American dominance over the supply of enriched uranium.

All the while, uranium kept its status as a politically sensitive resource—and still does. Hecht has recently argued that the uranium market cannot be construed as simply the result of "market" forces: participants exchange values and rules alongside technologies and raw materials. However, the uranium market *seems* wholly dependent upon these forces mainly because the exchanges typifying the politics of uranium are often made invisible by some of the actors involved in its acquisition. We have seen here that for a long time uranium was not a marketable commodity due to its military implications, but rather a resource stealthily dealt and monitored by intelligence agents, prospecting experts and (especially in the early postwar years) black market traders. In the circumstances of the Cold War, the potential of uranium to fuel military atomic projects meant that negotiations and agreements on uranium trade between countries were often kept hidden, while intelligence agencies carefully monitored new schemes and trading activities.

What does this paper make visible? First and foremost that since the end of World War II, the US administration, through its intelligence and atomic energy agencies, sought to put uranium deposits worldwide under surveillance in order to control their distribution in the foreseeable future. The United States's extensive reconnaissance and information-gathering activities are telling: the United

States was determined to secure Western Europe's uranium supplies or, at the very least, limit uranium acquisition by rivals in the region, including Italy, Spain, and France.

In the early 1950s, the recognition that there was too much uranium internationally to allow for an entirely "prohibitionist" approach led the AEC to offer bilateral agreements with many countries in Western Europe and elsewhere. Propaganda concerning the liberal trade of atomic energy served to distract America's partners from the fact that these agreements left these projects dependent on the AEC's provisions. And it kept hidden the continuation of intelligence activities focusing on prospecting and mining activities in foreign countries combined with diplomatic pressure to "cooperate" with US collaborative schemes. As we have seen, Spain and Italy capitulated, renouncing alternative partnerships with Western European nations. Only France's plans for an independent program were well enough advanced to keep its project's autonomy, and even France eventually accepted a bilateral deal with the United States.

However, the AEC could not prevent the further expansion of the uranium market and the availability of natural uranium at cheaper prices worldwide. In fact, this expansion was exactly what enabled the French to find at least some measures to limit US influence in Europe, especially by offering natural uranium reactors to Spain. Such a strategy even allowed another nuclear power, Britain, to sell a power reactor to Italy. But while these dealings could take place at the periphery of the atomic energy industry, the AEC remained the hegemon at the center.

In any case, to address attempts to subvert the production and circulation regime that the AEC desired, the US administration shifted decisively toward monitoring mid-stream and downstream activities rather than outsourcing processes in the atomic energy field. This was especially in order to certify their genuine peaceful intentions. From the late 1950s US intelligence agencies conducted less surveillance on uranium deposits and focused instead on monitoring the circulation of know-how and equipment needed to design as well as test nuclear weapons.[88] The US administration also promoted the establishment and consolidation of international agencies, such as the IAEA, to set international controls.

European resistance to US hegemony required scientific and diplomatic dexterity. Scientist-diplomats like Otero, Goldschimdt, and Ippolito working in their own, transnational environments, represented not only their respective national programs, but also constructed transnational networks throughout. And the scientist-diplomats were not alone. Geologists, engineers, and prospectors had their own interchanges, via cross-border visits and professional conferences, and almost always mediated through national atomic agencies. Viewing this from the transnational perspective and keeping uranium acquisition and exchange at the center of the story illuminates the role of these agencies role as instigators and gatekeepers.

Perhaps the most telling sign of the transnational nature of uranium acquisition is found in the development of the sciences involved: a little over a decade into the pursuit, some geologists were discussing the possible meaning of a European uraniferous province, while others attempted to establish nuclear geology or radiogeology to redefine their scientific effort to prospect uranium and create a new disciplinary identity. Just as the two world wars have respectively been labeled the chemists' war and the physicists' war, the Cold War could be considered the geoscientists' war, as thousands of experts worldwide were employed to chart many of the earth's features, including mapping its uranium deposits. Nuclear technoscience thus required transnational "tectonic" movements setting in motion crews of experts to study these features. The history of the search for uranium thus yields a new understanding of the interplay between geophysics and geopolitics.

Notes

1. From 507 tons in 1945 to 31,564 in 1965. Nuclear Energy Agency (ed.), *The Red Book Retrospective. Fourty Years of Uranium Resources, Production and Demand in Perspective* (Paris: OECD, 2006 [1965]), 89–91.

2. On the role of these missions see: Michael Goodman, *Spying on the Polar Bear* (Stanford: Stanford University Press, 2007), chap. 1; Simone Turchetti, *The Pontecorvo Affair* (Chicago: University of Chicago Press, 2012), chap. 2. For global intelligence see: Simon Schaffer, Lissa Roberts, Kapil Raj and James Delbourgo (eds.), *The Brokered World: Go-Betweens and Global Intelligence, 1770–1820* (Sagamore Beach, MA: Science History, 2009).

3. On hybridization: John Krige, "Hybrid Knowledge: the transnational co-production of the gas centrifuge for uranium enrichment in the 1960s," *British Journal for the History of Science* 45:3 (2012): 337–358. On technopolitics: Gabrielle Hecht (ed.), *Entangled Geographies. Empire and Technopolitics in the Global Cold War* (Cambridge, MA: MIT Press, 2011).

4. Including: Archivio General de la Administración, Alcalá de Henares, Spain (hereafter: AGA), Commissariat à l'Energie Atomique, Fontenay-aux-Roses, France (CEA); Jean Orcel Collection, Muséum National d'Histoire Naturelle, Paris, France (MNHN); Archives of the American Institute of Physics (AIP) and US National Archives (NARA), College Park, Maryland; Archive of the Istituto Nazionale di Geofisica, Roma, Italy (ING); Papers of Edoardo Amaldi, Istituto di Fisica, Università La Sapienza, Roma, Italy (AMA).

5. Jonathan E. Helmreich, *Gathering Rare Ores: The Diplomacy of Uranium Acquisition, 1943–1954* (Princeton: Princeton University Press: 1986) and G. Hecht, *Being Nuclear. Africans and the Uranium Trade* (Cambridge, MA: MIT Press, 2012).

6. On artificial scarcity, see Timothy Mitchell, *Carbon Democracy. Political Power in the Age of Oil* (London: Verso, 2011).

7. John Krige, *American Hegemony and the Postwar Reconstruction of Science in Europe* (Cambridge, MA: MIT Press, 2006).

8. Soraya Boudia, Néstor Herran and Simone Turchetti (eds.), Special Issue on the Transnational History of Science in the *British Journal for the History of Science* 45: 3 (2012).

9. As discussed in Helmreich, *Gathering Rare Ores*, 230. For an overview on Italy's nuclear program see: Leopoldo Nuti, *La Sfida Nucleare. La Politica Estera Italiana e le Armi Atomiche* (Bologna: Il Mulino, 2007). On France: Bertrand Goldschmidt, *The Atomic Complex: A Worldwide Political History of Nuclear Energy* (La Grange Park, IL: American Nuclear Society, 1982); Maurice Vaïsse (ed.), *La France et l'atome* (Bruxelles: Bruylant, 1994); Gabrielle Hecht, *The Radiance of France* (Cambridge, MA: MIT Press, 1998). On Spain see: Javier Ordóñez and José M. Sánchez-Ron, "Nuclear Energy in Spain: from Hiroshima to the Sixties", in Paul Forman and José M. Sánchez-Ron, *National Military*

Establishments and the Advancement of Science and Technology (Dordrecht: Kluwer Academic, 1996), 185–213; Albert Presas, "Science on the Periphery. The Spanish Reception of Nuclear Energy: An Attempt at Modernity?" *Minerva* 43 (2005): 197–218.

10. 16 Tons in all. Comité Scientifique, *Procès Verbal*, March 18, 1946, CEA.

11. Comité de l'Energie Atomique, PV, June 7, 1949 and September 19, 1949, CEA. Discussion with Robert Terrill, US Embassy in Paris on Status of the French CEA, April 28, 1952, Box 47, RG 59, NARA. See also Goldschmidt, *The Atomic Complex*, 311.

12. Antoine Paucard, *La mine et les mineurs de l'uranium français* (Paris: Editions Thierry Parquet, 1994), Vol. 2, 5–7, 25, 62, 153. See also: "Lettres à Joliot-Curie pour l'organisation des recherches d'uranium, 1945" in CEA, Rapport d'activité, 1946–1947, 11, MNHN.

13. Rénunion d'étude, Gif-sur-Yvette, September 13, 1951, in Kowarski Collection, S IV B4 F3, AIP.

14. Henry Lowenhaupt, Review of the French Atomic Energy Development, July 25, 1946, Box 173, RG 77, NARA. On Lowenhaupt see: Jeffrey T. Richelson, *The Wizards of Langley: Inside CIA's Directorate of Science and Technology* (Cambridge, MA: Westview, 2001), 2–4. See also Turchetti's article in this volume.

15. Arneson to Terrill, June 27, 1950, Box 47, RG59, NARA. See also Goldschmidt, *The Atomic Complex*, 250–251.

16. Procès Verbal du Comité d'Energie Atomique [the CEA steering committee], September 27, 1950, CEA. Founded in 1953, SOMAREM stood for the Société Marocaine de Recherches et d'Etudes Minières.

17. Procès Verbal du Comité d'Energie Atomique, October 8, 1953, CEA. The United States also quashed a Franco-Swiss collaborative deal. Procès Verbal du Comité d'Energie Atomique, Compte-Rendu de la Réunion Franco-Suisse, January 23, 1953, DRI F4/22.66, CEA. See also Peter Hug, "La Génèse de la technologie nucléaire en Suisse," *Relations internationales* 68 (1991): 325–344.

18. Felice Ippolito, *Intervista sulla Ricerca Scientifica* (Bari: Laterza, 1978), 23; Mario Silvestri, *Il Costo della Menzogna* (Turin: Einaudi, 1968), 30–40.

19. Antonio Carbonell, "Nota sobre los minerales de uranio," *Revista Ejército* 72 (1946); José L. Hernando and Rafael Hernando, "Descubrimiento, explotación y tratamiento de los minerales de uranio en Sierra de Albarrana, El Cabril (Córdoba)," *Boletín de la Real Academia de Córdoba* 143 (2002): 162–178. A similar reliance on uranium reserves typified the Portuguese nuclear program. See Júlia Gaspar, "The Two Iberian Nuclear Programmes: Post-War Scientific Endeavours in a Comparative Approach (1948–1973)", 7th STEP Meeting, June 17–20, 2010, Galway (Ireland).

20. Rafael Caro et al. (eds.), *Historia Nuclear de España* (Madrid: Sociedad Nuclear Española, 1955), 30–37. See also Gianni Battimelli (ed.), *L'Istituto Nazionale di Fisica Nucleare* (Bari: Laterza, 2001), 59–65.

21. José M. Sánchez-Ron, "International Relations in Spanish Physics from 1900 to the Cold War," *Historical Studies in the Physical and Biological Sciences* 33:1 (2002): 3–31, on 21.

22. R. Segovia and María A. Vigón, "Sobre la construcción de contadores Geiger y circuitos electronicos asociados," *Anales de la Real Sociedad Española de Física y Química* 44A (1948): 686–689.

23. Procès Verbal du Comité d'Energie Atomique, May 8, 1952, CEA. See also: Ana Romero de Pablos and José M. Sánchez-Ron, *Energía nuclear en España: de la JEN al CIEMAT* (Madrid: CIEMAT, 2001), 76.

24. Albert Presas, "Science on the Periphery,": 206, 209–213; Sánchez-Ron, "International Relations in Spanish Physics," 23.
25. *Herald Tribune*, September 27, 1949 cit. in Romero and Sánchez, *Energía nuclear en España*, 101.
26. Otero Navascués, "Desarrollo histórico de la minería del uranio desde 1945," *Energía Nuclear* 13 (1960): 92–99.
27. JEN General Director to JEN President, October 20, 1956, Box 71/8814, AGA.
28. C. A. Wendel to G. Arneson, January 5, 1953, Secret, Box 502, RG 59, NARA. On Arneson see Richard Hewlett and Francis Duncan, *Atomic Shield. A History of the AEC, 1947–1952* (Berkeley, CA: University of California Press, 1990), 426.
29. Wendel to Arneson, March 5, 1953, Secret, Box 503, RG59, NARA. On Lurisia mines: Robert Nininger, *Minerals for Atomic Energy* (New York: D. Van Nostrand, 1954), 82.
30. CISE/CNR agreement, May 3, 1952, Box 49, ING.
31. Pierre Guillaumat to High Commissioner Francis Perrin, October 22, 1952, and Guillaumat to Ministry of Foreign Affairs, November 5, 1952, CEA.
32. Arneson to Eldridge Durbrow (US Embassy, Rome), Confidential, March 12, 1953, Box 502, RG 59, NARA.
33. By 1955 global uranium production had increased to 20k tons of uranium. Nuclear Energy Agency (ed.), *The Red Book Retrospective*, 91.
34. Uranium recoverable in ores, or "natural uranium," contains 0.7 percent ca. of the fissile isotope U-235. Enrichment separates it from U-238 increasing the U-235 ratio; hence "enriched uranium."
35. Collaborative agreements are listed in James F. Keeley, *A List of Bilateral Civilian Nuclear Co-operation Agreements*, 5 Vols. (available at: http://dspace.ucalgary .ca/bitstream/1880/47373/10/Treaty_List_Volume_04.pdf, accessed May 8, 2012).
36. Manuel Alía Medina, Informe visita EEUU, November 21, 1956, Box 71/8794, AGA.
37. CoEA, PV, November 8, 1951, CEA. On Guillaumat and oil see Veneer and Cantoni in this volume.
38. CEA, Rapport d'activité, 1951, 9. J. A. Sarcia et al., "Geology of Uranium Vein Deposits in France," *Proceedings of the Second International Conference on the Peaceful Uses of Atomic Energy* (Geneva: 1958), Vol. 2, 592–611.
39. Robert Bodu, *Les secrets des cuves d'attaque: 40 ans de traitement des minerais d'uranium* (Cogéma, 1994).
40. Gabrielle Hecht, *The Radiance of France*, 60–61. On the British case see Margaret Gowing, *Independence and Deterrence*, 2 Vols. (London: Palgrave Macmillan, 1974).
41. Goldschmidt to S. Eklund, 22 January 1952, DRI F4/22.66, CEA. See also: Goldschmidt, *The Atomic Complex*, 286; Benyamin Pinkus, "Atomic Power to Israel's Rescue: French-Israeli Nuclear Cooperation, 1949–1957," *Israel Studies* 7:1 (2002): 104–138.
42. Otero to Renou, August 22, 1956, Folder Extrait de FAR 2008–22–36 (1954–1961), CEA; Manuel Alía Medina, "Plan de trabajo para el año 1956 de la sección de prospección geológica," February 7, 1956, Box 71/8814, AGA (13) 4.13.
43. "Convenio España-Estados Unidos relativos a los usos civiles de la energía atómica (1957)," *Forum Atómico Español, Boletín Informativo* 22 (1966): 3–13.
44. Otero to Renou, August 22, 1956. Folder Extrait de FAR 2008–22–36 (1954–1961), CEA.
45. CISE plan, May 9, 1952, Box 49, ING. CNRN Report 1, 20, AMA.
46. Third CNRN Meeting (minutes), September 29, 1952, Box 49, ING.

47. F. Ippolito, "Stato presente delle ricerche di uranio e torio in Italia," *Energia Nucleare* 17 (1955): 479–489. On prospecting studies at Pisa see S. Bonatti, "La Scuola mineralogica pisana Antonio e Giovanni D'Achiardi", Maggio 29, 1952 (Rotary International, Club di Pisa).

48. Frederich G. Houtermans and E. Picciotto, *Primo Convegno sulla Geologia Nucleare* (Rome: CNRN, 1955); Henry Faul (ed.), *Nuclear Geology. A Symposium on Nuclear Phenomena in The Earth Sciences* (New York: John Wiley, 1954).

49. Ippolito, "Stato presente delle ricerche di uranio e torio in Italia," *Energia Nucleare* 17 (1955): 479–489. On intelligence see: J. Krige, "Atoms for Peace, Scientific Internationalism, and Scientific Intelligence," in J. Krige and Kai-Henrik Barth (eds.), *Global Power Knowledge: Science and Technology in International Affairs – Osiris* 21 (2006): 161–181.

50. J. Freeman, Embassy to State Department, January 3, 1955, Secret in Box 402, RG59, NARA.

51. Italy/US Agreement for Co-operation Concerning Civil Uses of Atomic Energy, July 28, 1955 (UN document n. 3382, UN Treaty Series, 1956), 236–245.

52. Otero Navascués, "Necesidades españolas de elementos combustibles," *Energía nuclear* 5 (1958): 18–34, on 34.

53. Interview of Goldschmidt in Caro, *Historia nuclear*: 24–26.

54. Jefe División de Materiales to JEN Dirección General, May 12, 1962, Box 71/4045, AGA (13) 4.08.

55. Demetrio Santana, "Informe sobre un viaje a Suecia," June 30, 1955, "Minería del uranio en Canadá" (no date), "Informe visita Portugal," June 20, 1956, Box 71/8792, AGA (13) 4.13. See also JEN (ed.), *Minerales radiactivos en España. Cartilla del Prospector* (Madrid, 1956).

56. Manuel Alía Medina, "El servicio de investigación geológica," *Energía Nuclear* 6 (1958): 4–16.

57. Mabile to Santana, July 18, 1956 in Folder Extrait de FAR 2008–22–36 (1954–1961) and Couture to Otero, August 7, 1959, CEA. Armando Durán, "Las relaciones internacionales y la colaboración internacional," in JEN, *Coloquios de Información Geológica y Minera de los Materiales Radiactivos* (Madrid, 1959) Box 8816, AGA (13) 4. 13.

58. P. Baggio, F. Ippolito, S. Lorenzoni, G. Marinelli, and F. Silvestro, "Occurrence of Uranium in Italy," *Energia Nucleare* 4:3 (1957): 196–198.

59. CNEN Divisione Geo-Mineraria, Promemoria per il Presidente del Consiglio, 1956, AMA.

60. Jacques Geffroy and J. A. Sarcia, "La notion de 'gite épithermal uranifère' et les problèmes qu'elle pose," in EAES, *Compte-rendu du Colloque de Géologie des Gisements de Minerais d'Uranium et Méthodes de Prospection, 9–11 May 1957* (Madrid: JEN, 1957): 159–179.

61. André Lenoble and Jacques Geffroy, "Province uranifère en Europe: place occupée par la France," in Ibidem: 273–298, on 293–294.

62. Goldschmidt, *The Atomic Complex*, 285.

63. Nuclear Energy Agency (ed.), *The Red Book Retrospective*, 91. In 1967, production recovered, but it returned to 1962 levels only in 1977.

64. John Krige, "The Peaceful Atom as Political Weapon: Euratom and American Foreign Policy in the Late 1950s," *Historical Studies in the Natural Sciences* 38:1 (2008): 5–44.

65. Agreement for Cooperation Concerning the Civil Uses of Atomic Energy. Signed at Washington, July 3, 1957, UN Document n. 4462 (UN Treaty Series, 1958).

66. J. E. Helmreich, "The United States and the Formation of EURATOM," *Diplomatic History* 15:3 (1991): 387–410, on 394.

67. CIA, "Development of Nuclear Capabilities by Fourth Countries: Likelihood and Consequences," NIE 100–2–58 (July 1, 1958), Secret (available at http://www.foia.cia.gov).

68. "Euratom could also construct a plant to produce the enriched uranium needed [. . .]. But there is now no doubt that our countries can obtain enriched uranium from the United States in the necessary quantities, and at the low published prices," Three Wise Men, "A Target for Euratom," 8.

69. Procès Verbal du Comité d'Energie Atomique, "Les relations étrangères du Commissariat à l'Energie Atomique," February 8, 1958, CEA. Goldschmidt, *The Atomic Complex*, 296.

70. Ibidem, 281–282.

71. "Les relations étrangères du Commissariat à l'Energie Atomique."

72. Paucard, *La mine et les mineurs de l'uranium français,* Vol. 4, 21.

73. Amasa S. Bishop, US Embassy in Paris to John A. Hall, AEC, March 1, 1957, Box 407, RG 59. On the submarine: Maurice Vaïsse, "Une filière sans issue," *Relations Internationales* 59 (1989): 331–345.

74. Deputy Director of the Mining Division, "Relations avec l'Espagne," September 27, 1960, Folder Extrait de FAR 2008–22–36 (1954–1961), CEA.

75. "Resumen de las Jornadas Hispano-Francesas de Energía Atómica", *Forum Atómico Español, Boletín Informativo* 9 (1963): 17–23, on 18–19.

76. Otero Navascués, "Necesidades españolas de elementos combustibles": 18–34.

77. Francesc X. Barca, "Nuclear Power for Catalonia: the Role of the Official Chamber of Industry of Barcelona, 1953–1962," *Minerva, 43* (2005): 163–181.

78. On JEN-AEC negotiations see Ana Romero, "Energía nuclear e industria en la España de mediados del siglo XX. Zorita, Santa María de Garoña y Vandellós I", in Néstor Herran and Xavier Roqué (eds.), *La física en la dictadura. Físicos, cultura y poder en España, 1939–1975* (Bellaterra: UAB, 2012), 45–64, on 52–56.

79. On Vandellós I, see Esther M. Sánchez Sánchez, *Rumbo al Sur: Francia y la España del desarrollo, 1958–1969* (Madrid: CSIC, 2006): 375–394. See also: Gaston Palewski, *Mémoirse d'action, 1924–1974* (Paris: Plon, 1988): 281.

80. "Materiali," *Energia Nucleare* 5:2 (1958): 69; "Materiali," *Energia Nucleare* 5:2 (1958): 69; "La SOMIREN riaprirà la miniera di Novazza," *Energia Nucleare* 17:9 (1970): 514.

81. "Materiali," *Energia Nucleare* 6:2 (1959): 52; "Materiali," *Energia Nucleare* 10:4 (1963): 222.

82. F. Ippolito, "Dieci Anni di Ricerca Uranifera in Italia," *Notiziario CNRN* 9:7 (1963): 22–33.

83. His father's firm had secured funding for some CNEN projects which led to an accusation of conflicting interests. The new director of the Divisione Geo-Mineraria was also put on trial and found guilty. Andrea Barberi, "A Maggio il Processo Ippolito. Dieci a Giudizio per lo Scandalo CNEN," *L'Unità* April 5, 1964.

84. Previously unexplored areas such as glaciology and volcanology benefitted from this shift in the research agenda in Pisa and Rome. See, for instance, E. Tongiorgi et al., "Deep Drilling at Base Roi Baudouin, Dronning Maud Land, Antarctica," *Journal of Glaciology* 4:31 (1962): 101–110; E. Locardi, "Uranium and Thorium in the Volcanic Processes," *Bulletin Volcanologique* 31:1 (1967): 235–260.

85. E. Tongiorgi (ed.), *Stable Isotopes in Oceanographic Studies and Paleotemperatures. Proceedings of the Spoleto School, 1965* (Pisa: CNR, 1965).

86. Néstor Herran, *Aguas, semillas y radiaciones. El Laboratorio de Radiactividad de la Universidad de Madrid, 1904–1929* (Madrid: CSIC, 2008); Aitor Anduaga, *Geofísica, economía y sociedad en la España contemporánea* (Madrid: CSIC, 2009).

87. The term "radiogéologie" first appears in the Comptes Rendus des séances de l'Académie des Sciences in 1955. The principal focus of French radiogeology was the Centre des Recherches Radiogéologiques (University of Nancy), directed by Marcel Roubault, former head of the CEA's Mining Division.
88. See Herran and Turchetti, in this volume.

Chapter 2

Underground and Underwater: Oil Security in France and Britain during the Cold War

Roberto Cantoni and Leucha Veneer

In the years following World War II, global demand for oil increased continually, and Western European governments pursued various political and diplomatic strategies to obtain hydrocarbons as further reserves were revealed across the world. The tensions of the Cold War increased national concerns over energy security yet further, and in this chapter we shall discuss some aspects of the particular strategies employed by two leading Western European administrations to gain at least some control over the sources of their supply. These strategies included maneuvers such as stockpiling, encouraging diversification of supply, and, when the opportunity arose, controlling access to resources on home soil and abroad. This control required the mobilization of state and commercial geological surveying to obtain "geostrategic intelligence," that is, gathering information on what oil and gas reserves could be found underground; finding out what others (whether enemies or allies, co-producers, or business rivals) already knew about these reserves; and what acquisition strategies they had put in place. Surveillance in terms of both geophysical exploration and intelligence-gathering was therefore an essential element of oil security, an element often neglected in the existing literature on the history of oil exploration.[1] Oil surveillance operations also produced conflicts between diplomats, firm managers, government officials, and geoscientists of different countries. As Robert Jervis more generally shows, the bolstering of energy security through surveillance activities by one administration made its neighbors feel less reassured about their own security.[2]

The Cold War may have, broadly speaking, divided the world in two, but within that division was a somewhat fragmented system of alliances. Using archival material, this chapter focuses in particular on the traditional imperial powers of France and Britain because the international postwar order had forced both states into positions of lesser powers than before, their commercial and diplomatic footprints becoming accordingly smaller. Therefore, their oil security circumstances appear unique.[3] Moreover, both nations, due to that imperial past, were accustomed to maintaining significant numbers of scientific and intelligence personnel abroad, especially in colonies and countries that had recently gained independence. Here we reveal that these energy security urgencies prompted French geoscientific personnel to start exploring Algeria at the end of World War II. In Britain at a slightly later date, following the embarrassment of the Abadan and Suez Crises,[4] the discovery of gas in the North Sea started a rush for oil there as Britain sought self-sufficiency in fulfilling its energy needs.

The Secret Struggle for Algerian Oil

When World War II ended, the French government was not the only Western European administration concerned about how to source oil to fuel the nation. However, its officials were more alert to the plans of other countries asserting a presence in the world's oil-rich regions. Cold War divides notwithstanding, the plans of their allies worried the French government officials more than the distant Russian bear, especially as British and US oil firms held a dominant position in the French metropolitan oil market. The French oil industry was the largest in Europe before the conflict, but after 1945 it had to completely rebuild. Furthermore, its main source of supply in the Middle East was lost, an element that forced the French to be more vigilant both in terms of finding new deposits and controlling what other countries were doing.[5]

France's largest oil company, the Compagnie française des pétroles (CFP), had maintained a presence in the Middle East and owned shares in the Iraq Petroleum Company (IPC), a consortium that also included the Anglo-Dutch Royal Dutch Shell, the Anglo-Iranian Oil Company (controlled by the British government), and five American companies. But at the end of the conflict, British and US oil majors challenged the so-called Red Line agreement that ensured equal IPC shares within the territories of the former Ottoman Empire (see Figure 2.1). The dispute revealed to French government officials that the US administration had allowed American oil companies to challenge the status quo in the extraction and distribution of oil supplies from the Middle East. The war had quite evidently established new balances of power in the region. Moreover, the US companies— Standard Oil of New Jersey and Standard Oil of New York—now wanted to take a larger share of Saudi oil; in order to do so, they had to rid themselves of the constraints of the Red Line. Anglo-Iranian and Shell were willing to appease the Americans and did nothing to stop the Arabian-American Oil Company (ARAMCO), which was not originally part of the IPC and held concessions in Saudi Arabia, from "crossing" the Red Line. So in the spring of 1946, Jersey Standard publicly declared that it considered the agreement to have lapsed, owing to France's wartime status as an enemy power during the period of the Vichy regime.[6] During an autumn visit to Europe, representatives of the American oil majors further bolstered the US position by maintaining that the Sherman Antitrust Act forbade them to respect the restrictive provisions of the agreement, which would amount to cartelization.[7]

The secret deals between Americans and British convinced the French that they ought to reinforce their information-gathering activities on foreign oil agreements. At a CFP board meeting in December 1946, Director René de Montaigu confirmed press rumors that Jersey and New York Standards had acquired shares in ARAMCO, while President Victor de Metz informed the board of a new agreement between Anglo-Iranian and the two Standards.[8] The

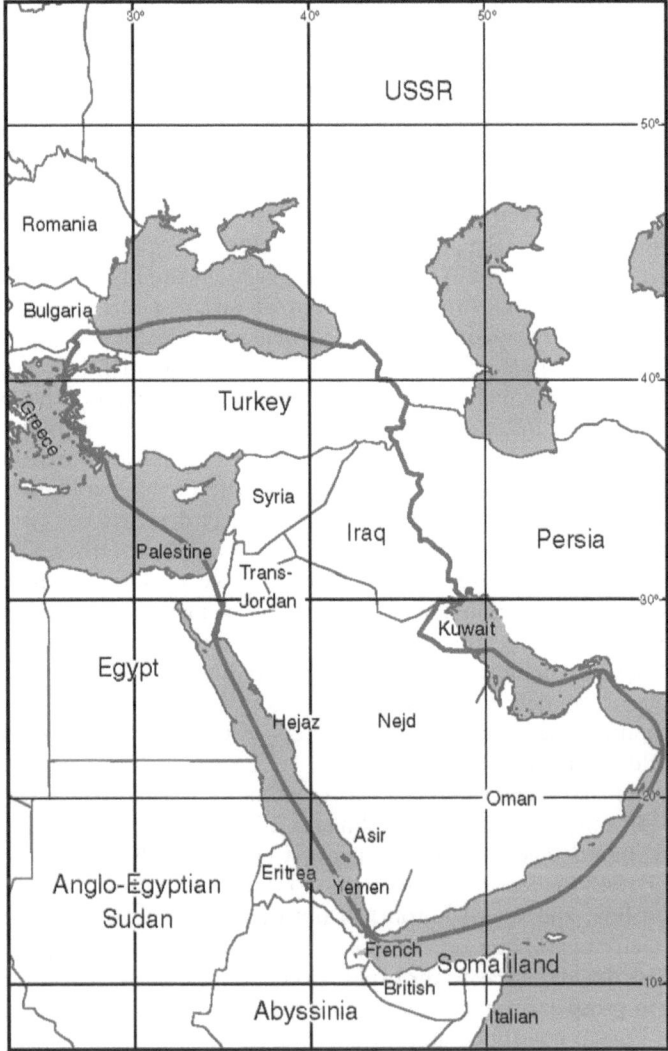

Figure 2.1 The area included in the Red Line Agreement. Map by Hans van der Maarel, Red Geographics

deal reassured British and US governments about controlling oil provisions in the Middle East at the expense of French energy security. During the next 20 years, the British government–controlled AIOC sold Jersey 160 million tons (Mt) of crude, and the two companies jointly built a pipeline linking the Persian Gulf to the Mediterranean.[9] The British used it for their Kuwaiti crude, which they exploited on a 50–50 basis with Gulf (another US firm). Shell's neutrality was also acquired through a very favorable contract, allowing the company free access to Kuwaiti oil: from May 1947, Gulf provided 30 percent of Shell's crude oil requirements in the Eastern Hemisphere.[10]

Naturally, the CFP managers understood that American and British governments had done nothing to prevent this situation (in fact, they had encouraged it), but it was especially the extension of British and US oil interests in Northern Africa that the French could not stomach. With the Treaty of Paris of 1947, Libya ceased to be an Italian colony: two of its three regions, Cyrenaica and Tripolitania, passed under British military control while the third, Fezzan, was controlled by the French. According to journalist Pierre Fontaine, a "secret battle" now ensued between British and French administrations to draw the borders between Cyrenaica and the potentially oil-rich Fezzan region. An army of scientific experts was mobilized to find out how much oil could be sourced out from this region and what the British were after. Thanks to pioneering prospecting activities carried out by geologist Conrad Kilian in the 1920s and 1930s, French oil authorities were in fact fully aware of the region's potential.[11] In 1942 Kilian had even been approached by ARAMCO and Shell consultants, who lured him into revealing them his Saharan secrets, but he had rejected their offers.[12]

Kilian now considered it imperative that the CFP establish a partnership with US oil concerns in prospecting work. In 1947 the French geologist met Jean Bédier, director of the powerful Banque de Paris et des Pays-Bas, and showed him his findings. With the help of the former director of the Office National des combustibles liquides, Bédier thus submitted a project calling for the establishment of a French-American consortium for the exploration and exploitation of Fezzan. French officials now understood that establishing a consortium would mollify American diplomacy in the dispute with the British for the definition of Libya's internal borders and encouraged the deal.[13] However, at the last minute, Kilian withdrew his support and the secrets of Fezzan's underground would remain hidden for some time.

Kilian had also urged the French government to prospect Algeria for hydrocarbons. His observations on the sediments of the Hoggar area, in the Sahara, suggested the existence of geological conditions apt to the presence of oil. In November 1948 his report was passed on to the French Académie des Sciences, sealed in a box, and stayed untouched until the geologist's death three years later.[14] Kilian's hypotheses on the area turned out to be only partially accurate, but his work paved the way to the exploration of the Sahara desert and instigated the CFP to prospect more.[15] The dispute between British and French governments on Fezzan continued and, by the end of the 1940s, it was informed by other factors. The Soviets sought to support Arab nationalism in the region, partly in an effort to gain control of oil resources. US and British oil companies now stealthily mobilized their own scientific monitoring networks to search for new oil fields out of Soviet and French reach. To this effect, the British even employed a geologist-turned-diplomat and military intelligence agent, Francis James Rennell Rodd. A specialist in the study of oil-bearing structures of the Fezzan region and a collaborator to the International Geological Map of Africa, Rodd exploited his knowledge of local territories and local elites in an attempt to gain control of its yet untapped oil deposits.[16]

Meanwhile, the information that the French had secretly acquired on the dealings of the Standards and ARAMCO in the Middle East was passed on to their government's legal departments with a view to launching proceedings in international courts. In 1946 de Metz urged that CFP's board take legal action at the British High Court of Justice against the American consortium's decision to terminate the Red Line agreement.[17] Two governmental commissars, the fuel director and president of the Bureau de recherches de pétrole (BRP), Pierre Guillaumat, and the president of the Bank of Algeria, Jacques Brunet, supported de Metz's proposition. It is important to note that during World War II, Guillaumat had worked together with the French intelligence in Tunisia and Algeria. Like Rodd, he had gathered information on foreign territories, their resources, and secretly managed details on oil deposits. He would play a decisive role in reorganizing the oil prospecting sector using this knowledge. Having to face the Anglo-American dealings in the Middle East excluding France, Guillaumat retaliated by seeking to reduce the influence of British and American oil interests in the French *Métropole*. He also instructed the French ambassador in Washington to deliver a letter of formal protest against the US denunciation of the Red Line agreement.[18]

On the eve of the first court hearing on this agreement in London, a Shell representative, John Boyle, proposed a compromise to de Metz and de Montaigu. The IPC would supply CFP as its managers wished, and a new pipeline would be built from Kirkuk to the Mediterranean Sea. The counterproposal was accepted but de Metz only agreed to postpone legal proceedings. The following February, the French filed the court petition again, hoping now to force the Americans to reopen negotiations.[19] In March, the French ambassador received a reply from George Marshall, the US acting secretary of state, so the legal challenge at least helped the French to force the Americans to compromise. By the end of May 1947, a settlement was reached. The CFP would withdraw its objections in exchange for an increase in its share of IPC production, which would also be increased considerably to accommodate French oil demand.[20] The Heads of Agreement were signed by all the major IPC partners a year later.[21] British and Americans diplomats thus realized that the French government was prepared to make use of its experts, intelligence agents, and lawyers in defense of oil security, or in order to force them to renegotiate existing agreements to increase French oil supplies in the wake of the Cold War. However, more security concerns soon arose in Paris and soon forced the French to reconsider their position in the Middle East consortium with a plan to invest more in North African resources.

The Scramble for Oil in North Africa

The new IPC deal appeared to be short-lived. The superpowers' influence in the Middle East increased quickly and dramatically, making diplomatic and oil relations more volatile. In 1946, the Soviet government had urged Iran to start up an oil exploration company, though the Iranians had later cancelled the deal and struck a military agreement with the US government. The Arab–Israeli war of 1948 led to the permanent closure of one of IPC's terminals. And both Iraqi and Iranian officials sought to obtain 50–50 contracts from the oil majors modeled on the one conceded by ARAMCO to Saudi Arabia. The new arrangements would make the two contracting parties equal partners, thus ending the exploiter–exploited relation that characterized the previous contracts. The majors' refusal to agree on this request produced tensions and contributed to instability in the area. The French government now decided to partly disengage from the Middle East. The decision was taken as a consequence of the reorganization of the oil administration at government level, and also because of the presence of Anglo-American interests within CFP; only 35 percent of its shares now belonged to the French state.

The French provisional government led by Charles de Gaulle instigated the foundation of new public agencies responsible for exploring for oil in the French Union, as well as for the technical development of French know-how in oil exploration. It then urged these agencies to shift exploration from the Middle East to Africa, especially the Sahara desert and the Guinea Gulf. Threatening French oil security in the Middle East thus stimulated sweeping surveys in the French colonies. The creation of the oil exploration agency, Bureau de recherches de pétrole (BRP), and the presence of former intelligence agent Guillaumat at its helm, marked the beginning of a new era in the history of French oil. BRP managers came from similar educational backgrounds as most of them had been trained at the Parisian military academy, École Polytechnique, and at the Corps des Mines, institutes that offered the most prestigious technical and engineering training in France. Soon a small group of experts, characterized by strong personal and political links, took control of the French oil agencies. Many *corpsards*—like Guillaumat—had entered the cadres of the French intelligence services during the war and now engaged in bringing together intelligence and geological expertise. Guillaumat had also kept many informants in North Africa. André Rauscher, a Shell engineer in Tunisia and a fellow *polytechnicien*, had helped Guillaumat by spying on the Italian army's prospecting activities in Libya. Pierre Taranger of the Compagnie générale de géophysique (CGG), and Léon Kaplan of Shell were also close to Guillaumat and directly involved in French oil exploration.[22]

The French began to explore North Africa by using all techniques made available by CGG, especially through a grand gravimetry reconnaissance campaign in 1948, and through seismology from 1951. However, gravimetry was slow and its interpretation in the region proved hard; reflection seismology produced deceptive results, while photogeology could not be applied outside the Saharan Atlas Mountains, where the mass of Mesozoic layers hindered surface geology. Because of these problems, the chief geologist of BRP's Algerian affiliate SN REPAL, Igor Ortynski, suggested that CGG apply seismic refraction, a method that had been out of fashion for decades, but which seemed to be more suitable to the geological characteristics of the Sahara desert.[23] From 1952 CGG started a new campaign that produced more successful results. Besides eliminating the problem of multiple reflections, refraction seismology allowed penetrating younger geological layers and forming a picture of deeper layers.[24]

The accumulation of this knowledge on local underground resources helped the French to focus on specific areas to explore. Those in the French Union—as the area formerly included in the French Empire was called from 1946—could now be used without previous negotiations or the establishment of new consortia.[25] The union was French territory and could therefore be treated as a private ground. In the Algerian Sahara, the BRP sponsored novel explorations through SN REPAL and with the collaboration of CFP. French experts thus gained a refined understanding of the geology of the Saharan region, which would soon prove of capital importance in the oil discoveries that took place between 1952 and 1954.

US and British oil concerns also developed an interest in Algerian oil and tried to get their share of territory to explore. Guillaumat soon realized that Kilian's early findings and recently acquired geophysical knowledge should not be divulged, but this was not enough to grant safe and quick oil supplies from Algeria. The French needed to prospect more if they wanted energy autonomy within a few years. They were therefore faced with two main options: let the Anglo-American enterprises into the Sahara, and gain in efficiency, financial backing, and technological knowledge; or continue their path independently at the risk of having to carry a colossal prospecting burden over many years. This second choice would be a dangerous scheme for a country heavily struggling with inflation. In June 1947, the French director of general affairs, Pierre Maisonneuve, organized a conference at the Under-Directorate of Algeria to discuss the prospecting plans of three foreign companies (Caltex, Gulf, and AIOC) that had shown an interest in Algeria.[26] Several ministries and oil company representatives (including that of SN REPAL) attended. Guillaumat argued that collaboration with foreign companies would be extremely profitable to the French economy, due to the shaky state of French finances.[27] But his proposal to develop joint participations with these companies was met with resistance. The representative of the Algerian governor general, Henri Urbani, challenged Guillaumat's favorable attitude toward foreign companies, expressing his serious concern over a too permissive stance.[28] What worried Urbani was especially lack of knowledge of what US crews were doing in Algeria and how much they knew about French operations:

First of all, every day we see Americans coming back and forth to Algeria. We don't know much about what they come to do, but what we do know is that they are interested in oil. [...] Once we will have given the Americans exploration permits, we will see them arriving in Algeria *en masse* and, from that moment on, what will their action be in the country?[29]

Urbani's reservations were understandable: in June 1947 France was still very weak, both financially and politically, whereas the role of the United States as a superpower had been made clear by the enunciation of the Truman Doctrine only three months earlier. There was little doubt that France would be forced to give in if the Americans decided to deploy all their influence in North Africa; even more so if they opted for a major prospecting effort, something that the French could not match. Urbani's point was thus that the French ought not to make concessions in order to have the upper hand in the region. Guillaumat, however, disagreed. He believed that a few American companies could work outside the United States with the same proficiency they had at home. Furthermore, he was not at all convinced that such frantic foreign activity had taken place from 1942 to 1945 in Algeria. As a former intelligence officer, and thanks to his relations with people such as Taranger, Guillaumat had access to restricted information that Urbani simply lacked.[30] Lucien Bonneau, plenipotentiary minister and director of Africa and Near East at the Foreign Ministry, also downplayed the extent of American influence in North Africa but showed wariness. If the

Americans were determined to access North Africa, they would use their powerful transport or radio companies—and, undoubtedly, their secret services.

The notable difference of viewpoints between Urbani and Guillaumat was also discussed at SN REPAL's board meetings, where it was concluded that foreign assistance was needed only in the supply of materials and specialized drilling staff. SN REPAL took the lead in general operations, and only collaborated with foreign enterprises in associations where it held a majority of shares.[31] Guillaumat succeeded in convincing his colleagues that collaboration in Algeria would not threaten French interests in the region. Was he really aware of what the British and Americans were really doing? While the French had been reconnoitering the Algerian underground, the Americans had been busy reconnoitering it from the sky.

British and American Attempts to Enter Algeria

The surveillance of potential oil-bearing areas in Algeria was a decisive element in establishing whether or not the oil majors would try to enter Algeria and whether the French could stop them. Guillaumat knew that during World War II the US Air Force had taken aerial photographs in Algeria and that the photos contained details of geological structures that revealed the presence of oil deposits. After the war, the French government had agreed with US diplomats that photographic material ought not to be shared without prior French consent. On the other hand, the French did not hold copies of these photographs either.[32] In 1947 French officials authorized Jersey Standard experts to see the photographs and the following year an agreement signed at the French embassy in Washington enabled the French Air Force's Chiefs of Staff to obtain copies of the photos. It is likely that SN REPAL obtained access to these copies as well.[33]

Caltex, Gulf, Anglo-Iranian, and Shell had been attracted by the French exploratory activities in Algerian territories. Should oil be found, these companies were ready to wield the power they enjoyed as a result of the war in major exploratory campaigns. Yet, as they did not have access to the photos, they could not know enough about the real Algerian potential. So before setting foot in Algeria, they used their lawyers to sound out French reactions, and to determine from these reactions if the French had found oil deposits. In September 1947, one of Gulf's lawyers sent a letter to Yves Chataigneau, governor general of Algeria, through the French embassy in Washington.[34] After that, Gulf received useful data on Algerian geology through the French embassy and decided to begin large-scale works, provided the French government agreed. Gulf representatives now approached the officials of the Direction des carburants and BRP in Paris, and met Guillaumat and BRP's delegate-general, Paul Moch. Gulf was ready to carry out prospecting works for over $1 million, including surface geological and seismic works, and photogeology.[35] We have seen that while Guillaumat saw favorably the collaboration with foreign interests, he wanted to retain absolute control of geological data. But Gulf demanded instead a series of guarantees, including the availability to Gulf of documentation kept by the Mine Service, SN REPAL, the Hydrography Service, as far as geology and oil exploration was concerned. Thus, a conflict between different departments in the French government ensued and when Gulf applied for an exploration license, it was refused.[36]

The same strenuous opposition, however, did not characterize SN REPAL's board, which welcomed a collaborative project with an affiliate of the US oil firm, Jersey. In 1947 Jersey had shown some interest for searching oil in Algeria and had obtained authorization to send a team of geologists and carry out a study on the oil potential of the Saharan region, provided an account of the team's activities was transmitted to SN REPAL. Jersey could even use the set of aerial photographs taken during the war by the US Air Force for its exploration. But the collaboration was short-lived. When the Americans realized that the only area deemed to have serious commercial oil possibilities was one that would be assigned to SN REPAL, they pulled out.[37]

One reason for Guillaumat to encourage collaboration with foreign enterprises in Algeria was that he hoped to gain some influence in oil exploration projects in other areas of the world. In October 1951 Shell, through its affiliate Shell Française, informed the new governor general of Algeria, Roger Léonard, of its intention to ask for a vast exploration permit. Guillaumat reckoned that room for Shell could be found in the Sahara if they accepted French participation in their exploration activities in Canada and Venezuela. Representatives of the Ministry of Finances and Economic Affairs stressed the difficulties of this solution, so eventually the French administration decided to participate in Shell's works, even without getting any substantial quid pro quo in terms of permits.[38]

That Guillaumat and his collaborators were willing to facilitate relations with foreign companies did not mean all the oil executives in French companies embraced his viewpoint, especially when yielding to those companies could jeopardize their primacy in the area. An episode regarding Shell and SN REPAL clearly shows this point. In February 1952 the president of SN REPAL, Roger Goetze, forwarded to Moch (the BRP delegate general) two letters to urge him to consider the consequences of allocating foreign companies permits in areas bordering those requested by CFP-SN REPAL (see Figure 2.2). Goetze stressed the existence of a clause contained in SN REPAL's permit allowing the company to prospect outside its permit zone. The French had requested their permits in August 1950, earlier than Shell, but these had not yet been awarded. Furthermore, since geological knowledge about the area was less detailed at the time that SN REPAL had requested its permits, Shell could now ask licenses for more promising areas. Goetze pointed out that in the light of the new geological data about the Saharan basin, it would be preferable to allow SN REPAL priority over Shell or other companies on unexplored areas. In order to prevent Shell from gaining uncontrolled access to the desired area, Goetze even proposed that SN REPAL take a financial stake in all companies engaged in the Sahara, especially in prospecting activities, so as not to miss any opportunity that might accrue.[39] Eventually, although the requests jointly made by Shell and by the Régie autonome des pétroles (RAP, a French public oil agency) were approved, SN REPAL received the governor's support to obtain a counselor seat with no financial stake, or a small stake (up to 5 percent) in the companies to be formed by Shell and RAP.[40]

The results of collaborative prospecting activities would eventually prove Guillaumat's strategy right, as they enabled the first major oil discoveries in Algeria. Up until the outbreak of the Algerian war in 1954, the Algerian Sahara remained firmly in French hands and provided France with the supply of oil it

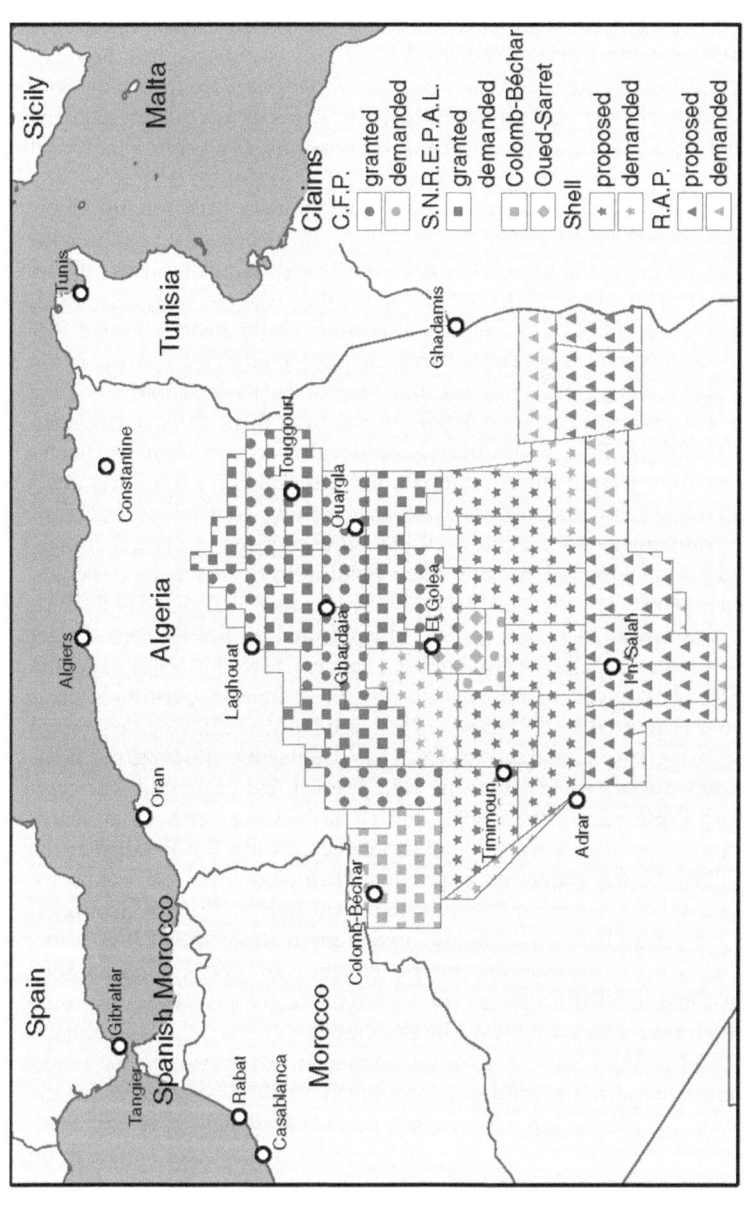

Figure 2.2 Assigned and demanded permits for oil exploration in the Sahara by 1952. Map by Hans van der Maarel, Red Geographics re-adapted from original map in Affaires algériennes (1873/1964), b. 81F/2060, fd. Recherche pétrolière, correspondence, 1949–55, ANOM–FM

badly needed to cope with national demands. In 1953, the reopening of Anglo-French negotiations on joint collaborative work led to the constitution of two companies, the Compagnie de recherche et d'exploitation de pétrole au Sahara (65 percent RAP; 35 percent Shell), and the Compagnie des pétroles d'Algérie (65 percent Shell; 35 percent RAP).[41] In 1954 the first of these companies discovered the first gas field of commercial value in the area, and two years later also discovered the Edjeleh oilfield.

Thanks to the geophysical knowledge put compiled during the past 10 years, the French administration could thus address its energy security needs. SN REPAL and BRP discovered in 1956 the two largest Algerian oil and gas fields, Hassi Messaoud and Hassi R'Mel, and kept control of them. Since the main American concern regarding the conflict was to keep the Soviets out of North Africa, the early Cold War tensions on oil supplies between American and British oil companies and the France administration relaxed somewhat. However, this situation was complicated by events in the late 1950s, principally the conflict for Algerian independence. While this chapter does not discuss the impact of this conflict on French energy security in detail, it is worth mentioning that it partly upset the system of selective collaborations with foreign concerns that Guillaumat had put in place. In particular, both Italian and US firms now used the conflict as a lever to gain more influence in the exploitation of Algerian hydrocarbons. Meanwhile, the French and the British joined forces in the Suez Canal crisis, thus overcoming their traditional enmity in oil affairs. This leads us to consider another case, that of the North Sea, where the gathering of knowledge on oil and gas fields was decisive in shaping the relations between another former imperial power, Britain, and its neighboring countries.

The Race for the North Sea

The British case presents various points of contrast and comparison with the French one, especially following the discovery of gas in the North Sea in 1959 (with the concomitant realization of the long-held expectation that oil would also be found). The oil disputes in the Middle East and North Africa between key Cold War allies highlighted to the British government the urgency of reaching agreements on oil security through supranational organizations formed along the political fault lines drawn by the Cold War. For instance, from the early 1950s Britain encouraged NATO to take oil stockpiling seriously for both military and energy security purposes, and mustered support for a Petroleum Planning Committee to examine supply lines and storage of military and civilian oil across the alliance. The committee was even made responsible for developing strategies to protect these supplies in the event of World War III.[42]

Another element producing anxiety in the British government was the diplomatic victory of Egyptian president Gamal Abdel Nasser in the Suez Crisis. This was bad for British oil security, but even worse for the French colonial interests, as, so the French claimed, Nasser was acting as a proxy between the Soviets and the Algerians, and was the main channel through which Algerian independence fighters received weapons. The outcome of the Suez expedition strengthened the Algerian liberation movement and made Nasser the champion of Arab nationalism. Britain now found itself isolated in its quest to secure more oil. Similar to the experience of the administrators at the French BRP, their British colleagues at the Ministry of Power understood that the solution to the oil supply problem rested with the ability to know more about oil deposits underground. But in contrast with the French, the British found that the solution to their problems was much closer to home and required no intervention in colonies or former colonies.

Although underlying security concerns in Britain were similar to those in France, the diplomatic and scientific pressures were different, and so were the strategies and fuel policies that the Ministry of Power adopted in response. The most important was to assemble essential information regarding North Sea gas fields so as to quickly establish control on its output through licensing. This was imperative given that Britain had no state oil company, and was therefore reliant on commercial enterprise. The British government held a controlling stake in AIOC, but the company had been greatly changed by the Iranian government's decision to nationalize the Iranian oil industry in 1951, which had led to the Abadan crisis. AIOC became the British Petroleum Company (BP) in 1954.[43] Before the discovery of gas in the North Sea, the British government encouraged companies such as BP, Shell, and Esso to stockpile crude oil and diversify their sources of supply as much as possible.[44] Nationalization never became part of the ministry's strategy. While the UK gas industry had been nationalized in 1948, no state oil company was founded, even in the wake of the North Sea discoveries, for fear it would damage existing relations with the major oil companies.[45] Nor were these arguments ever simply about business methods and oil revenues: in the international Cold War context, there was the ever-looming possibility that Soviet military action might affect both civilian and military oil supplies from the Middle East, and the great risk inherent in British dependence on imported oil became starkly clear as domestic demand continued to rise.

In the early 1960s, the British government felt the potential threat to its supplies was serious and quickly put into action a policy that not merely encouraged but actively forced rapid exploitation of North Sea oil through commercial and scientific avenues. Although oil and gas had been found just before World War II, from 1959 the location of the Slochteren gas field in the Groningen province produced a rush toward prospecting and finding more hydrocarbons.[46] The UK Parliament soon passed a bill claiming sovereignty over all submarine resources on its continental shelf and in 1962 BP established the first offshore prospecting site in Weymouth Bay. However, BP experts knew that sites in British waters had limited prospects, whereas the North Sea basin (which geologically resembled northern Netherlands) held much more promise.[47] Once the surveys—geological sampling, gravity, and seismic—revealed that the richest field were located on the Norwegian shelf, the British government sought to prevent other European countries from staking a claim on these fields by quickly reaching an agreement with Norway and licensing exploitation soon afterward (see Figure 2.3).

The agreement, signed in 1964, was far more advantageous to the Norwegian administration since it would increase the amount of tax revenues from oil and gas extraction, but would enable oil companies supplying Britain to gain permits to pump gas out quickly and virtually unchallenged, offering "oil security" for the foreseeable future. Agreement was rapidly reached with Norway over the division of all waters too far north for continental European countries to

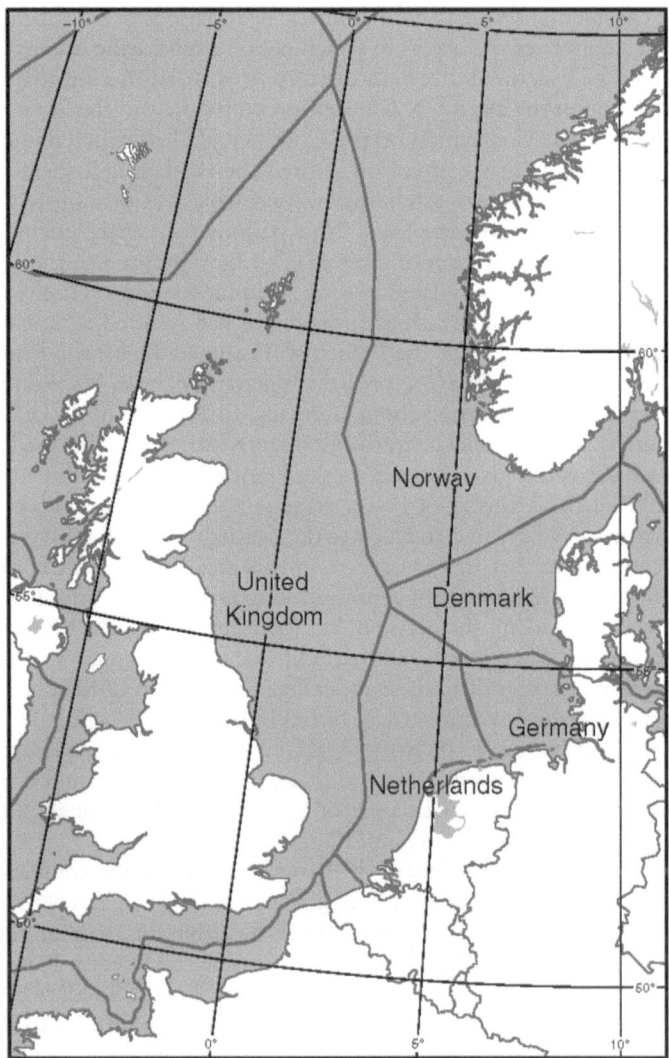

Figure 2.3 International Boundaries in the North Sea. Map by Hans van der Maarel, Red Geographics

claim them,[48] Britain agreeing to a slight reduction in territory in order to begin extraction at once and forestall potential rivals. Exploration licenses were issued already in 1964, though a final agreement on the division of the North Sea was not concluded until 1972. By 1967, when minister of power Richard Marsh reported to the House of Commons on fuel policy, 54 exploration wells had been drilled in the North Sea by at least 15 different companies.[49]

The deal with Norway compelled British authorities to constantly monitor the prospecting efforts of other countries and firms in the North Sea. The UK and Norway were, naturally, not the only countries hurrying to exploit resources on their continental shelves, and diplomatic negotiations over which companies from which countries had access to which nation's submarine resources became complicated and required a certain delicacy. It is worth noting that following the completion of the 1958 UN Convention on the Law of the Sea a new international treaty, the Convention on the Continental Shelf, entered into force on June 10, 1964. The new convention ratified the right of individual nations to exploit portions of the sea extending beyond what was previously considered international waters and compelled these nations to prepare documents providing a geological definition of their shelf to be submitted to the UN.[50] The impending convention stimulated a secret scramble for new oil licenses.

Although French and British administration had restored amicable relations on oil matters by the 1960s, the French still competed with the British in the North Sea and kept a close eye on British activities there. Ministry of Power officials were far more worried about Germany. In 1963 BP was excluded from a consortium that was to operate in the German North Sea and learned that West Germany had instead given licenses to the French BRP. The following year, BP requested that the Ministry of Power arrange a diplomatic approach to enable BP's German subsidiary to gain access to the German North Sea. BP was excluded from a consortium that was to operate in this area because, the company was told, their interest had not been expressed soon enough. BP officials were reluctantly prepared to accept the decision, but then learned that the French BRP was admitted after BP was refused, and that BRP had expressed interest later than BP had. Ministry of Power officials suggested to the Foreign Office that if BP was discriminated against, reciprocal actions might be taken regarding the issuing of licenses for British areas of the North Sea. In the end, BP became resigned to the situation when the Germans promised that the consortium would operate only in the German North Sea, with areas of German continental shelf in the Baltic remaining open to other companies, BP included.[51]

The attitudes of both Ministry of Power officials and the companies themselves could change very quickly though, especially when restricted information on other countries' intentions was made available. In 1966 another area of continental shelf came under negotiation, this time the Swedish Baltic, and the Ministry of Power learned from the Foreign Office that the Swedish government was believed to be forming a consortium of Swedish companies to explore its own waters. Though the Swedes would have to import equipment and expertise, the British government—and BP—felt the formation of the consortium indicated that foreign companies would probably not be permitted direct access.[52] Fears also lingered concerning the Soviet Union potentially gaining access to Swedish Baltic oil fields. Though the earlier case had not resulted in either French or German companies being treated differently by the minister of power, Frederick Lee, when he allocated licenses for the North Sea, the possibility of reciprocal action was now raised again. Since British companies had no urgent need to gain access to the Swedish continental shelf, British diplomats eventually agreed that there was no reason to openly challenge the decision. It seemed better to "adopt a liberal attitude ourselves and then take what credit we can for it rather than threaten reciprocal restrictions," as one ministry official put it.[53] Furthermore, there was another potential security issue with the Swedish Baltic Sea: if an open invitation to Western companies to prospect was demanded by

Western governments, it would not be unreasonable to suppose that the USSR would begin to take an interest as well.

Not unlike the Algerian case, the accumulation and circulation of new geophysical knowledge on potential oil deposits played a key role in disputes on the North Sea licenses. The first set of licenses for British North Sea oil explorations had been issued by the minister of power between 1964 and 1965 and expired in 1970. By 1967 some companies had been gathering more accurate geological data for two years, and all of them would continue to do so for another three. Ministry officials now realized that they would not be able to make sound evaluations on the reissuing of those licenses in the next round unless they possessed the same information,[54] and since the urgency to explore the North Sea as quickly as possible remained, those licenses would have to be immediately reissued. Although the British government did not have an equivalent of the BRP, it too compelled the companies to which it had issued licenses to share data with the Ministry of Power, so that information could be made readily available to its officials. The lack of scientific capacity within the Ministry of Power to process the raw data created anxiety. The licensed oil firms would share the data, but not their scientific interpretations, thus placing the Ministry at a clear disadvantage when evaluating the new round of licenses.

The Ministry of Power therefore turned to the newly established Institute of Geological Sciences (IGS, formerly the British Geological Survey) and its new director, Kingsley Charles Dunham.[55] Although recently appointed, Dunham had considerable academic and commercial experience, having served as professor of geology at the University of Durham since 1950 and consulted for several mining and oil companies, including Iraq Petroleum in Oman and South Yemen.[56] One specific duty that made him eminently suitable for the tasks ahead was that he had served on the committee chaired by Sir Frederick Brundrett in 1963, which, along with the Trend report on the reorganization of government science, had recommended the formation of the IGS under the new research councils in 1965. Unlike Guillaumat, Dunham was never involved in intelligence work. But serving in the Brundrett committee exposed him to the management style of one of Britain's pioneers of scientific intelligence. During the war Brundrett had developed the Royal Naval Scientific Service, subsequently rising to become head of the Atomic Energy Intelligence Committee and then science adviser to the Ministry of Defence.[57] His connections with both scientific policy and the intelligence services were thus well established, and his influence had not waned with his retirement in 1959; indeed he was still responsible for the Defence Research Policy Committee, the chief planning organization for defense research in post-1945 Britain.[58]

Brundrett's policy review recommended that geological work previously carried out separately by the Overseas Geological Survey and the geological survey unit of the Atomic Energy Division be centralized under the IGS.[59] It also more clearly directed IGS work toward oil security needs. Following a recommendation from fisheries scientist Ray Beverton, assistant secretary of the newly established Natural Environment Research Council (NERC), Dunham agreed to establish a Mineral Resources Consultative Committee (MRCC) under the aegis of the Department of Education and Science (DES). This enabled information (and concerns about data acquisition) to be shared between various government agencies, including the DES, the Ministry of Power, and the Ministry of Technology.[60] As director of the IGS, Dunham was therefore aware of the overall implications of controlling information on natural resources, especially oil, and not merely the geological details. He was willing to meet the Ministry of Power's demands but also cautious: he did not have sufficient staff; they did not yet have the data they needed from the operators; and the time available would not be enough.[61]

The ministry agreed to Dunham's request for funds, but by mid-1968 the ministry officials were concerned that the IGS would not have any material to work with, as the information they needed would only really begin to flow in early 1969.[62] The key problem for Dunham was that North Sea operators were being deliberately sluggish in sending their data, and there was little that either the ministry or the IGS could do about it, especially as the physical well core samples were large, unwieldy items requiring careful handling and storage. The ministry needed at least some information, and as quickly as possible. Most of the commercial operators were willing to assist, at least in principle, and the UK North Sea Operators' Committee recommended to its members that they cooperate with any approaches for data from the IGS.[63] On the one hand, they were legally obliged under the licenses to provide at least their raw data, and since the ministry issued the licenses, they could not afford to appear uncooperative. On the other hand, some of the operators knew that the Ministry of Power, following the IGS's advice, would use their data to advance the interest of the state rather than that of the oil firms. Elements within the governing Labour Party even pressed for the establishment of a National Hydrocarbon Corporation to centralize oil exploration and development, although Prime Minister Harold Wilson and his cabinet never followed through on the proposal.[64]

In these difficult circumstances, Dunham skillfully used the power and knowledge he had gained from the Brundrett committee. Firstly, he sought to reorient the existing geology programs so as to bring to the fore oil security matters. His predecessor at the Overseas Geological Survey had already outlined a survey of the British continental shelf. Now Dunham used NERC and Ministry of Power's support to put forward an ambitious plan to prospect the North Sea basins and correlate geophysical and geological results in order to gain greater knowledge of oil-bearing structures.[65] To counter the oil firms' sluggishness, the IGS began to carry out its own surveys. Continental Shelf Unit I (CSU I) was established at the Leeds office to survey the central North Sea and the Irish Sea, along with the Mineral Assessment Unit (MAU) in London to analyze the findings. Continental Shelf Unit II (CSU II) was formed in Edinburgh to survey the northern North Sea. The MAU worked closely with the MRCC and was heavily involved in assessing North Sea resources, though it also had broader strategic concerns. In the North Sea, the first task was to establish the stratigraphy, so both the immediate and the long-term program for the continental shelf work were quickly established, beginning with the "interpretation of data from commercial exploration of North Sea and Irish Sea" and "offshore geophysical and geological investigations in the northern Irish Sea, Humber Estuary and North Sea off Lincolnshire."[66] The IGS received about half the data the operators then held by the end of 1967, and the flow continued in 1968.

Although there was no nationalized oil company, the operators still feared that the ministry might use confidential data they had produced to justify issuing licenses to British Gas (which *was* a state-owned company), prompting ministry officials to issue reassurances.[67] The lack of a national company meant that from the state's perspective, the greatest threat to Britain's oil security appeared to be that it could not guarantee sufficient investments in geophysical research, a situation that contrasted with other Western European states. In case of conflict or due to sudden lack of supply, it was unclear if these companies would protect national or commercial interests. Dunham also had to face the consequence of a white paper that advocated a different allocation of funds to NERC and IGS, thus putting in jeopardy both the Continental Shelf Survey and the IGS Mineral Intelligence program.[68] Nevertheless, the threat of establishing an independent national survey unit ultimately convinced private oil firms operating in the North Sea to align their operations more closely to the chief imperative of the British government—to secure a constant supply of oil to Britain.

By 1971 a number of major oil fields had been discovered in the North Sea, and the rate of commercial exploration and discovery was rapidly increasing. A 110-mile marine pipeline was being planned from the Forties field to the coast near Peterhead, and Dunham reported that "official statements suggest that by the end of the decade a substantial proportion of the United Kingdom's oil requirements may be available from the North Sea," adding that this was "the most important geology-based development in Britain since the opening up of the coal-fields."[69] The pressure began to tell on the IGS, however, in the 1970s, as basic survey work suffered while projects associated with the North Sea became paramount—not only analysis of the operators' data, but also extended seabed surveys, pipeline projects, and growing transport links. By 1975, when Dunham left office, he had effectively put in place a comprehensive plan of explorations prioritizing British oil security by aligning oil firms to IGS plans, since—as Dunham stressed—"the fossil fuels come first."[70] Notwithstanding the oil crisis of 1973, Britain's concerns for oil security had been successfully addressed.

Conclusions

Oil supply was at the center of national security strategies in both Cold War Britain and France, due to the critical role of oil in the military-industrial complex and an ever-increasing domestic demand. Guided by similar fortunes, namely the degree of autonomy fostered by the discovery of oil reserves in nearby and colonial territories and finding their interests in the Middle East threatened by increasing Soviet and American influence, the French and British governments attempted to secure control of Algeria and the North Sea respectively. While the French had relatively little difficulty in achieving quasi-exclusive control of exploration in Algeria, a territory over which France had complete political control, the British strategy was made more complicated by the very nature of the area of exploration, bordered by a number of independent European countries with their own energy and security concerns.

In the late 1940s, French oil security was shattered both by ploys to evict CFP from the Middle East and by external political factors that France and CFP could not control. To rebuild their security, the French mobilized their army of exploration geophysicists and intelligence agents to know more about what could be found in the French Union and what other countries intended to do in these regions. They moved their focus to the French Union, in an attempt to guarantee national supplies from an area they could use as their own private ground. The possession and management of confidential geological information shaped the beginning of Algeria's oil era, and gave a marked advantage to French companies, which were able to use their practical monopoly rights for their exploration activities without having to worry too much about competitors. But foreign oil companies sought to establish their presence in the Sahara too and take advantage of their influence internationally. The repeated requests of British and US oil firms to this end caused a long-lasting quarrel between French institutional sectors, resulting in the strictly government-moderated entrance of Shell into Algeria, in a joint-venture with French public agencies, or in making limited concessions to gain collaborative deals in oil exploration ventures elsewhere. The rest of Algeria, however, would be safely in French hands. At least for a few years.

The British government, in contrast with the French, could count already on a fairly constant flow of oil from the Middle East. So it maintained an interest in entering other areas outside Britain, but this waned somewhat when the North Sea revealed to be the best source available of hydrocarbons. At this point the chief strategic urgency for the British minister of power became to administer North Sea licenses effectively, controlling the underlying knowledge that enabled private firms to extract the oil. Yet, geostrategic knowledge both in the form of geological studies and an understanding of what other countries wished to do on their own territory was equally important. British security concerns stimulated continual urgency of both exploration and policy development within the Ministry of Power, speeding exploitation of the North Sea by encouraging both the operators and the IGS to map, survey, and drill as much as possible as quickly as possible, to the extent that by the early 1970s the possibility of British near self-sufficiency was being mooted. In essence, The British government's system of supply was based on control of licenses and thus on geophysical knowledge that would allow preventing access to other nations and a mechanism of distribution to companies allowing to secure a regular flow of oil to home.

At the same time, oil exploration and geological knowledge had become another kind of intelligence for nations long accustomed to intelligence gathering, partially through the influence of officials such as Kinglsey Dunham and Guillaumat, who had backgrounds in the two sides of geostrategic intelligence, namely information on the underground resources and the plans of those nations who had an interest in them.

Notes

1. This is, for example, the case for: Daniel Yergin, *The Prize: The Epic Quest for Oil, Money, and Power* (New York: Simon & Schuster, 2009); André Nouschi, *La France et le pétrole: de 1924 à nos jours* (Paris: Picard, 2001); Jean Prouvost (ed.) *La recherche pétrolière française* (Paris: Éditions du CTHS, 1994).
2. Robert Jervis, "Cooperation Under the Security Dilemma," *World Politics*, 30:2 (1978): 167–214; R. Jervis, "Was the Cold War a Security Dilemma?" *Journal of Cold War Studies*, 3 (2001): 36–60.
3. Amongst other: the archives of the French oil firm TOTAL (Archives Historiques du Groupe TOTAL, AH TOTAL), the US National Archives and Records Administration (NARA), and the British National Archive (TNA).

4. James H. Bamberg, *The History of the British Petroleum Company, volume 2: the Anglo-Iranian Years, 1928–1954* (Cambridge: Cambridge University Press, 1994).

5. On the postwar reconstruction of France from a political and diplomatic point of view, see: Irwin M. Wall, *The United States and the Making of Postwar France, 1945–1954* (Cambridge: Cambridge University Press, 1991); William I. Hitchcock, *France Restored: Cold War Diplomacy and the Quest for Leadership in Europe, 1944–1954* (Chapel Hill/London: University of North Carolina Press, 1998); Jean-Pierre Rioux, *The Fourth Republic, 1944–1958* (Cambridge: Cambridge University Press, 1987).

6. CFP, Minutes of the Meeting of the Board of Directors (MBD, henceforth), June 5, 1946, pp. 2–3, TOTAL-CFP, b. 92.10/1, AH TOTAL.

7. CFP—Minutes of MBD— November 6, 1946, p. 3, TOTAL-CFP, b. 92.10/1, AH TOTAL.

8. CFP—Minutes of MBD— December 27, 1946, pp. 2–4, TOTAL-CFP, b. 92.10/1, AH TOTAL.

9. Irvine H. Anderson, *ARAMCO. The United States and Saudi Arabia. A Study of the Dynamics of Foreign Oil Policy, 1933–1950* (Princeton: Princeton University Press, 1981), 153–154. Also in: Nouschi, *La France et le pétrole*, 198. See also: A. Nouschi, "Un tournant de la politique pétrolière française: les *Heads of Agreement* de novembre 1948," *Relations Internationales* 44 (1985): 379–389; Burton I. Kaufman, *The Oil Cartel Case: A Documentary Study of Antitrust Activity in the Cold War Era* (Westport: Greenwood Press, 1978), 123–36.

10. Jean Rondot, *La Compagnie Française des Pétroles. Du franc-or au pétrole-franc* (Paris: Plon, 1962), 93–95; Michael B. Stoff, *Oil, War, and American Security. The Search for a National Policy on Foreign Oil, 1941–1947* (New Haven, CT: Yale University Press, 1980), 200.

11. Pierre Fontaine, *La morte étrange de Conrad Kilian, inventeur du pétrole saharien* (Paris: Les Sept Couleurs, 1959), Chaps. 6–7.

12. Ibidem, 23.

13. Ibidem, 65–66.

14. Alain Perrodon, "Historique des recherches pétrolières en Algérie," in J. Prouvost (ed.), *La recherche pétrolière française*, 323–340, on 325.

15. Ibid., 326.

16. Fontaine, *La morte étrange de Conrad Kilian*, 64–69.

17. CFP, Minutes of MBD, December 27, 1946, p. 2–4 in TOTAL-CFP, b. 92.10/1, AH TOTAL.

18. Henri Bonnet to the US Under-Secretary of State for Economic Affairs William L. Clayton, January 4, 1947, *Foreign Relations of the United States* (1947), V, 627–9, NARA.

19. Emmanuel Catta, *Victor De Metz. De la CFP au Groupe TOTAL* (Paris: Total Edition Presse, 1990). See also Rondot, *La Compagnie Française des Pétroles*, 97.

20. CFP, Minutes of MBD, March 5, 1947, pp. 3–5 in TOTAL-CFP, b. 92.10/1, AH TOTAL. See also: Anderson, *ARAMCO*, 158.

21. CFP, Minutes of MBD, June 4, 1947, pp. 3–5 in TOTAL-CFP, b. 92.10/1, AH TOTAL. See also: Nouschi, *La France et le pétrole*, 200.

22. Pierre Péan in George-Henri Soutou and Alain Beltran (eds.), *Pierre Gullaumat, la passion des grands projets industriels* (Paris: Rive Droite, 1995), 12–13; P. Péan and Jean-Pierre Séréni, *Les émirs de la République* (Paris: Seuil, 1982), 28.

23. C. Layat, A. Clement, G. Pommier, and A. Buffet, "Some Technical Aspects of Refraction Seismic Prospecting in the Sahara," *Geophysics* 26:4 (1961): 437–46. See also: A. Beltran and Sophie Chauveau, *Elf Aquitaine, des origines a 1989* (Paris: Fayard, 1998), 59.

24. Perrodon, "Historique des recherches pétrolières en Algérie," 328.
25. Unlike, for example, French colonies in sub-Saharan Africa, the three *départements* of Algeria were considered integral parts of the French political unit.
26. Minutes of the Conference held on June 19, 1947 at the Under-Directorate of Algeria, p. 4 in ELF-ERAP, b. 07AH0168–6, Papers SN REPAL, AH TOTAL.
27. On the Monnet Plan see Gérard Bossuat, *La France, l'aide américaine et la construction européenne, 1944–1954* (Paris: Comité pour l'histoire économique et financière de la France, 1997).
28. The Algerian Government General, established in July 1834, was responsible for the administration of French North African territories.
29. Minutes of the Conference held on June 19, 1947 at the Under-Directorate of Algeria, p. 4 in ELF-ERAP, b.07AH0168–6, Papers SN REPAL, AH TOTAL [my own translation].
30. Ibidem, 8 [my translation].
31. SN REPAL, Minutes of the 4th MBD, June 24, 1947, ELF-ERAP, b. 07AH0168–28, Papers SN REPAL, AH TOTAL.
32. SN REPAL, Minutes of the 6th MBD, December 30, 1947, in ELF-ERAP, b. 07AH0168–28, papers of SN REPAL, AH TOTAL.
33. Giovanni Buccianti, *Enrico Mattei: Assalto al potere petrolifero mondiale* (Milan: Giuffrè, 2005), 151–152.
34. Gulf's Legal Department [unsigned] to Yves Chataigneau [via French Embassy in Washington], September 10, 1947, in ELF-ERAP, b. 07AH0168–6, Papers SN REPAL, AH TOTAL.
35. On Paul Moch see Péan and Séréni, *Les émirs de la République.*
36. P. Guillaumat to Inspector General of Mines in Alger Georges Betier, April 9, 1948, in b.07AH0168–28, ELF-ERAP.
37. Paul Moch to P. Guillaumat, October 18, 1950, and Note by Governor General of Algeria Marcel-Edmond Naegelen, "Recherches d'hydrocarbures dans le Sahara," ELF-ERAP, b. 07AH0168–6, Papers SN REPAL, AH TOTAL.
38. BRP, Minutes of the 29th MBD, October 30, 1951, pp. 6–9 in b. 10AH0832–26, ELF-ERAP and SN REPAL, Minutes of the 29th MBD, November 24, 1951, pp. 16–17, in b.07AH0168–6, AH TOTAL.
39. Goetze to Moch, February 27, 1952 in ELF-ERAP, b.07AH0168–6, SN REPAL, AH TOTAL.
40. Goetze to Moch, September 17, 1952 in ELF-ERAP, b.07AH0168–6, SN REPAL, AH TOTAL.
41. A. Morange, A. Perrodon, and F. Héritier, *Les grandes heures de l'exploration pétrolière du groupe ELF Aquitaine* (Boussens: Elf Aquitaine Éditions, 1992), 97. See also Perrodon, "Historique des recherches pétrolières en Algérie," 327.
42. Leucha Veneer, "Oil Security and the North Sea: British Explorations in the 1960s," unpublished paper. On British attitudes at NATO, see also Roberto Cantoni, "Oily Deals. Exploration, Diplomacy and Security in early Cold War France and Italy," Ph.D. thesis, Manchester: University of Manchester, 2014, chapter 5.
43. The British government held a controlling stake in AIOC, but the company had been greatly changed by the Iranian government's decision to nationalize the Iranian oil industry in 1951. AIOC became the British Petroleum Company (BP) in 1954. See Bamberg, *The History of the British Petroleum Company*, vol. 2, 495–504.
44. Ministry of Power report, February 22, 1966, in "Security of Oil Supplies; and UK Stockpiling Policy," POWE 63/75, TNA.
45. Ibidem.

46. P. E. Kent, "North Sea Exploration—A Case History," *The Geographical Journal* 133:3 (1967), 289–301, on 290. See also P. E. Kent, "The North Sea—Evolution of a Major Oil and Gas Play," *Facts and Principles of World Petroleum Occurrence* 6 (1980): 633–652.

47. Kent, "North Sea Exploration—A Case History," 291–292.

48. Øystein Noreng, *Oil Industry and Government Strategy* (Boulder, CO: ICEED, 1980), 40; see also Helge Ryggvik, *The Norwegian Oil Experience: A Toolbox for Managing Resources?* (Oslo: University of Oslo, 2010), 11.

49. Ministry of Power, *Fuel Policy. Presented to Parliament by the Minister of Power by Command of Her Majesty, November 1967* (London: HMSO, 1967), 5.

50. For an overview see: UN Division for Ocean Affairs and the Law of the Sea, "The United Nations Convention on the Law of the Sea (A Historical Perspective)," 2012. Available at: http://www.un.org/Depts/los/convention_agreements /convention_historical_perspective.htm, accessed January 29, 2014. See also D. C. Watt, "Britain and North Sea Oil: Policies Past and Present," *The Political Quarterly* 47:2 (1976), 377–397, on 378.

51. "North Sea Oil," General Correspondence, Economic Relations (U): Supplies, FO 371/178156, TNA.

52. "North Sea Oil and Gas," General Correspondence, International Oil (IOD), FO 371/187603, TNA.

53. Hubert Scholes to John T. Fearnley, April 18, 1966, in "North Sea Oil and Gas," FO 371/187603, TNA.

54. Untitled Memorandum, June 21, 1967, in "Investigation by the IGS of the Geophysics of the North Sea from Information Given by Offshore Licencees," POWE 63/201, TNA.

55. Ibidem. For an overview on the institution's history see: Harold E. Wilson, *Down to Earth: One Hundred and Fifty Years of the British Geological Survey* (Edinburgh: Scottish Academic Press, 1985). Post-1967 developments are examined in Dennis Hackett, "Our Corporate History: Key Events Affecting the British Geological Survey, 1967–1998," British Geological Survey Technical Report WQ/99/1 (available at: www.bgs.ac.uk, accessed August 19 2013).

56. G. A. L. Johnson, "Sir Kingsley Charles Dunham, 1910–2001," *Biographical Memoirs of Fellows of the Royal Society* 49 (2003): 147–162, on 155.

57. Richard Aldrich, *Espionage, Security, and Intelligence in Britain, 1945–1970* (Manchester: Manchester University Press, 1998), 60.

58. On this committee, see Jon Agar and Brian Balmer, "British Scientists and the Cold War: the Defence Research Policy Committee and Information Networks, 1947–1963," *Historical Studies in the Physical Sciences* 28:2 (1998): 209–252.

59. K. Dunham, "IGS Final Report, 1967–1975," Appendix 1 in Hackett, "Our Corporate History: Key Events Affecting the British Geological Survey, 1967–1998," 9–11.

60. Ibidem, 10.

61. "Investigation by the IGS of the Geophysics of the North Sea," POWE 63/201, TNA.

62. "IGS: Continental Shelf Programme," POWE 63/465, TNA.

63. Ian McCartney to N. E. Martin, January 11, 1968, in "Investigation by the IGS of the Geophysics of the North Sea," POWE 63/201, TNA.

64. Dunham to R. J. H. Beverton (received December 21, 1967), "Investigation by the IGS of the Geophysics of the North Sea," POWE 63/201, TNA. On the Labour proposal, see Fuel Study Group, "A National Hydrocarbons Corporation," London: Labour Party, 1968. See also Watts, "Britain and North Sea Oil," 383.

65. K. Dunham, "IGS Final Report, 1967–1975," 9.

66. IGS, Annual Report 1967 (London: HMSO, 1968), 63.
67. Ian McCartney to N. E. Martin, January 11, 1968, in "Investigation by the IGS of the Geophysics of the North Sea," POWE 63/201, TNA.
68. K. Dunham, "IGS Final Report, 1967–1975," 10.
69. IGS Annual Report 1971 (London: HMSO, 1972), 2.
70. K. Dunham, "IGS Final Report, 1967–1975," 11.

Section II

Monitoring the Earth: Nuclear Weapon Programs

Chapter 3

"Unscare" and Conceal: The United Nations Scientific Committee on the Effects of Atomic Radiation and the Origin of International Radiation Monitoring

Néstor Herran

On December 3, 1956, the General Assembly of the United Nations established the Scientific Committee on the Effects of Atomic Radiation (UNSCEAR) to collect and evaluate information on the worldwide levels and the effects of ionizing radiations. The committee, which was compelled to present a complete report to the UN General Assembly by late 1958, became a key space for international scientific exchange, setting of standards in radiological protection, and establishment of transnational networks of radiological monitoring. Nonetheless, its very creation, design, and operation were surrounded by controversy. Diplomatic tensions regarding nuclear disarmament, surveillance ambitions, and the interest in playing down a transnational collaborative project seeking to find out more about radioactive fallout shaped the structure and inner dealings of the committee.

The constitution and early activities of the UNSCEAR were a result of two complementary and sometimes contradictory activities of nuclear powers: the downplaying of the effects of radiation exposure vis-a-vis the findings of activists and the concealing some of the methods to detect atmospheric radiation used to gain information on foreign nuclear tests. These activities involved the development of transnational networks of radiation monitoring to assess the health hazards of radioactivity for human populations that, additionally, would conceal the preexisting military monitoring infrastructure.[1]

In order to reveal this logic of "playing down and conceal," I first present the state of military radiological surveillance networks that were developed in the late 1940s and 1950s by the nuclear powers to monitor foreign nuclear activities, as well as the emergence of nonmilitary programs for the study of environmental radioactivity in the wake of the global controversy regarding fallout from nuclear tests. Then, I analyze the composition of UNSCEAR and discuss its relevance to diplomatic and intelligence-gathering concerns, and how evidence on the effects of radiation was managed according to different national interests. In this point, I stress the role of the United States and the United Kingdom as the most influential nations in the shaping of the committee and its increasing relevance at the expense of other UN institutions such as the UNESCO and the IAEA, or preexisting international institutions dealing with the regulation of radioactivity, such as the International Commission on Radiological Protection (ICRP).[2]

Finally, I study the contents and reception of the first UNSCEAR report in 1958, as well as the effect of this institution's activities in the establishment of further projects of radioactive monitoring and the setting up of international standards in the collection and measurement procedures. Backed by its international and intergovernmental character, UNSCEAR's approach—based on the idea of maximum permissible dose—spread globally in the following years. Indeed, the institution would acquire a dynamic of its own, which helped to create an international community of experts in radiation language, tied by a common set of and approaches and practices.

Early Military Radiological Monitoring

Military radiological monitoring can be defined as the set of techniques and practices created and used to detect nuclear activities, especially those activities related to the development and deployment of nuclear weapons. The origin of these techniques and practices can be traced back to American efforts to assess the extent of the German nuclear program at the end of World War II. General Leslie Groves, head of the Manhattan Engineering District—and also responsible for nuclear intelligence operations since the fall of 1943—considered obtaining information about the development of nuclear weapons in Germany a priority. Groves adopted an innovative approach to intelligence-gathering, based on the detection of nuclear operations at distance by measuring the radioactivity such operations released into the environment.[3] Luis Alvarez, a Massachusetts Institute of Technology Radiation Laboratory–trained physicist, was charged with the task of developing a method. He came out a procedure involving the collection of air samples by US Air Force planes and the determination of the presence of xenon-133, a rare gas produced during the operation of nuclear reactors. This monitoring schema was first used in the fall of 1944 over sites of suspected German nuclear activities, with negative results. A parallel intelligence operation also devised by Groves, the Alsos Project, confirmed that the German nuclear program had been unsuccessful.[4]

Radioactive emissions from American nuclear tests and facilities were also monitored in order to assess the reliability of the method employed. Manhattan District scientists unsuccessfully tried to detect the Trinity test (July 16, 1945) by seismic and radiological methods, and flights over plutonium-producing facilities at the Hanford nuclear production complex in Washington state revealed that the xenon-based system was only reliable for detecting nuclear activities at a short distance. However, scientists alien to the Manhattan project were able to retrospectively detect the Trinity test, suggesting that it was possible to detect nuclear explosions as far as 1,000 kilometers away by sonic and seismological methods, and up to 2,000 kilometers by measuring radioactive debris.

After the war, the prospect of nuclear proliferation furthered nuclear intelligence-gathering. American intelligence operations were reorganized, with monitoring operations expanding under the impulse of different branches of the US military. One of the first initiatives in this direction was Project Mogul, established in 1945. Geophysicist Maurice Ewing, working at the Woods Hole Oceanographic Institution, had studied the transmission of sound underwater and discovered the existence of "sound channels" able to carry sounds at long distance in deep layers of ocean water. The suggestion that a similar channel could exist in the stratosphere raised the interest of officials in the US Army and US Air Force, which appointed Ewing as head of a research group at Watson Laboratories.[5] At the same time, the US Navy established a monitoring program as a part of the Operation Crossroads, a series of nuclear tests conceived to assess the effects of nuclear blasts on ships. Geophysicist and oceanographer Roger Revelle participated in this effort, which led to inconclusive results. But, as had happened in the Trinity test, scientists not affiliated with the military announced that they could detect nuclear explosions using either seismographs or Geiger-Müller counters. This seemed to confirm again that it was possible to detect signals of distant nuclear tests against the background of natural radiation once the time of the explosion was known.

By 1946, while sonic and seismological monitoring stagnated due to the limitations of the available high-altitude balloons or seismographs, the radiological method consolidated as the most promising technique for long-range detection.[6] This idea was reinforced by the results of Operation Fitzwilliam, carried out to monitor the Sandstone nuclear tests in spring 1948 and compare the different monitoring methods.[7] The private company Tracerlab, a firm that specialized in manufacturing radioactivity-measuring equipment, played an important role in these studies, which were based on the radiochemical analysis of air filters from ground stations and monitoring flights. The success of these experiences paved the way for the participation of Tracerlab technicians in subsequent monitoring operations.

The Central Intelligence Group, the institutional predecessor of the Central Intelligence Agency (CIA), fostered the development of a centralized body in charge of monitoring based on a preexisting Air Force group, AFMSW-1 (for Air Force deputy chief of staff for Materiel, Special Weapons Group, Section One). By July 1948 this group was under the direct control of the AEC under the name AFOAT-1 (Air Force deputy chief staff of operations, Atomic Energy Office, Section One).[8] In the following years, AFOAT-1 expanded its mission from the detection of nuclear tests to the monitoring of all aspects of the nuclear cycle, from uranium mining to the stockpiling of fissionable materials. In the radiological domain, this led to the improvement of methods based on the detection of krypton-85, and the development of a compact cryogenic collection unit for the remote monitoring of tests, reactor operations, and plutonium production.[9]

By 1949, AFOAT-1 had expanded to include two central laboratories (operated by Tracerlab in Berkeley and Boston), four dedicated squadrons of BW-29 air-sampling bombers, and 24 ground stations, which were operated primarily as a backup for the more effective monitoring carried out by aircraft. An agreement with the United Kingdom had been reached on the monitoring of Soviet nuclear activities, which was reflected in the geographical distribution of the stations. American stations were mostly scattered in the Pacific, from Northern Alaska to the Philippines, and, in minor, extent, in the Atlantic (Washington DC, Bermuda, and the Azores), while the British were in charge of monitoring the northern regions of the Atlantic Ocean.[10] In particular, the United Kingdom had established ground stations at air bases in Scotland, Northern Ireland, and Gibraltar, equipped with Geiger counters to analyze filters coming from surveys in Greenland and Western Atlantic (Operation Bismuth) and the Mediterranean (Operation Nocturnal). Samples with high readings were sent to the UKAEA laboratories at Harwell to be analyzed in detail.[11]

In August 1949, the British-American monitoring system was able to detect the first Soviet nuclear test after finding radioactive dust in filters exposed in flights carried out over the Pacific some days after the explosion. Collaboration between British laboratories and the US Weather Bureau Special Projects Section, headed by meteorologist Lester Machta, allowed for the determination of the bomb type and the location of the explosion in Central Asia. The precision of the monitoring system was improved in the following years by implementing a complete network of seismic, sonic, and radiological stations, whose reliability was regularly checked by following American nuclear tests in the Pacific. In this schema, the different methods of detection combined to provide a detailed picture of Soviet explosions: the radiological was used to make out the composition of the bomb, the electromagnetic pulse to provide the exact time of explosion, the sonic to help ascertain the yield, and the seismic to determine the location of the test.[12]

While those British and American monitoring projects were carried out, the Soviet Union also established its own intelligence operations in this field. They involved not only traditional spying activities that made them aware of American monitoring practices,[13] but also regular monitoring of environmental radioactivity. In particular, analysis of the radioactive debris of American tests seem to have been instrumental in the design of the Soviet hydrogen bombs based on the Teller–Ulam design, first detonated in November 1955, only three years after the first American thermonuclear test.[14] But this could also apply to the British nuclear weapons program, which was fostered by the knowledge provided by its monitoring programs, and the French program that by late 1957 involved a parallel network of 15 monitoring stations in continental France and at least one more station in Tahiti.[15]

The Fallout Controversy and Its Consequences

On March 1, 1954, a nuclear test codenamed "Castle Bravo" detonated a 15-kiloton hydrogen bomb in the Marshall Islands. The explosion, exceeding the predicted yield, spread nuclear fallout over regions hundreds of kilometers away from the test site. The population of nearby atolls had to be eventually evacuated after receiving high doses of radiation, and a secret study was launched to assess the effects.[16] However, public concern would possibly have remained limited if not for the fallout affected some Japanese fishing boats. After returning to Japan, the crew of one of the ships showed signs of radiation sickness. A controversy escalated when measurements of tuna catches—which had been already marketed—revealed them to be radioactive and unfit for human consumption. In the following months, more reports on contaminated fish and radioactive rain coming from nuclear tests featured frequently in Japanese media. From May to July 1954, a team including marine biologists, meteorologists, food scientists, and radiologists performed an oceanographic survey of the waters in the Pacific

Ocean. Considered the first environmental radiation survey outside the nuclear test sites, it challenged claims by the US Atomic Energy Commission (AEC) that no radioactive contamination could be found outside the danger zone established around the Bikini atoll. These findings were contested by the results of a new AEC survey, codenamed "Operation Troll," which indicated that the measured activities remained below the maximum permissible dose. However, this clash of claims and counterclaims did not help to reassure Japanese public, fostering the emergence of a strong grassroots Japanese antinuclear movement. This involved, for example, the creation of a unified Japanese Council against Atomic and Hydrogen Bombs gathered more than 35 million signatures calling for a ban on nuclear weapons.[17]

Concern over fallout was not restricted to Japan, and soon acquired a global scope. News about radioactive rain was reproduced by newspapers in all continents, fostering worldwide awareness on the risks of nuclear testing. In the United States, a public controversy originated around a specific isotope present in fallout, strontium-90. In 1953, the AEC had launched Project Sunshine, a secret project to collect samples of air, water, soil, milk. and human bones in the United States and abroad, in order to measure the presence of strontium-90. Led by atomic chemist and AEC commissioner Willard F. Libby, Project Sunshine was the first large-scale survey of radioactive environmental contamination.[18] Its preliminary results, released to appease public anxiety, were moderately reassuring, but the controversy was not over and some associations tried to undertake independent measurements. For example, the Consumers Union conducted a national study of strontium-90 concentrations in milk, which was published in the magazine *Consumer Records*, and the Greater St. Louis Citizens' Committee for Nuclear Information started a survey of strontium-90 in children's teeth.[19]

In the diplomatic arena, the fallout controversy soon became intertwined with decolonization issues. India's first prime minister Jawaharlal Nehru was one of the most outspoken critics of nuclear tests, requesting in April 1954 a "standstill agreement" on nuclear testing as a first step toward disarmament. His appeal referred directly the effects of Pacific atomic tests in Japan, pointing out that "Asia and her peoples appear to be always nearer these occurrences and experiments, and their fearsome consequences, actual and potential."[20] One year later, he was instrumental in the organization of the Bandung Conference, which launched the movement of nonaligned countries.

A transnational movement of scientists, under the auspices of the UN Educational, Scientific and Cultural Organization (UNESCO), began to mobilize to assess the dangers of radioactive fallout for human health. The International Council of Scientific Unions (ICSU), meeting in Oslo in August 1955, resolved to ask its members to conduct studies of radiation effects independently of the United States and to sponsor a scientific study independent of governments. This initiative was regarded with suspicion by nuclear powers, which preferred a committee of scientists designated by national governments.[21] In December 1955, the issue was addressed in the United Nations General Assembly, leading to a resolution, approved unanimously, which requested the establishment of a committee to study the effects of radiation on human health in general. Under this formulation, it would not only be a consultative body on the question of fallout, but also provide a standard reference on radioprotection to countries interested on developing civil nuclear programs in nations but lacking radioprotection agencies.[22]

The United Nations Scientific Committee on the Effects of Atomic Radiations (UNSCEAR) was constituted by representatives of 15 states, including the three nuclear powers at this time (the United States, the Soviet Union, and the United Kingdom), countries with advanced nuclear programs such as France and Canada, strategic providers of uranium such as Czechoslovakia, Belgium, and Australia, and six nonaligned countries (India, Brazil, Egypt, Argentina, Mexico, and Sweden). Japan, epicenter of the fallout controversy, completed the panel. However, the committee's composition was in line with that of the early United Nations, which favored the United States. Only two of the countries participating in the commission were in the Soviet Union's sphere of influence.

UNSCEAR was born in this way as a quite exceptional offspring of international diplomacy, as the only scientific committee in the UN system and as a precedent to other international institutions devoted to the regulation of nuclear affairs, such as the International Atomic Energy Agency (IAEA), which was established in 1957. Its origin, however, was rather serendipitous and has to be seen as a response of national governments, and especially of the United States, to an incipient transnational movement linking initiatives at different levels: grassroots antinuclear movements, nonaligned countries aiming at nuclear disarmament, and scientists acting as moral crusaders on behalf of populations sacrificed to the atomic cause.[23]

UNSCEAR at Work

In early 1956 the members of the committee were appointed by their respective governments, with each country presenting at least one scientific representative and one or more substitutes, as well as accompanying diplomats. Among them were some of the leading experts of the rapidly growing discipline of medical physics, such as British representative William Mayneord, American representative Warren Shields, and Swedish representative Rolf Sievert, and scientists with experience in advising governments and international institutions such as the Mexican representative Manuel Martínez-Baez. The committee was thus typified by disciplinary diversity, since it included directors of medical and public health institutions, biophysicists, members of radioprotection institutions, and geneticists (see Table 3.1).

The main idea inspiring the committee was to reproduce the spirit of the recent Geneva "Atoms for Peace" Conference, which was remembered as the first instance of open international discussion in nuclear matters since the beginning of the Cold War. The feeling that an open discussion could also be attained in the UNSCEAR was explicitly written down in the committee minutes:

> The session has merely confirmed what the Conference on the Peaceful Uses of Atomic Energy had already shown in August 1955: namely that, despite all differences in background and language, scientists everywhere had the fundamental approach to their work, reached the same results, and always welcomed the opportunity to compare their own results with those of others and to achieve standardization.[24]

Table 3.1 Heads of national delegations at the UNSCEAR

Country	Representative	Background
Argentina	Constantino Nuñez	Director of the Department of Medicine and Biology, National Commission of Atomic Energy
Australia	C. E. Eddy / D. J. Stevens	Director of the Commonwealth X-Ray and Radium Laboratory
Belgium	Zenon Bacq	Professor of Radiobiology, University of Liège
Brazil	Carlos Chagas	Professor of Biological Physics, University of Rio de Janeiro
Canada	E. A. Watkinson	Department of Health and Welfare of Canada
Czechoslovakia	Ferdinand Hercik	Director of the Biophysics Institute, Czechoslovak Academy of Sciences.
Egypt	A. Halawani	Director of Department of Research and Endemic Diseases, Ministry of Public Health
France	Louis Bugnard	Director of the National Institute of Hygiene, Ministry of Public Health
India	V. R. Khanolkar	Director of the Indian Cancer Research Centre, Chairman of Biological and Medical Advisory Committee, Department of Atomic Energy
Japan	Masao Tsuzuki	Director of Japan Red Cross Central Hospital, Professor Tokyo University
Mexico	Manuel Martínez Baez	Founder of Institute of Tropical Disease
Sweden	Rolf Sievert	Professor of Radiophysics, Royal Carolinska Institute, Stockholm
United Kingdom	William V. Mayneord	Professor of Physics, University of London; Director of Medical Physical Department, Institute of Cancer Research, Royal Naraden Hospital
United States	Shields Warren	Scientific Director of the New England Cancer Research Institute,
USSR	Andrei V. Lebedinsky	Professor of Physiology

The committee met five times before the submission of its first report: twice in 1956 (from March 14 to 23 and from October 22 to November 2),[25] once in 1957 (from April 8 to 18, 1957) and twice again in 1958 (from January 27 to February 28, and from June 9 to 13). The final report was approved in the fifth session and submitted to the General Assembly. In addition to regular meetings, which were recorded in the committee minutes, the scientific committee also established informal—and unrecorded—discussion groups focusing on the main issues examined: genetic effects of radiation, the difference between internal and externally absorbed irradiation, natural radiation levels, exposures during medical procedures and occupational exposure, and environmental contamination.

These meetings, of more technical character, were organized around the discussion and evaluation of documents submitted by national governments under strict formal procedures. The need to classify and evaluate a huge amount of information was evident since the first meeting, and the committee requested the hiring of two young scientists, the Swedish radiophysicist Bo Lindell and the Argentinean biophysicist Dan Beninson, to present data in a form suitable for the committee's discussions.[26]

Despite the apparently technical format in which UNSCEAR meetings took place, discussions in the committee concurrently dealt with diplomatic issues related to the framing of discussions, the selection of sources of information, and the dissemination of its results. Scientific representatives to the commission were aware of the diplomatic implications of their task, and when they forgot, the accompanying diplomats duly reminded them. Previously classified documents on the committee held in the UK national archives reveal, for example, that the British delegation received explicit instructions from the Foreign Office warning them that "the question of halting or banning nuclear weapon tests lays outside the competence of the Scientific Committee and should be discussed only in the context of disarmament" and that "the UK representatives should resist any attempt to get the Committee to make recommendations on this or other aspects of the nuclear tests problem."[27] This instruction was consistent with the British (and American) policy to separate disarmament negotiations from a multilateral debate on fallout.

UNSCEAR meetings previous to its first report were strongly conditioned by the publication of scientific reports on the effects of ionizing radiations produced by the United States National Academy of Sciences (NAS) and the United Kingdom Medical Research Council (MRC) in June 1956, a few months after the establishment of the committee. Both institutions had engaged in "independent" surveys under the supervision of their respective national atomic energy authorities. As Jacob Hamblin and Soraya Boudia have shown, the conclusions of both reports report were a product of a negotiation between MRC and NAS-appointed experts and the AEC administration, resulting in a pretended scientific consensus on safe levels of radiation.[28] In relation to the effects of nuclear testing, the reports aimed at dispelling fears, stressing that radiation released was negligible in comparison to the natural background and playing down concern for genetic damage. On this point, the timely publication of the first comprehensive review of the data obtained by the Atomic Bomb Casualty Commission on the survivors of Hiroshima was also instrumental. This report, published in 1955, provided some evidence of a minimal dose under which no genetic effects could be observed in humans and was widely used in the media to dispel fears of health hazards from fallout.[29]

The MRC and NAS reports were taken as a basis for discussion in UNSCEAR, which generally avoided contributing to public scare on fallout. In March 1956, for example, radiologists William V. Mayneord and Rolf Sievert presented Swedish and British surveys, suggesting that exposure to radiation due to diagnostic procedures was higher than previously expected, and chairman Carlos Chagas had to remind them that "the committee duty (was) to inform, not to alarm."[30] All at once, rhetorical strategies were put in place to give legitimacy to the commission, such as the replacement of the word "experts" by "scientists," or the emphasis on the fact that commission was the result of scientific cooperation between nations.[31]

The recollections of first UNSCEAR secretary, the Canadian Ray Appleyard, also reveal that the UN press corps established rigorous procedures to avoid scientists making off-the-record remarks, instructing delegates about how to address the press with strictures such as "never answer hypothetical questions." [32] These recommendations were strictly followed by the committee members, as shown by records of press conferences:

> *Question*: (...) On the basis of what you have found so far, are you in a position to say that there should not be any more testing of nuclear weapons or bombs in view of the fact that, some time ago, a large number of Nobel prize winners, among them scientists of great repute, have asked for the banning of tests? ...
>
> *Prof. Bacq (Belgium)*: I must say that this question is outside the scope of our work. We were not asked by the Assembly of the United Nations to answer this question. We shall give to the Assembly all the data evaluated by us and put it in a readable form. It will be the responsibility of the Assembly of the United Nations to take such decisions as the one you have suggested.[33]

On July 13, 1958 UNSCEAR presented its first report to the General Assembly of the United Nations. The 232-page report drew heavily on the material presented by the United States and the United Kingdom, which counted together not only the 53 percent of the pages examined, but also provided the most detailed synthetic reports.[34] Given the composition of the committee and the selection of material examined, it is not surprising that the report's contents and conclusions closely followed the path of its British and American predecessors. That is, it too downplayed the hazards of radioactive fallout, minimizing its genetic impact by comparing with natural levels of radiation, and relied specifically on studies about the distribution of strontium-90.[35] This isotope had been the focus of the Project Sunshine and also of the American debate, which revolved around the danger this element posed by entering the human food chain through milk, an important element of Americans' diet. The only mention to the idea that genetic effects of radiation might have no threshold, which was the basic argument of critics of atmospheric testing, was included as a particular vote of the Soviet, Czech, and Egyptian delegations in the final report.

It is also remarkable that other isotopes produced in nuclear tests, particularly carbon-14, were excluded from the first report. Despite carbon-14 being one of the main sources of radiation in living organisms, the contribution of fallout to its global increase was widely ignored in the study. This discrepancy was a centerpiece in the arguments of the two main opponents of nuclear testing at both sides of the Iron Curtain. Andrei Sakharov, who published his results in June 1958, estimated 10,000 deaths and other health injuries from the low-dose radiation effects from each megaton of nuclear weapons exploded.[36] Linus Pauling, who knew of Sakharov's findings though his network of scientific activists against nuclear proliferation, estimated in November 1958 that one year of atmospheric nuclear testing would cause "55,000 children with gross physical or mental defects, 170,000 stillbirths and childhood deaths, and 425,000 embryonic and neonatal deaths."[37] However, the lack of emphasis on this element can be understood by the lack of available studies on the biological effects of this isotope and the reluctance to disclose data on its production by thermonuclear bombs, which could give details about the bombs' components.

Towards a Transnational System of Radiation Monitoring Networks

British and American representatives had the upper hand in UNSCEAR negotiations mainly because, in contrast with other countries, they had been working on radiological monitoring during the previous 10 years, and had begun establishing a network of stations that ensured global coverage. A large worldwide network of 122 stations had been put in place after the Castle Bravo tests under the direction of Lester Machta, who coordinated measurements obtained from 39 US Weather Bureau stations in the continental United States and 14 at overseas locations, 23 overseas stations operated by the Air Weather Service, 31 stations from the State Department, three operated by the Navy and the Coast Guard, and two by the Atomic Bomb Casualty Commission. The Canadian Meteorological Service and the Canadian Atomic Energy Commission cooperated with respectively ten more stations. In addition, daily measurements based on gummed film stands were performed by most ships of the Military Sea Transport Service on routes in the Pacific Ocean.[38]

This scientific hegemony is evident in the first UNSCEAR report, which provides one of the first graphical accounts of the distribution of radiation monitoring stations worldwide. According to the first UNSCEAR report, the committee had had access to measurements from about 350 stations. This number seems rather impressive, but their authors noted that the uneven distribution of stations meant that "large areas of the earth are insufficiently covered by the network of stations collecting data," while also noting that "the different stations and laboratories do not all operate with comparable collection and evaluation methods."[39] A disproportionately large number of the stations were located in the United States in comparison with the lack of measurements in large areas of Asia and Africa. Indeed, most measurements were obtained from terrestrial stations, leaving out of the picture most of the Atlantic and Indian oceans, and large parts of the Pacific. In global terms, the Northern Hemisphere was overrepresented in comparison to the Southern, making the calculation of global doses difficult and uncertain.

In order to get a better picture, the US Air Force launched in 1957 the High Altitude Sampling Program (HASP), aiming at the direct measurement of stratospheric concentrations of debris from tests of nuclear weapons (and especially of Sr-90), and the estimation of their stratospheric residence times and mechanisms and rates of transfer within the stratosphere and from the stratosphere to the troposphere. Between August 1957 and June 1960, about 3,700 samples of air were collected by aircraft at more than 70,000 feet (around 20,000 meters) of altitude in a meridian sampling corridor.

The interest of the United States to build a comprehensive map of the distribution of atomic debris can be related to the beginning of the Geneva talks with the Soviet Union about the establishment of a treaty banning nuclear tests in

1958. One of the most important problems faced by negotiators was the unreliability of methods for the verification of violations of the agreement (see chapter 4 by Simone Turchetti). Indeed, these methods could also be used to track nuclear tests by nations engaged in the development of nuclear weapons, such as France, in the scenario of widespread nuclear proliferation.

UNSCEAR proved to be an ideal space to further the collaboration of other nations in the global collection of data on environmental radioactivity and to strengthen bilateral relationships in the nuclear domain in the same spirit as Eisenhower's Atoms for Peace campaign. And, as John Krige has shown in relation to this later initiative, American hegemony was constructed by means of technical instruments, such as the establishment of standards and procedures.[40] But while these monitoring procedures spread globally, more sophisticated methods utilized to garner intelligence on Soviet tests were kept away from these public debates and international negotiations on nuclear fallout in order not to undermine their effectiveness.

The very methods established to collect fallout samples provide a good example of this construction of scientific hegemony by controlling standardization procedures. The fundamental technique used for measuring fallout was the "gummed film," a method initially developed to trace the pattern of fallout distribution of nuclear tests performed in Nevada since fall 1951. It was based on putting a gummed film, made from cellulose acetate, on a platform situated at about one meter from the ground. The film, which supposedly caught fallout as it descended, was subsequently burned and its ashes analyzed for the presence of radioisotopes by means of a beta-radiation detector. This method worked relatively well and produced simple test patterns when fission devices were concerned, allowing for estimations of individual long-lived nuclides such as strontium-90. However, when thermonuclear devices were first used, they delivered the test debris into the stratosphere, producing a more rapid dilution and making them more difficult to follow. Accordingly, mathematical models had to be developed to allow computation of the doses produced from the available data. Techniques to compute the actual strontium-90 level included comparison with other techniques, such as the collection of radioactive debris in soil pots, but they were not stabilized at the moment of the UNSCEAR reports. In 1960, for example, an internal memorandum of the US AEC showed that measurements computed from gummed films between 1956 and 1958 underestimated by a factor of 20 the presence of strontium-90 in comparison with direct measurement in soil pots.[41]

These important disagreements did not prevent the United States from suggesting to UNSCEAR that the global radiation monitoring network be based on gummed films, proposing to provide other countries with standardized foils and formularies and to collect them for analyzing the data and process the data automatically to produce global radioactivity maps. Standardization under US norms was not accepted straight away. An example of this is the opposition of the Japanese representative to the imposition of American method in one of the first meetings of the commission, stressing the differences existing between different measurement methods.[42] However, even if the report pointed out the uncertainty related to the gummed film method, the system was approved thanks to the extensive American network and the abundance of data.

Conclusion

From its military origin in the late 1940s to its transformation in an international endeavor under the auspices of the UNSCEAR, radiological monitoring was an important instrument for the surveillance of nuclear activities. It is impossible to disentangle public health concerns from diplomatic and intelligence activities, as all these dimensions mattered and were fundamental to the establishment, extension, and coordination of the network. The participation of the United Nations in the development of radiological monitoring can be understood as a movement consistent with the scenario of international relationships in the late 1950s, when the process of decolonization added a new terrain of confrontation between the two ideological blocks.

As Soraya Boudia has shown, and this chapter corroborates, the UNSCEAR was used by the nuclear powers in order to counteract public criticism of nuclear testing or, in the case of the Soviet Union, to further it. At the same time, a parallel effort was made to reveal details on the availability of advances methods for analyzing the composition of bombs from nuclear test fallout, such as those related to carbon-14. This concealment can also be related to the standardization of measurements around the gummed film technique, a method considered as unreliable for determining radiation doses from fallout, but not for analyzing bomb composition.

In relation to its scientific impact, UNSCEAR followed the model of other international institutions such as the World Meteorological Organization. It exemplifies the adaptation of scientists to a the Cold War model of international relationships, and the transition from nongovernmental internationalism—which in the UNSCEAR case could be represented by the International Commission on Radiation Protection (ICRP)—to an institution based upon nationally appointed representatives. And, as with the WMO, its legitimization was based on the establishment of a network of collection, analysis, and standardization of data, which generally implied the coordination of scientific centers in countries arising from the decolonization process with the more developed standards and networks developed by former colonial powers, especially in the United States. As this chapter shows, this dependency, based on an asymmetry in the development of military monitoring networks in the early 1950s, and more generally in American diplomatic soft power and scientific hegemony, shaped scientific consensus and objectives to fit American geopolitical goals. By putting the surveillance imperative as the motor of the radiological monitoring, the "infrastructural globalization" led by UNSCEAR can be considered as an instrument of the global hegemony of the United States.[43] Given the chronological limits of this case study, we cannot conclude if this function was still prevalent in the following decades, an interesting question that could be addressed by studying how UNSCEAR developed since the 1960s in relation to the development of civil nuclear industry and the subsequent local and transnational opposition to it.

Notes

1. On the use of international institutions as spaces susceptible of transnational historiographical approaches, see the special issue on transnational history of science of the *British Journal for the History of Science* (45:3, 2012), and specially

the introduction: Simone Turchetti, Néstor Herran, and Soraya Boudia, "Have We Ever Been Transnational? Towards a History of Science across and Beyond Borders," *British Journal for the History of Science* 45:3 (2012): 319–336.

2. On the UNESCO and radiological monitoring, see Jacob D. Hamblin, "Exorcising Ghosts in the Age of Automation: United Nations Experts and Atoms for Peace," *Technology and Culture* 47:4 (2006): 734–756. On the establishment of an international regulatory system, see Soraya Boudia, "Global Regulation: Controlling and Accepting Radioactivity Risks," *History and Technology* 23:4 (2007): 389–406, in which she suggests that the scientific expertise embodied in the committee and other international organizations for the global regulation of radioactivity were essentially instruments used by political leaders "for restoring and strengthening public trust."

3. For a comprehensive study of early US nuclear monitoring activities, see Charles A. Ziegler and David Jacobson, *Spying without Spies. Origin of America's Secret Nuclear Intelligence Surveillance System* (Westport, CT: Praeger, 1995).

4. On the German nuclear bomb project, see Mark Walker, *Nazi Science: Myth, Truth, and the German Atomic Bomb* (Cambridge, MA: Perseus, 1995).

5. Maurice Ewing, "Long Range Sound Transmission in the Atmosphere," October 14, 1945, p. 1, General LeMay Collection, microfilm roll No. 1760, frames 1978–1983, Air Force Historical Research Center. Quoted in Ziegler and Jacobson, *Spying without Spies*, 57.

6. An article analyzing different monitoring methods and stressing the superiority of the radiological method was published in the October 1946 issue of the *Intelligence Review*, a classified journal distributed to government officials involved in intelligence activities to keep them informed about recent developments in intelligence.

7. "Report of Operation Fitzwilliam," Vol. I, Tab A. Period 1948, Microfilm Roll A15941, Frames 4–495, Air Force Historical Research Center. Quoted in Ziegler and Jacobson, *Spying without Spies*, 138–139.

8. This did not imply the extinction of all monitoring programs in other branches of the US military. An example was the US Navy project *Rainbarrel*, aiming at the monitoring of radiation in rain water.

9. This last aspect of the nuclear cycle revealed as the most difficult to ascertain, as it implied not only the collection of air samples from different geographical locations—a task that involved the installation of portable stations in military and commercial planes—but also a better understanding of the global mixing of gases in the atmosphere. The difficulties associated with fully implementing this system of global data collection prevented correct estimations of plutonium production until the mid-1950s.

10. This collaboration had been forged despite the limitations imposed by the 1946 McMahon Bill, which prevented the exchange of American nuclear information with foreign countries. The need of gathering intelligence about Soviet nuclear activities implied, however that as early as 1948 the AEC allowed collaboration with Canada and the United Kingdom in the fields of geophysics and meteorology. On the US–UK collaboration in surveillance of the Soviet Program, see Michael S. Goodman, *Spying on the Nuclear Bear: Anglo-American Intelligence and the Soviet Bomb* (Stanford: Stanford University Press, 2007), 43–46.

11. Ibidem, 44.

12. Jeffrey T. Richelson, *Spying on the Bomb: American Nuclear Intelligence from Nazi Germany to Iran and North Korea* (New York: Norton, 2006), 113.

13. The participation of British spy Harold "Kim" Philby seems to have been essential in providing the Soviet Union information about American monitoring programs.

See Verne W. Newton, *The Cambridge Spies: The Untold Story of Maclean, Philby, and Burgess in America* (Lanham: Madison Books, 1999), on 336–341.

14. G. A. Goncharov, "Thermonuclear Milestones," *Physics Today* 49:11 (1996): 44–61.

15. In 1958, this network allowed them to detect and determine the causes of Sellafield accident, and to survey Bristish nuclear tests in the Pacific. Division of International Affairs, *Memorandum of Conversation. Discussion with Dr. Yves Rocard, French Physicist*, February 26, 1958. NARA archives, box 490, folder 21.33.

16. A complete report on the Castle Bravo test, entitled *CASTLE BRAVO: Fifty Years of Legend and Lore. A Guide to Off-Site Radiation Exposures* has been recently released by the Defense Threat Reduction Agency, a Department of Defense body charged with the analysis on nuclear and conventional weapons–related topics (http://www.stormingmedia.us/87/8722/A872275.html, accessed on June 20, 2013).

17. The origin and consequences of the fallout controversy in Japan has been studied by Toshihiro Higuchi in a series of paper: Toshihiro Higuchi, "An Environmental Origin of Antinuclear Activism in Japan, 1954–1963: The Politics of Risk, the Government, and the Grassroots Movement," *Peace & Change* 33:3 (2008): 333–366; "Atmospheric Nuclear Weapons Testing and the Debate on Risk Knowledge in Cold War America, 1945–1963," in J. R. McNeill and Corinna R. Unger, eds., *Environmental Histories of the Cold War* (Cambridge: Cambridge University Press, 2010), 301–322; "Tipping the Scale of Justice: the Fallout Suit of 1958 and the Environmental Legal Dimension of Nuclear Pacifism," *Peace & Change*, 38:1 (2013): 33–55.

18. US AEC, *Worldwide Effects of Atomic Weapons. Project Sunshine, August 6, 1953* (Santa Monica: Rand Corporation, 1956).

19. Ralph H. Lutts, "Chemical Fallout: Rachel Carson's Silent Spring, Radioactive Fallout, and the Environmental Movement," *Environmental Review* 9:3 (1985): 210–225.

20. "Nehru Proposes Atom 'Standstill' Pending UN Curb," *New York Times*, April 3, 1954. Quoted in Matthew Jones, *After Hiroshima: The United States, Race and Nuclear Weapons in Asia, 1945–1965* (Cambridge: Cambridge University Press, 2010), on 202.

21. Jacob D. Hamblin, "'A Dispassionate and Objective Effort: Negotiating the First Study on the Biological Effects of Atomic Radiation," *Journal of the History of Biology* 40:1 (2007): 147–177.

22. The idea of promoting international cooperation in radioprotection by means of the UNSCEAR fitted perfectly with the objectives of the Atoms for Peace program, launched by American president Eisenhower in the United Nations three years before. On the UN and the Atoms for Peace see, for example, Hamblin, "Exorcising Ghosts in the Age of Automation: United Nations Experts and Atoms for Peace," 734–756.

23. The argument that UNSCEAR was to delay the implementation of more radical solutions endorsed by activists (in this case, the test ban) has been raised in Boudia, "Global Regulation: Controlling and Accepting Radioactivity Risks," 389–406. A similar example of this dynamics can be found in the establishment of the Intergovernmental Panel on the Climate Change in the 1980s, as discussed by Shardul Agrawala, "Context and Early Origins of the Intergovernmental Panel on Climate Change," *Climatic Change* 39:4 (1998): 605–620.

24. Minutes of the Eight Meeting of the UNSCEAR, March 25, 1956.

25. The second meeting resulted in a yearly progress report, which was submitted to the General Assembly.

26. Minutes of the Seventh Meeting of the UNSCEAR, March 23, 1956. See also David Sowby, "ICRP and UNSCEAR: Some Distant Memories," *Journal of*

Radiological Protection 21 (2001): 57–62; and David Sowby, "Some Recollections of UNSCEAR," *Journal of Radiological Protection* 28 (2008): 271–276.

27. Brief for UK Delegation to United Nations Scientific Committee on Radiation, UK National Archives, folder AB 12–313.

28. Boudia, "Global Regulation: Controlling and Accepting Radioactivity Risks," 389–406 and Hamblin, "'A Dispassionate and Objective Effort: Negotiating the First Study on the Biological Effects of Atomic Radiation," 159–65. As seen by Hamblin, classified correspondence between the MRC and the NAS show that both institutions had regular exchanges in order to assure the convergence of results and conclusions. Despite their scientific character, these exchanges were coherent with both countries nuclear cooperation policy, aimed at the exchange of information while restricting other countries access to it.

29. James V. Neel and William J. Schull, *The Effect of Exposure to the Atomic Bomb on Pregnancy Termination in Hiroshima and Nagasaki* (Washington, DC: National Academy of Sciences National Research Council, 1956 [Publ. No. 461]). According to Susan Lindee, this report was a result of much compromise regarding "information inclusion and emphasis… [reflecting] the authors' sensitivity to the public impact and institutional meaning of their work." Susan Lindee, *Suffering Made Real, American Science and the Survivors of Hiroshima* (Chicago: University of Chicago Press, 1994), on 219. In particular, this could have meant incertitude the about the continuation of the program, which was on the verge of being cancelled because of its cost and uncertain scientific results.

30. Minutes of the Eight Meeting of the UNSCEAR, March 25, 1956. UN Archives.

31. Minutes of the Seventh Meeting of the UNSCEAR, March 23, 1956. UN Archives.

32. Ray K. Appleyard, "The Birth of UNSCEAR—the Midwife's Tale," *Journal of Radiological Protection* 30 (2010): 621–626.

33. Press conference held by Officers of the UNSCEAR held on January 30, 1958. Note No 1718, January 30, 1958. UN Archives, box S262/16.

34. Bibliography included in the final report made reference to 213 reports submitted by 30 countries and five international institutions, such as the Food and Agriculture Organization (FAO), the World Health Organization (WHO), and the World Meteorological Organization (WMO). That is, around 7,500 pages of reports. Of them, almost 3,400 pages had been submitted by the United States and almost 600 by the United Kingdom, which is a contribution of a similar order than the Soviet one (around 700 pages) and the Swedish one (around 550 pages), and slightly more than the Japanese one (around 450 pages) or the contribution from international organizations (around 400 pages).

35. A semantic analysis of the report shows that the most employed words were the terms "radiation/irradiation," object of study of the report, and "dose," implying the persistence of the toxicological notion of dose, taken from the MRC and NAS reports. Indeed, the verb "may" features in third place, reflecting the insistence on the incertitude about health effects in humans. Other widely used terms were "bones," marking the focus on Sr-90, and "mutation," indicator of the interest on genetic effects. Despite the effort to isolate discussions from the fallout controversy, the word "fallout" was also prominent, but in the same proportion than "natural (radiation)," putting at the same level the two factors by which exposure to radiations was interpreted.

36. Andrei D. Sakharov, "Radioactive Carbon from Nuclear Explosions and Non-threshold Biological Effects," *Soviet Journal of Atomic Energy* 4:6 (1958): 159–169.

37. Linus Pauling, "Genetic and Somatic Effects of Carbon-14," *Science* 3333 (1958): 1183–1186.

38. Robert J. List, Worldwide fallout from Operation Castle, May 17, 1955. US Weather Bureau, Washington, DC.

39. UNSCEAR, *Report to the General Assembly of the United Nations* (New York: UN, 1958), on 112.

40. John Krige, "Atoms for Peace, Scientific Internationalism, and Scientific Intelligence," *Osiris* 21 (2006): 161–181.

41. John H. Harley, Naomi A. Hallden, and Long D. Y. Ong, *Summary of Gummed Film Results through December 1959* (New York: U.S. Atomic Energy Commission's Health and Safety Laboratory, 1960), 14–15.

42. UNSCEAR Minutes of the seventh meeting, March 23, 1956.

43. On "infrastructural globalism," see Paul N. Edwards, *A Vast Machine. Computer Models, Climate Data, and the Politics of Global Warming* (Cambridge, MA: MIT Press, 2010), and Paul N. Edwards, "Meteorology as Infrastructural Globalism," *Osiris* 21 (2006): 229–250.

Chapter 4

"In God We Trust, All Others We Monitor": Seismology, Surveillance, and the Test Ban Negotiations

Simone Turchetti

During the second half of the twentieth century, seismologists gained a greater understanding of how waves produced by seismic events travel across the inner earth and reach distant places. Stations devoted to detection and recording of these events grew in number and the sensitivity of monitoring instrumentation like seismometers increased considerably. Yet the object of enquiring that allowed this considerable expansion was not the seismologist's traditional focus of research, the earthquake, but rather new and problematic: the nuclear weapons test (see Figure 4.1). Gazing through seismograms, experts now analyzed the characteristics of seismic movements produced by nuclear weapons, thus gaining new knowledge on their yield and location. Seismological studies thus featured in surveillance operations distinctive of the Cold War conflict and coupled with the gathering of "atomic" intelligence (intelligence on foreign atomic weapons programs).[1]

Seismology also played a key role in the unfolding of the Geneva nuclear test ban negotiations. The talks famously began on July 1, 1958, and saw British, American, and Russian delegates at the United Nations consider the possibility of abolishing nuclear tests. Since decisions on whether or not to sign a comprehensive treaty were made conditional upon the ability to policing a ban, especially by using seismic detection and verification, the talks allowed seismological research to thrive. As Kai-Henrik Barth has shown, the Geneva talks allowed seismologists to determine key areas to attack, thus prompting critical reviews of existing research programs. The result was seismology's "transformation [. . .] from a small academic discipline to a large academic-military-industrial enterprise."[2]

But the recent declassification of restricted information suggests that seismology's hidden ambitions also played a part in the proceedings and offers a different narrative for this important chapter in the history of nuclear disarmament.[3] The documents reveal that while negotiating as independent delegations, US and British teams played a secret game with their Communist discussants. Western intelligence had by then united in the effort of spying on the Soviet atomic program through the establishment of the "I" committee. So the stances

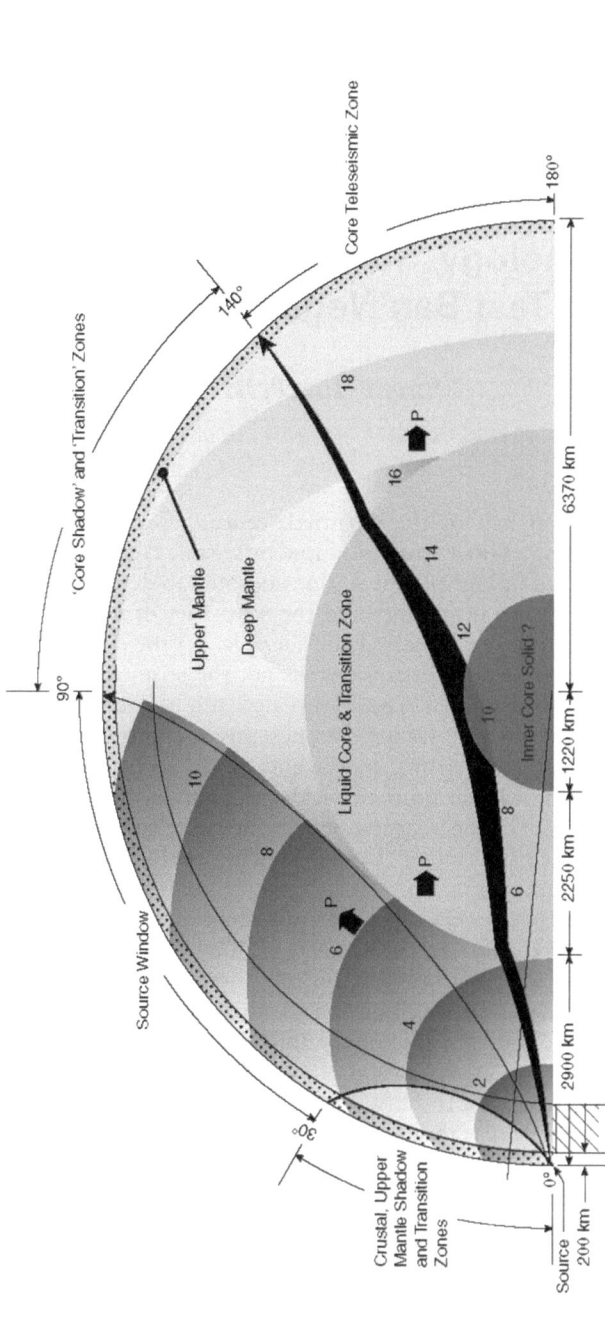

Figure 4.1 Propagation of seismic signals (P-waves)

Source: UKAEA, *The Detection and Recognition of Underground Explosions*, London: 1965, p. 19.

of Western delegations in Geneva did not derive exclusively from an evaluation on the possibility to police a test ban, but also from considerations on how the talks could be exploited to garner intelligence.[4] While it would be wrong to suggest that the talks were only a cover for attempts to extract information, so it is to say that their only ambition was nuclear disarmament.

This chapter charts the I committee's history in connection with the test ban negotiations and the advancement of seismology. I first illustrate how the committee's establishment followed the completion of the 1958 Conference of Experts. I then look at the history of the 1959–1961 Conference on the Discontinuance of Nuclear Weapons Tests, when intelligence gathering was prioritized by Western and Soviet officials alike in light of the prevailing pessimism about the possibility of reaching an agreement. And I finally consider how intelligence urgencies informed the patronage of seismology nationally and internationally when the Geneva talks reached an unsuccessful conclusion.

Atomic Intelligence and the Test Ban Talks

The 1958 nuclear test ban negotiations emerged at a key point in the history of nuclear disarmament characterized by the mounting public pressure to ratify an agreement and the growth of intelligence activities focusing on nuclear weapons. Proposals to ban nuclear tests were first put forward at the United Nations (UN) in 1948 and became a focus of international policymaking when the UN Disarmament Commission was established. In 1954 the Indian prime minister, Jawaharlal Nehru, called for testing to be halted in order to defuse global tensions. The incident of Japanese ship *Lucky Dragon*, exposed to the nuclear fallout of the Bravo test (Castle series), produced more protests. The American chemist Linus Pauling investigated the public health implications of nuclear fallout and in January 1958 he presented a petition to the UN signed by 9,000 scientists.[5]

Pauling's initiative persuaded decision makers. In March 1958 Soviet premier Nikita Khrushchev decided to unilaterally suspend weapons testing and this forced US president Dwight Eisenhower to look for a political solution. A panel of US experts led by physicist Hans Bethe concluded that a test ban treaty needed to be successfully policed and this could only be done through the realization of a worldwide network of stations equipped with advanced instrumentation. Eisenhower thus countered Khrushchev's initiative with a new proposal: experts should meet to plan how to monitor a test ban. Further discussions between diplomacies helped to complete the preparations for the so-called Conference of Experts. Two delegations comprising representatives of four countries for each bloc (the Soviet Union, Czechoslovakia, Poland and Romania, the United States, Britain, France, and Canada) met between July 1 and August 21, 1958.[6]

These initiatives received public coverage but by the time experts met in Geneva, several American, British, and Soviet government agencies were already, in great secrecy, busy monitoring nuclear explosions. Herbert (Pete) Scoville Jr., who was appointed in 1955 as Central Intelligence Agency (CIA) assistant director of scientific intelligence, referred to weapons testing as the "most reliable window" to collect information on a foreign nuclear program.[7] And from the late 1940s the collection of radioactive debris, the recording of seismic and acoustic waves, and the analysis of electromagnetic pulses generated by a nuclear blast helped ascertaining location, size, and composition of atomic devices tested by the two superpowers.

The collection of scientific information useful to intelligence personnel grew considerably during these years and in 1947 the US Air Force set up an agency specialized in the detection of nuclear explosions. Two years later, the Air Office of Atomic Energy (AFOAT-1) successfully detected the first Soviet nuclear test. Later on the agency set up a network, the Atomic Energy Detection System (AEDS), including stations in Turkey, Australia, and Germany. Information collected through the AEDS was processed in collaboration with the US Atomic Energy Commission (AEC) Sandia and Livermore laboratories.[8] In 1956 the CIA director instigated the establishment of a Joint Atomic Energy Intelligence Committee to oversee dissemination of information regarding foreign nuclear explosions obtained through the AEDS. It included representatives of the Joint Chiefs of Staff, State Department, Army, Navy, Air Force, the AEC, the CIA, and the Federal Bureau of Investigation (FBI).[9]

The British government sponsored atomic intelligence too. From 1955 a purposefully designed committee brought together the directors of the Atomic Weapons and Atomic Energy Research Establishments (AWRE, William Penney; AERE, John Cockcroft), the chairman of the Joint Intelligence Committee (Patrick Dean) and the Ministry of Defence (MoD) scientific adviser, Frederick Brundrett.[10] The intelligence setup superintended a group, the Technical Research Unit (TRU), liaising with AWRE experts on analysis and techniques, with MI6 for intelligence gathering, and with the Royal Air Force for surveillance operations and debris collection. US and UK atomic intelligence organizations worked together on an ad hoc basis on several occasions.[11]

And although Kristie Mackrakis has recently argued that the Soviets privileged gathering intelligence through human agents, surveillance of foreign nuclear tests by seismological detection was not neglected in Soviet Russia.[12] In 1947 seismologist Mikhail Aleksandrovich Sadovsky established a detection group at the Institute of Chemical Physics. And from 1960 he formed a much larger group devoted to seismic monitoring at Moscow's Institute of Physics of the Earth. Meanwhile, the Soviet Ministry of Defense established the Special Monitoring Service, whose activities were entirely covered by secrecy and coordinated with GRU's spymasters.[13]

So when the Conference of Experts was envisaged for the first time, these atomic intelligence operatives thought about the talks mainly as a valuable opportunity for gathering otherwise unavailable information on these rivaling monitoring systems. Actually, by the time the conference was organized, US and British intelligence agencies had already coordinated their efforts to jointly spy on the Soviet nuclear program through a new echelon: the I committee.

The I Committee and the Conference of Experts

Six months before the test ban talks, the AWRE research director William Penney suggested reviewing the ad hoc mechanism that had typified US–UK collaboration in the atomic intelligence field until then. In December 1957, he put forward the proposal for establishing a joint committee to oversee information sharing on seismic data and data from debris collection in order to more profitably examine them and share the deriving assessments on the state of advancement of the Soviet nuclear program.[14] The TRU officers transmitted Penney's proposal to Scoville and the AEC Military Liaison with the Department of Defense, General Herbert B. Loper. Their comments on the proposal were returned to Brundrett and Dean.[15] Yet, the Americans hedged on the scheme. Britain had yet to test the H-bomb and information sharing may have accelerated its weapons program. This resistance continued up until August 1958 when the Grapple test series reached successful completion: the AWRE could now produce H-bombs.[16]

Meanwhile atomic intelligence officers attended the Conference of Experts either as delegates or observers. Penney played a key role during the proceedings while keeping his role within his country's atomic intelligence organization hidden. Yet he passed on to the Foreign Office details that had intelligence implications, also arguing that one of the British delegation's secondary objectives was exactly gaining knowledge on the Soviet monitoring system.[17] The TRU director Robert Press also visited Switzerland. And Scoville's colleagues at the CIA sought to obtain intelligence shadowing the Eastern delegates. They learnt for instance that the Russian Igor Tamm had "considerable fondness of pretty girls" and that the Czech Alois Zatopek was unhappy with travel restrictions.[18] On August 21, 1958, the meeting ended, but intelligence-gathering continued since the Second Conference on the Peaceful Uses of Atomic Energy was about to take place in the same location (the UN *Palais des Nations*). Now the CIA had laid out "extensive plans" for the "intelligence exploitation" of the conference.[19] And agent Henry S. Lowenhaupt, who had previously contributed to the localization of the Soviet atomic polygon in Semipalatinsk through a seismological computation method, supervised a number of intelligence operations with the help of the AEC director of intelligence, Charles Reichardt.

The I committee was established at the end of these intelligence activities and following the signing, on August 4, 1958, of the US–UK Mutual Defense Agreement. The treaty catered to collaboration on the uses of atomic energy for mutual defense purposes and entailed sharing classified information on atomic weapons. On September 30, a letter from TRU personnel to Penney confirmed that the Americans were now willing to hold intelligence talks on the Soviet program and had organized a meeting in Washington, DC, for this purpose, due to take place the following October.[20] So the I committee brought together key players in the US and British atomic intelligence communities and became a key feature in the defense partnership across the Atlantic. In 1959 CIA director Allen Dulles argued that the Mutual Defense Agreement represented "a meaningful frame of reference for the exchange of corresponding Restricted Data information on Soviet nuclear weapons" and that the intelligence agencies of both countries "have a keen interest in this vital area."[21]

The minutes of the first I committee meeting are unavailable, but Scoville filed an intelligence report on the Conference of Experts in the week that followed the meeting, thus suggesting that the same assessment was discussed with the British.[22] Scoville argued that a ban treaty could be extremely beneficial to Western intelligence. The experts in Geneva had signed a joint statement claiming that a test ban treaty could be policed by 160–170 stations; some located in Soviet Union. These installations had "great intelligence potential" and Scoville valued especially Soviet concessions on flights over Russia's territory.[23] He concluded that Western intelligence had to gain from a test ban. The Eastern delegation had proven to be less skilled in the scientific analysis of detection problems. The Russian experts relied on theories rather than the statistical approach that typified Western science.

British atomic intelligence officers came to similar conclusions, but disagreed with the US colleagues on Soviet monitoring. Penney stressed that the Soviets were skilled in the use of acoustic methods and was unconvinced about their expertise in the electromagnetic technique that, conversely, had impressed Scoville.[24] Moreover Scoville's optimistic forecast failed to impress British delegates. A ban might have been favorably perceived at the CIA headquarters also because it was bound to have a restraining effect on other countries' nuclear ambitions. A disarmament agreement would thus "bring into play strong public pressure against testing [...] even though such countries might not initially be parties to the agreement."[25] This made the intelligence and defense partnership with Britain even more important. Britain's role in the NATO alliance would work as a deterrent to efforts by other European nations.

President Eisenhower took these considerations back into the domain of policymaking. He now indicated that while detection of nuclear tests was technically feasible, only those countries that already possessed nuclear weapons should undertake future talks.[26] He thus capitalized on the intelligence partnership with Britain, knowing that while British, American, and Soviet delegations would meet again as independent parties, two of them shared defense information and, more significantly, atomic intelligence.

False Intentions at the Discontinuance Conference

While Eisenhower was busy advocating new talks, testing resumed. In October 1958 a new series of nuclear tests (Hardtack II) took place at the Nevada site. The British government ordered three more tests. Khrushchev could not let these explosions go unchallenged and approved 14 more. This prelude to the Conference on Discontinuance of Nuclear Weapon Tests admittedly did not cast a positive light on the talks. The proceedings opened on October 31, 1958, with three delegations, led by the US representative on the UN Security Council, James Wadsworth; his Soviet counterpart Semyon S. Tsarapkin; and the UK minister of state for foreign affairs, David Ormsby-Gore. The talks confirmed that a comprehensive test ban appeared a distant target. The three nuclear powers agreed on a self-imposed two-year moratorium but the delegations quarreled on everything else.

John Walker has recently claimed that political opportunism was the main reason why talks resumed given that Eisenhower and Harold Macmillan, the British prime minister, had agreed during a number of meetings that they couldn't renounce to test further.[27] A test ban would have compromised further nuclear weapons development including the ongoing US miniaturization program.[28] The opposition of key players in this program, such as nuclear scientist Edward Teller, was well known. In January 1959 Penney advised Macmillan that it was in the British interest to agree only to an atmospheric tests ban, in order to continue testing underground.[29]

But the recently disclosed documentation tells us is that talks continued not just because of political opportunism, but also because of the intelligence gains that could be derived from these meetings. In particular, US and British delegations took important decisions on what to divulge during the talks in order to force the Russian to inadvertently give away information.

The focus on underground testing entailed growing attention toward seismic detection, making it the chief means to monitor foreign nuclear explosions. Underground tests, in contrast with atmospheric tests, released less or no radioactive debris above ground and were undetectable through acoustic and electromagnetic means. This is the reason why seismology took center stage in the proceedings. On January 5, 1959, Wadsworth presented the document "New Seismic Data," based on the recent Hardtack II series, to his British and Soviet counterparts. The series, he argued, offered experimental evidence that as much as 10 times the number of low yield explosions (below 5 kilotons—kt hereafter) estimated during the Conference of Experts would go undetected if the policing system the experts had suggested was adopted.[30]

As sufficient information was provided on how these conclusions had been reached, the Soviets refused to accept them and the negotiations came to a deadlock. It did not help that on January 20, 1959, in a farcical twist of events, the existence of AFOAT-1 was mistakenly revealed when his former director, Doyle Northrup, received a presidential award. The revelation persuaded the US Air Force to rename AFOAT-1 as Technical Applications Center (AFTAC). AFTAC's motto: "In God We Trust, All Others We Monitor" was (and still is) a somewhat unsubtle declaration of its surveillance ambitions.

In light of the disagreement with the Soviets, the US delegation could have easily walked out. But, in fact, in March 1959 US officials kept the talks drag on. They even suggested restoring scientific debating by arranging technical working groups, a proposal that Tsarapkin eventually accepted. Technical Group 1 (TWG1) met in Geneva between June 23 and July 15, 1959. The recommendations put before the conference at the end of TWG1 showed agreement on ways to police atmospheric testing, but left the vexed issues of underground testing and the ability of seismologists to monitor illegal tests unanswered.[31]

Now the US experts released information selectively to avoid letting their counterparts know more about their seismic monitoring capability and, in so doing, understand more about that of their adversaries. Eisenhower had by then appointed a Panel on Seismic Improvement (chaired by US geophysicist and science administrator Lloyd Viel Berkner) to identify new requirements for seismic monitoring. The panel eventually concluded that more funding should be made available to US institutions to carry out basic and applied seismological research.[32] Meanwhile US seismologists had found out that a nuclear explosion could be "muffled" by setting it off in a large cavity. But the bulk of information on *decoupling* (the muffling effect) was divulged after the report by the Berkner Panel and a paper on the subject by scientists Ernest Martinelli, Edward Teller, and Al Latter were both published. And this was only days before the beginning of TWG1.[33]

Moreover details on nuclear tests that had enabled to reach the analysis contained in New Seismic Data were not made available. The last US test series, Hardtack II, had comprised a number of explosions in the kiloton and subkiloton range. It was a series of 37 tests (some underground) but only four had been officially announced (Figure 4.2).[34] Omissions on test data had a significant impact on the talks, especially because some of the unannounced tests (like

Figure 4.2 Atmospheric test De Baca, Hardtack II series, Nevada test site, 26 October 1958
Source: National Nuclear Security Administration / Nevada Site Office.

the 6kt *Socorro*) had helped US seismologists to understand the correlation between amplitude of seismic waves produced by atomic explosions and yield of tested weapons.[35] Notably, other scientific papers discussing the implications of these underground tests for seismic detection were embargoed; a paper by seismologists Frank Press and D. T. Griggs could not be published before 1961.[36]

If TWG1 represented an opportunity for the US delegation to release information selectively, then Technical Working Group 2 (TWG2) was typified by attempts to exploit this selective release to induce the Soviet delegates to involuntarily offer precious information on their own monitoring program, as we shall now see.

Looking for Revelations during TWG2

On occasion of the meetings of the TWG2, experts from both sides appear to have acted like skilled intelligence agents, presumably because they were briefed by atomic intelligence officers before meeting up. In August 1959 Scoville and AWRE director Nyman Levin agreed on information-sharing procedures in the I committee, indicating that priority should be given to "exchanges [...] in respect of research work in connection with control systems for monitoring the Discontinuance of nuclear tests."[37] The agreement anticipated the organization of the second I committee meeting due to take place in London between October 5 and 9, 1959. Further exchanges between Levin and Loper anticipated the meeting, which was attended by 32 delegates from both countries.[38] These prominent administrators from scientific and intelligence agencies discussed key features of the Soviet nuclear program. Brundrett and Penney led the British delegation, which also included TRU and AWRE personnel. The US delegation included Martinelli, whose work on decoupling had just been published, and high-level officials in the US atomic intelligence establishment like Reichardt, Loper, AFTAC's new director Jermain F. Rodenhouser, and CIA officer Irl D'Arcy Brent (who replaced Scoville on that occasion).[39]

The list of questions that anticipated the meeting show that it entailed exchanging knowledge on Soviet thermonuclear tests, characteristics of warheads, testing at high altitude and underground, and "subkiloton shots." Recent intelligence data put together on the occasion of the last series of tests carried out in the Soviet Union were also analyzed.[40] These items were discussed mainly during the first day of the conference, but the second day was devoted instead to the conclusions reached at TWG1. The delegates thus considered the intelligence implications of the control system outlined in Geneva and the groundwork for an intelligence exploitation of the forthcoming TWG2 (November 24, 1959–December 19, 1959).[41]

A key target for both intelligence groups was to find out more about Soviet monitoring efforts and there was no clear indication that a Russian network exclusively devoted to seismic detection existed. Western seismologists knew that Soviet monitoring stations near the Arctic such as those in Mikhnevo and Kuldar (Sakhalin Island) had importance because of their geographical location. But it was especially the establishment of Antarctic seismic stations in preparation for the International Geophysical Year (IGY, 1957–58) that had provided evidence of Soviet ambitions to improve their seismic capability. In 1956, a new Soviet seismic station was set up at Mirny (East Antarctica).

The chairman of the Panel on Seismic Improvement, Lloyd Berkner, had played a key role in the IGY organization and prepared reports for the US State Department (and CIA) on the intelligence gains to be derived from promoting international collaboration in geophysical research, including seismology.[42] And before the IGY, the US National Committee, responsible for its organization, had transmitted to the CIA technical data on other countries' programs. The report documented Soviet seismic work in the polar regions, the characteristics of Soviet instruments to be utilized and the location of new seismic stations.[43]

The polar regions were critical to long-range detection as seismic signals propagated through polar "pipelines" thousands of miles away from the epicenter of a nuclear explosion. US seismologists knew they could travel very far, as Nevada tests were recorded at the Canadian station of Mould Bay (above the Arctic Circle). These waves, however, crossed the Arctic and went through the old geological structures of continental Russia, also reaching Soviet seismic stations such as that of Borovoy (Kirkov Oblast) and thus offering to Soviet seismologists a wealth of data on US tests.

Thus US seismologists attending the TWG2, such as AFTAC's Carl Romney, wanted to know more about what their Soviet colleagues had managed to detect. Romney had been a US delegate at the Conference of Experts as well and he had tried to figure out if the Soviets possessed an operational monitoring system when the Russian seismologist Ivan Pasechnik had enquired about location and time of the 1955 US test, Wigwam.[44] During TWG2 he seized the opportunity to know more about Soviet seismology when, due to the lack of information made available by the US delegation, the chief scientific delegate Yevgeni K. Federov asked impatiently if all the US data had been presented. Romney now "pressed him for any data the Soviets might have," forcing Federov to confirm that Soviet seismologists had detected the Hardtack II shot *Blanca* (22kt) at the stations of Mirny and Kuldar.[45] That the much lower yield *Logan* (5kt) was not detected confirmed the lack of sensitivity typifying the Soviet monitoring system.

Russian experts had an equal interest in withdrawing seismic data and acquiring details on the Western control systems. They refused to discuss verification procedures to avoid entering a "nitty-gritty" area in which sensitive information could inadvertently be divulged. They were also careful enough to avoid showing too many seismograms. And they were cautious in discussing their professional affiliation. For instance, Sadovsky kept hidden his responsibility as leader of the guarded seismic detection center based in Moscow.

Unsurprisingly, the TWG2 ended unsuccessfully as delegations could not agree on the scientific assessment of how to police underground explosions. In particular, there was "disagreement regarding the interpretation of the new data from the Hardtack experiment [...] and the question of de-coupling." Actually, the Soviet delegation filed separate conclusions stating that their US colleagues were "on the brink of absurdity."[46] Perhaps even more absurd was the British statement released at the end of TWG2. The UK delegation concluded that US delegates were right in claiming that "the 1958 Experts were too optimistic" even if the evidence proving it, the Hardtack II data, had yet to be fully divulged.[47]

But looking at the conclusions reached in Geneva exclusively from a scientific or political viewpoint would be misleading. TWG2 was designed to help Western delegates to gain a better understanding of the Soviet monitoring program. The attempt to assemble scientific intelligence had been prepared well in advance through the activities of the secret I committee. Thus British experts knew much more about the Hardtack II series than what was officially released at the end of the TWG2. And, as we shall now see, this intelligence groundwork was decisive in instigating a scientific race to improve methods of seismic detection in the three countries whose delegations had featured so prominently in the Geneva talks.

The Rise of Seismology in Britain

On January 12, 1960, talks in Geneva resumed once again but proposals and counterproposals did not lead to an agreement. It did not help that on May 1, 1960, a U-2 reconnaissance plane was shot down in Russian territory. The incident suggested that the US administration had put surveillance before policymaking. By the end of 1960, Soviet and US officials still disagreed on inspections of test sites. And the beginning on that year of French nuclear testing in the Sahara desert cast a negative light on the talks.[48] In January 1961, the new US president, John F. Kennedy, argued that he hoped to continue talks, but two months later he announced new tests. In June, shortly after the beginning of the Berlin crisis, Tsarapkin indicated that the Geneva talks were a waste of time. The Soviet Union would also resume testing the following September.[49] The now mounting disagreement on a test ban made it even more urgent to advance seismological research. Whatever world leaders would agree upon, the three administrations that had promoted the Geneva negotiations now fast-tracked the setting up of a seismic monitoring complex, allowing to collect fresh data on the tests recently announced.

In 1960 Penney advised the prime minister to sponsor more seismological work in Britain. This was especially in the knowledge that although the I committee granted data sharing with US seismologists, this may end unless Britain's ally saw advantages in the collaboration, especially in terms of refining seismic techniques. Already in December 1959 the AWRE Field Experiments Division had carried out decoupling experiments with nonnuclear explosives.[50] After that, seismological detection work moved to Blacknest (near Aldermaston) and the head of the AWRE division, Ieuan Maddock, was appointed director of the new Seismic Detection and Verification Unit.[51] Hal Thirlaway, formerly a student at the Department of Geodesy and Geophysics (University of Cambridge) and a MI6 consultant in Pakistan, was now given responsibility for overseeing the seismic program.[52] Thirlaway and his co-workers pioneered a new method of detection based on sets of seismometers forming electronically steerable arrays (see Figure 4.3).[53]

Blacknest was established mainly because of the need to continue sharing atomic intelligence with the Americans. Prime Minister Macmillan believed it necessary to invest as little as possible in seismology, just the amount needed to allow the UK government to play some role in the Geneva negotiations. And

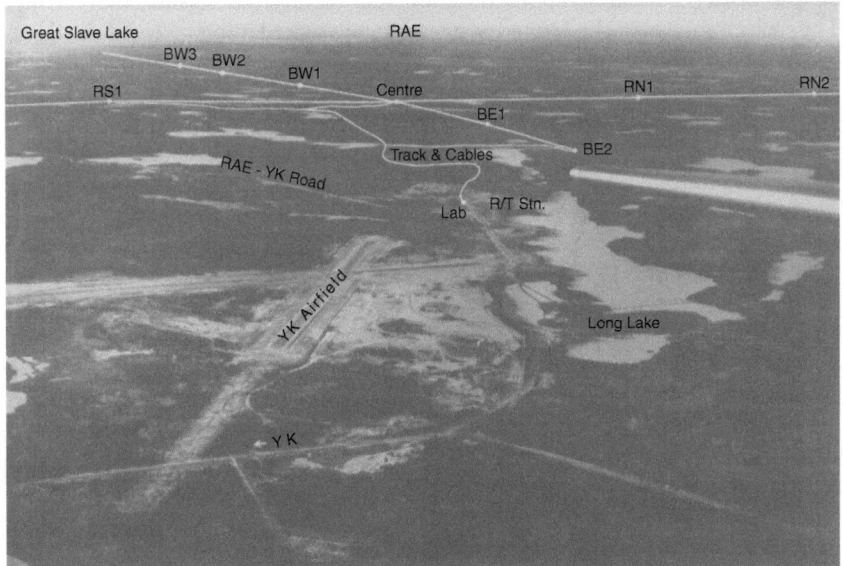

Figure 4.3 Aerial view of the Yellowknife array (Canada)

Source: UKAEA, *The Detection and Recognition of Underground Explosions*, London: 1965, p. 35.

when the talks were about to end unsuccessfully, the prime minister indicated that funding for seismological work should be kept "to the absolute minimum."[54] But Macmillan was eventually persuaded to reconsider his decisions by the Foreign Secretary Alec Douglas-Home (prime minister from 1963). Levin had recently informed the foreign minister about the consequences of withdrawing from seismology work and, in particular, about the nature of the high-level secret encounters that had taken place in the context of the I committee.[55] So Douglas-Home promptly informed the prime minister that, "If we were to maintain *full* co-operation with the US [Italic mine, . . .] it would be important that we should have behind us the backing of our own research program."[56] Macmillan was eventually persuaded to fund Blacknest for another year and Douglas-Home convinced the Treasury to make Blacknest's budget part of the "self-generated" military R&D expenditure. In this way he effectively removed its funding from those sections of the budget that required the prime minister's approval.[57]

The promotion of coordination in seismic detection and intelligence work was also decisive to the trajectory of Britain's atomic testing program as it now became possible to for British specialists to conduct join tests with US colleagues at Nevada test site (such as on occasion of the 1962 trial *Pampas*). This meant furthering knowledge on weapons and detection systems as well as making it possible for Britain to save considerably on testing costs by closing down the site of Maralinga (Australia).[58]

The "Iron Curtain" of Seismology: US and Soviet Detection Plans

The end of the Geneva talks confirmed that the only way to gain new knowledge on the Soviet nuclear weapons program was to improve seismic monitoring. Thus seismologists in the United States and the Soviet Union received the funding needed to complete a number of research programs that on many occasions had the covert ambition to put foreign nuclear tests under surveillance.

The 1959 Berkner panel's report and the conclusions reached at the I committee meetings both had an impact on the US program of seismological research. This was a program typified by large-scale research projects and although experts from the United Kingdom and the United States continued meeting regularly and shared ideas, soon their understanding of seismic detection differed substantially.[59] Romney criticized the seismic array technique even when his British colleagues claimed to have detected low-yield French, Soviet, and US tests.[60] In 1962 bilateral talks at the State Department confirmed the existence of different opinions across the Atlantic about how to make seismic detection more effective.[61]

This divergence can be explained in terms of different approaches to monitoring foreign explosions. British seismologists believed that a method based on data collection from a few dispersed (but very sensitive) arrays targeting specific areas sufficed to the task of securing a comprehensive monitoring of foreign nuclear tests. Conversely American seismologists wished to increase the "density" of seismic data by augmenting the number of monitoring stations across the globe because they believed that an efficient seismic detection system needed to spread out globally.[62]

Vela Uniform, a $10 million project managed by the DoD Advanced Research Projects Agency (ARPA), allowed the growth of US seismology providing funding to both the military and academic departments.[63] The program was officially announced on May 7, 1960, and entailed offering equipment for seismic observatories to foreign countries. The US Coast and Geodetic Survey developed new instrumentation for this purpose and by 1963 the World-Wide Standardized Seismic Network (WWSSN) consisted of 76 operative stations and reached a remarkable number of 125 by the time of completion.[64] Although the WWSSN produced open data, correlation with information from the AEDS (whose stations had also increased by then) offered otherwise unavailable details on Soviet tests to US experts.[65]

Scoville moved out of the CIA's scientific intelligence office to take a similar position within the US Arms Control Development Agency (ACDA), established on September 26, 1961.[66] The ACDA sponsored the placing of Temporary Mobile Seismic Observatories abroad, while downplaying (or concealing) their significance for detection of nuclear test. When in February 1963 seismic equipment was transferred to Lagos (Nigeria), one ACDA officer informed the local embassy that in presenting the research program emphasis should be put on its "aims in the realm of science" such as the "improved prediction of destructive earthquakes" rather than increasing "capability for detection and identification of underground nuclear tests."[67] ACDA schemes entailed avoiding the sharing of sensitive information with the host countries. The placing of equipment in Bolivia followed the recommendation "to avoid attention to our mobile unit" and indicating that "if enquiries or public comments appeared in the press, details about the project should be provided by placing some emphasis [...] to the advantages deriving from seismological research to pure science."[68] US offers of seismological equipment to Yugoslavia's government were promptly refused as it "would only be a 'cover' for American detection of Soviet atomic tests."[69] A renewed offer was made after the Skopje earthquake of July 26, 1963, and rejected again.

The Soviet Union responded to these efforts by sponsoring an independent seismic program, which entailed primarily the extension of its already existing network: the Uniform System of Seismic Observations (USSO). The 106 stations located in Soviet territory in 1959 almost doubled by 1965. Meanwhile the Soviet Union established new seismic stations in Antarctica and in a number of allied countries such as Cuba and Vietnam. Moreover, Soviet seismic bulletins such as the *Byulleten's eti seysmicheskikh stantsiy SSSR* introduced censorship procedures to avoid letting Western seismologists know about seismic events in Russian territory that offered data on nuclear tests.[70] The reorganization, growth, and development of competing seismic surveillance networks set up by the two superpowers allowed a sort of Iron Curtain to descend over the international seismological community. The division entailed promoting alternative methods of standardization and equipment and curtailing access to seismic data, while also affecting the international dialogue between researchers. This division also materialized in the location of new seismic stations along the spheres of influence that Cold War alliances produced.

Conclusions

On August 5, 1963, the United States, the USSR and Britain jointly signed a Partial Test Ban Treaty (PTBT) prohibiting all nuclear tests aside from those underground. The new agreement was welcomed as a means to promote détente, especially after the Cuban Missile Crisis. The treaty was, however, the only political outcome of five years of negotiations, which had been typified by political disagreement and scientific quarrels. American, Russians, and British negotiators signed a partial rather than a comprehensive ban because of the deadlock on how to set up a system to verify compliance. US delegates reiterated Romney's argument that seismologists could not successfully detect low-yield nuclear explosion and thus could not sign a treaty extending to underground testing.

So in July 1963 Khrushchev proposed a ban that did not include testing underground and encompassed exclusively nuclear tests in the atmosphere, space, and underwater. This was exactly the proposal that Macmillan and Eisenhower had discussed back in 1959 and since the AEC had already in place plans to move its testing program underground, Khrushchev found its interlocutors sympathetic. In this way, however, the PTBT effectively sanctioned (and therefore increased the number of) underground tests, thus making seismic detection an even more compelling means to garner atomic intelligence. And indeed, the considerable growth of seismology as a discipline was mainly a result of this growing interest in putting foreign nuclear programs under surveillance by detecting earthquakes generated by underground nuclear explosions. Thus for quite some time seismology did not feature in international negotiations on a nuclear test ban either. And no internationally recognized organization adopted seismological means to verify nuclear explosions up until 1997 when, with the Cold War already ended, an International Monitoring System was set up by Comprehensive Test Ban Treaty Organization in order to finally enable to bring the prohibition to test nuclear weapons into force.

This chapter showed that these developments cannot be explained exclusively in light of policymaking or not even political expediency. The Conference of Experts was the first opportunity for a number of agencies devoted to surveillance to figure out how much information on methods to detect nuclear explosion could be gained in scientific meetings. It is for this reason that exchanges of information on what should be discussed in these talks was agreed upon beforehand by US and UK intelligence groups in the newly established I committee. This led to two important developments. At the TWG1 information was released selectively so as to prevent Soviet delegates from gaining information on recent American tests. At the TWG2 the selective release of this information was decisive in knowing more about Soviet detection methods and results. Thus Scoville was certainly correct in suggesting that testing was the main window of opportunity to collect intelligence on the advancement of a foreign nuclear program. But it is also true that talking about a test ban was the most suitable way for US and UK atomic intelligence groups to find out about how reliable this window really was.

Unfortunately, the heavily redacted files on the I committee do not allow us to fully explain the role that scientific intelligence played in the conference proceedings. But the new evidence suggests that scientific and political goals openly laid out in Geneva overlapped covert intelligence ambitions. Key experts involved in the talks (Penney, Sadovsky, and Romney) also played key roles either in the I committee or similar atomic intelligence agencies in the Soviet Union. The documents also reveal that they agreed to discuss with key intelligence officers (Scoville, Press, and Reichardt) attending Geneva and designing plans for the intelligence exploitation of the talks. Key pronouncements at the conference were eventually reviewed in the context of the I committee meetings.

The new evidence thus helps us to reconsider the history of seismology during the Cold War. True as it is that the Geneva negotiations boosted the search for new means for seismic detection, they did so mainly because of the trajectories of intelligence cooperation that the meetings instigated. Britain's decision to continue a seismic program derived from the wish to exchange more intelligence with the US community. US seismological research nationally and internationally thrived partly because of the conclusions that Berkner's Panel of Seismic Improvement arrived at, but also because of the implications of seismic work for surveillance of atomic tests. The ACDA research activities demonstrate that the goals of seismic detection were at times hidden from the public because of their intelligence connotations. During the Cold War, these entangled open and secret dimensions of seismological research sanctioned the existence of invisible detection networks and set intelligence and research communities together in the gathering of restricted information and expert advice.

Notes

1. On atomic intelligence see: Jeffrey Richelson, *Spying on the Bomb. American Nuclear Intelligence from Nazi Germany to Iran and North Korea* (New York: W. W. Norton & Co., 2006); Charles A. Ziegler and David Jacobson, *Spying without Spies. Origins of America's Nuclear Surveillance System* (Westport, CT: Praeger, 1995); Michael S. Goodman, *Spying on the Nuclear Bear: Anglo-American Intelligence and the Soviet Bomb* (Stanford: Stanford University Press, 2007). On scientific intelligence see also: Ronald Doel, "Scientists as Policymakers, Advisors and Intelligence Agents," in T. Söderqvist (ed.), *The Historiography of Contemporary Science and Technology* (Amsterdam: Harwood, 1997), 215–244; and Kristie Mackrakis, "Technophilic Hubris and Espionage Styles during the Cold War," *Isis* 101 (2010): 378–385.

2. Kai-Henrik Barth, "The Politics of Seismology," *Social Studies of Science* 33 (2003): 743–781, on 743 and 758. See also K. Barth, "Science and Politics in Early Nuclear Arms Control Negotiations," *Physics Today* 51 (1998): 34–39; and Kai-Henrik Barth, "Catalysts of Change: Scientists as Transnational Arms Control Advocates in the 1980s." In *Global Power Knowledge: Science and Technology in International Affairs (Osiris* 21), edited by J. Krige and Kai-Henrik Barth, 161–181. Chicago: University of Chicago Press, 2006. On seismology during the Cold War, see also: Bruce Bolt, *Nuclear Explosions and Earthquakes. The Parted Veil* (San Francisco: W. H. Freeman & Co., 1976) and Carl Romney, *Detecting the Bomb. The Role of Seismology in the Cold War* (Washington, DC: New Academia, 2008).

3. The FOI application asked for the release of intelligence files produced in the British Atomic Weapons Research Establishment (AWRE) in the 1950s. Previous accounts include: Harold K. Jacobson and Eric Stein, *Diplomats, Scientists, and Politicians: The United States and the Nuclear Test Ban Negotiations* (Ann Arbor: University of Michigan Press, 1966); Robert Divine, *Blowing on the Wind/ The Nuclear Test Ban Debate* (New York: Oxford University Press, 1978); John Walker, *British Nuclear Weapons and the Test Ban, 1954–1973* (London: Ashgate, 2010). See also Disarmament Administration, *Geneva Conference on the Discontinuance of Nuclear Weapon Tests. History and Analysis of Negotiations* (US State Department, 1961).

4. For a similar case see John Krige, "Atoms for Peace, Scientific Internationalism, and Scientific Intelligence," in Krige and Barth (eds.), *Global Power Knowledge: Science and Technology in International Affairs*, 161–181.

5. Divine, *Blowing on the Wind*, chap. 5.

6. Ibid, 143–146.

7. Scoville cit. in Goodman, *Spying on the Nuclear Bear*, 86. On his background and role see also: J. Richelson, *The Wizards of Langley. Inside the CIA's Directorate of Science and Technology* (Cambridge, MA: Westview, 2001), 8–9.

8. On the AFOAT-1 see Mary Welch, "AFTAC [ex AFOAT-1] Celebrates 50 Years of Long Range Detection," *The Monitor*, October 1997, 8–32. See also: Romney, *Detecting the Bomb*, 6–11; Ziegler and Jacobson, *Spying without Spies*, 189.

9. A. W. Dulles, Control of Initial Information Regarding Foreign Nuclear Explosions, CIA Director Directive 1/6, August 5, 1959, Secret [accessible in document 0001216234 at www.foia.ucia.gov].

10. Between 1954 and 1959 Brundrett was also chairman of the Defence Research Policy Committee, the government's group of scientists responsible for the overall planning of military research in Britain. On this see: Jon Agar and Brian Balmer, "British Scientists and the Cold War: The Defence Research Policy Committee and Information Networks, 1947–1963," *Historical Studies in the Physical and Biological Sciences* 28:2 (1998): 209–252. See Cantoni and Veneer, this volume.

11. TRU's structure went through several revisions over time as shown in Goodman, *Spying on the Nuclear Bear*, 171–174 and 210–211.

12. Mackrakis, "Technophilic Hubris and Espionage Styles During the Cold War," 379.

13. *Glavnoye Razvedyvatel'noye Upravleniye* (GRU) was Soviet Union's foreign military intelligence agency (and is still an agency of the Russian Federation's armed forces). On the Soviet Monitoring Service see: Vitaly Khalturin et al., "A Review of Nuclear Testing by the Soviet Union at Novaya Zemlya, 1955–1990," *Science and Global Security* 13:1 (2005): 1–42, on 5–7. See also A. P. Vasiliev et al., *The Service born in the Atomic Century: The History of the Soviet Special Monitoring Service* (Moscow: SSK, 1998, in Russian).

14. Undated, List of Questions—Draft, September 28, 1959, Secret. FOI application 12–03–10–133704–001.

15. TRU to Penney, December 16, 1957, Secret and TRU to Penney, January 10, 1958, Secret. Loper was responsible for AEC relations with DoD. FOI application 12–03–10–133704–001.

16. On the series, see Lorna Arnold (with Katherine Pine), *Britain and the H-Bomb* (London: Palgrave Macmillan, 2001), 166–170.

17. Goodman, *Spying on the Nuclear Bear*, 114.

18. H. Scoville, Intelligence Analysis on the Geneva Conference to Study the Methods to Detect Violations of a Possible Agreement on the Suspension of Nuclear Tests, CIA/SI 205–58, October 28, 1958, Secret [accessible in document 0000818707 at www.foia.ucia.gov].

19. The episode is discussed in Richelson, *Spying on the Bomb*, 116 and 127 on the basis of Lowenhaupt's account.

20. XXXXX, TRU to Penney, Secret, September 30, 1958. FOI application 12–03–10–133704–001.

21. Cit. in Michael Goodman, "With a Little Help from My Friends: The Anglo-American Intelligence Partnership, 1945–1958," *Diplomacy and Statecraft* 18:1 (2007): 155–153, on 176.

22. A. W. Dulles, Control of Initial Information Regarding Foreign Nuclear Explosions, CIA Director Directive, August 5, 1959, Secret [accessible in document 0001216234 at www.foia.ucia.gov].

23. H. Scoville, Intelligence Analysis on the Geneva Conference [accessible at www.foia.ucia.gov].

24. *Ibidem.*

25. CIA, Development of Nuclear Capabilities by Fourth Countries: Likelyhood and Possibilities, NIE 100–2-58, July 1, 1958, Secret [accessible in document 0001108555 at www.foia.ucia.gov].

26. Divine, *Blowing on the Wind*, 229.

27. Walker, *British Nuclear Weapons and the Test Ban, 1954–1973*, 135.

28. US atomic weapons laboratories were by then testing devices in the subkiloton range to improve flexibility in the deployment of nuclear devices to be used as tactical nuclear weapons or as clusters in delivery launchers.

29. Walker, *British Nuclear Weapons and the Test Ban, 1954–1973*, 120–124.

30. US Disarmament Administration, *Geneva Conference on the Discontinuance of Nuclear Weapon Tests*, 30.

31. *Ibidem*, 56–57.

32. Barth, "Politics of Seismology," 751–752.

33. The research was carried out at the AEC Lawrence Livermore Laboratory. See A. L. Latter, E. A. Martinelli, and E. Teller, "Seismic Scaling Law for Underground Explosions," *Physics of Fluids* 280 (1959): 280–282.

34. Blanca (22kt), Logan (5kt), Tamalpais (0.072kt), and Evans (0.055kt). Bolt, *The Parted Veil*, 252. For an updated list of the tests see US Department of Energy, *United States Nuclear Tests, July 1945 through September 1992* (Nevada: Nevada Operations Office, December 2000).

35. Romney, *Detecting the Bomb*, 106. On lack of transparency in data release see Chuck Hansen, "Open Secrets, Closed Minds," *Bulletin of the Atomic Scientists* 51:4 (1995): 16–17.

36. D. T. Griggs and F. Press, "Probing the Earth with Nuclear Explosions," *Journal of Geophysical Research* 66:1 (1961): 237–258.

37. XXXX, TRU to Levin, August 18, 1959, Secret. FOI application 12–03–10–133704–001.

38. SR/CR, TRU to Levin, July 29, 1959, Confidential. FOI application 12–03–10–133704–001.

39. F. H. Panton, TRU to Levin, September 17, 1959, Secret. FOI application 12–03–10–133704–001.

40. SRCR, TRU to Levin, 28.9.59, Secret. FOI application 12–03–10–133704–001.

41. Second US-UK Scientific Evaluation of Soviet Nuclear Weapons Tests, September 28, 1959, Secret. FOI application 12–03–10–133704–001.

42. On this see Alan Needell, *Science, Cold War and the American State* (Washington DC: Smithsonian/Harwood, 2000), 146 and 357.

43. US IGY Committee, "Combined Report on Program of Other Countries Participating in the IGY," in Technical Data Forwarded to CIA, Box 14, Record Group 59, National Archive and Records Administration, College Park, Maryland [RG59, NARA from now on]. Soviet IGY publications include: A. Sytinskiy, "Seismic Observations at Mirny [in 1956]," in M. Somov (Ed.) *Pervaya Kontinentalnaya Expeditsia 1955–1957. Nauchnye rezultaty* (Leningrad, 1960), 153–156; A. Lazareva and A. Sytinskiy, "Seismic Observations at Mirny in 1960," in E. Korotkevich (ed.), *Pyataya Kontinentalnaya Expeditsia 1959–1961. Nauchnye rezultaty* (Leningrad, 1967), 280–284. I am thankful to Irina Gan for pointing me toward these works.

44. Romney, *Detecting the Bomb*, 89.

45. *Ibidem*, 136.

46. Disarmament Administration, *Geneva Conference on the Discontinuance of Nuclear Weapon Tests*, 76–77.

47. "Report of the United Kingdom Delegation from the Proceedings of Technical Group II," *Bulletin of the Atomic Scientists* 16:2 (1960), 44–45.

48. Disarmament Administration, *Geneva Conference on the Discontinuance of Nuclear Weapon Tests*, 119.

49. On Kruschev's decisions see: Aleksandr Fursenko and Timothy Naftali, *Kruschev's Cold War* (New York: W. W. Norton & Co., 2006), 369.

50. Walker, *British Nuclear Weapons and the Test Ban, 1954–1973*, 169.

51. W. E. Duckworth, "Ieuan Maddock, 1917–1988," *Biographic Memoir of the Fellows of the Royal Society* 37 (1991): 323–340.

52. David Davies, "Hal Thirlaway (obituary)," *The Guardian*, January 19, 2010. On Thirlaway and Bullard, see also Solly Zuckerman, *Men, Monkeys and Missiles: An Autobiography, 1946–88* (London: Collins, 1988), 333.

53. Eric Carpenter, "An Historical Review of Seismometer Array Development," *IEEE Proceedings* 53 (1965): 1816–1821.

54. Oliver Wright, FO, to Philip de Zulueta, PM's Private Secretary, August 25, 1960, Secret, in ES13/43, The National Archive, Kew Gardens, London [TNA]. It is beyond the scope of this chapter to examine Macmillan's science policy stance. From 1957 Macmillan famously embraced a policy of defense research cuts (with Duncan Sandys as minister of defence), believing it necessary to make Britain more reliant on defense coordination with its allies. He thus understood seismological detection research as duplicative of American efforts and advocated that Britain be wholly reliant on US detection systems. See: Agar and Balmer, "British Scientists and the Cold War," 240. See also Wyn Rees, "The 1957 Sandys White Paper: New Priorities in British Defence Policy?" *Journal of Strategic Studies* 12:2 (1989): 215–229.

55. N. Levin, Seismic Research Program, Secret, May 15, 1961, in ES13/43, TNA.

56. R. Makins to Hugh Stephenson, FO, May 18, 1961, Secret and Lord Hume, 'Seismic Research Program', June 2, 1961 in ES13/43, TNA.

57. De Zulueta to Claude Pelly, June 7, 1961 in ES13/43, TNA.

58. Lorna Arnold, *A Very Special Relationship: British Weapon Trials in Australia* (London, HMSO, 1987), 220–225.

59. Walker, *British Nuclear Weapons and the Test Ban, 1954–1973*, 128.

60. AWRE, *The Detection and Recognition of Underground Explosions. A Special UKAEA Report* (London: AWRE/HMSO, 1965).

61. H. I. S. Thirlaway, "Earthquake or Explosions?" *New Scientist* 18 (1963). On the controversy see: Romney, *Detecting the Bomb*, 187.

62. A similar difference typified US and British approaches to sea surveillance. See Robinson's chapter in this volume.

63. Charles Bates, "Vela Uniform. The Nation's Quest for Better Detection of Underground Nuclear Explosions," *Geophysics* 26:4 (1961): 499–507. See also Barth, "The Politics of Seismology," 752.

64. J. Oliver and L. Murphy, "WWNSS: Seismology's Global Network of Observing Stations," *Science* 174 (1971): 254–261.

65. On data correlation see Romney, *Detecting the Bomb*, 12.

66. ACDA, *Arms Control and Disarmament Agreements: Texts and History of Negotiations* (Washington, DC: US Government Printing Office, 1977). In 1963 Scoville resigned from the CIA because of his unhappiness with the launch of new reconnaissance projects. Richelson, *The Wizards of Langley*, 57–58.

67. ACDA to Amembassy Lagos, Airgram, February 22, 1963, in Box 4181, RG59, NARA.

68. ACDA to Amembassy, Bonn and Amembassy, La Paz, May 21, 1963 in Box 4181, RG59, NARA.

69. Amconsul, Zagreb to Department of State, August 9, 1963 in Box 4182, RG59, NARA.

70. N.V.Kondorskaya and Z. I. Aronovich, "The Uniform System of Seismic Observations of the U.S.S.R. and Prospects of Its Development," *Physics of The Earth and Planetary Interiors* 18:2 (1979): 78–86.

Section III

Seeing the Sea—From Above and Below

Chapter 5

Stormy Seas: Anglo-American Negotiations on Ocean Surveillance

Sam Robinson

The origin of this project [...] emanated from the Americans on the basis of a philoso-phy of the need for world-wide surveillance with which we ourselves do not agree.[1]

On April 14, 1986, 18 US F-111 fighter-bombers passed unhindered over the Strait of Gibraltar on a mission intended to assassinate Libyan president Muammar Gaddafi. The mission failed, and Gaddafi would not be ousted from power until the Arab Spring and subsequent uprising of 2011. Yet the episode was important for another reason: This was the first time that the US administration invoked the UN International Convention on the Law of the Sea (UNCLOS III) to assert free passage through sea straits for ships and aircraft in order to launch an air attack without violating the airspace of neighboring European countries. This was a decisive move, especially because the Strait of Gibraltar has long been the site of various tensions, even between "special" allies. Gibraltar has been a British bastion for centuries, but the Spanish government has always contested the right of free passage through the Strait, and the United States and France have also vied for control of these strategically vital waters.

This chapter focuses on competing visions for understanding and policing this important waterway and another, the so-called Greenland-Iceland-UK gap, in an era when physical oceanography had begun to render the oceans controllable spaces but before spy satellites were ubiquitous and everything that moved by sea, in the air, or on land was traceable from a computer screen. In particular, using archival documents from British and Spanish repositories, it looks at one dispute between the US and British authorities on different approaches to sea reconnaissance that took place at the height of the Cold War.[2] The controversy originated in the late 1950s when British and American naval officers elaborated different plans for surveillance operations. The conflict that unfolded during the following decade was played out through the sponsorship of competing projects in vital sea passages for the Cold War conflict (see Figure 5.1).

These differences were eventually put aside and from the mid-1970s a comprehensive maritime surveillance system based on sonar technologies was implemented by the US Navy with British assistance. The establishment of a US

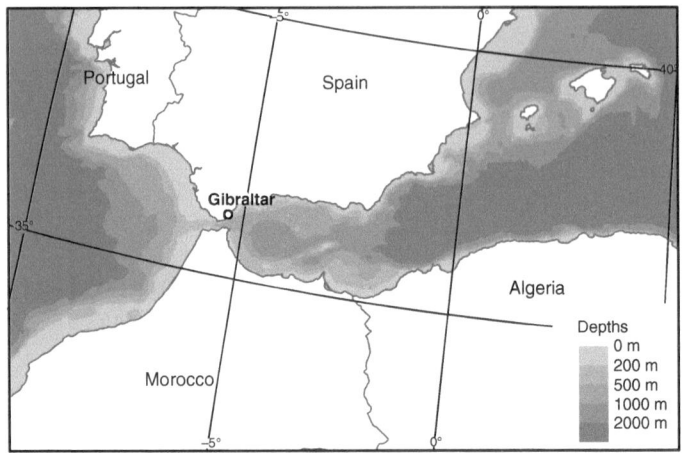

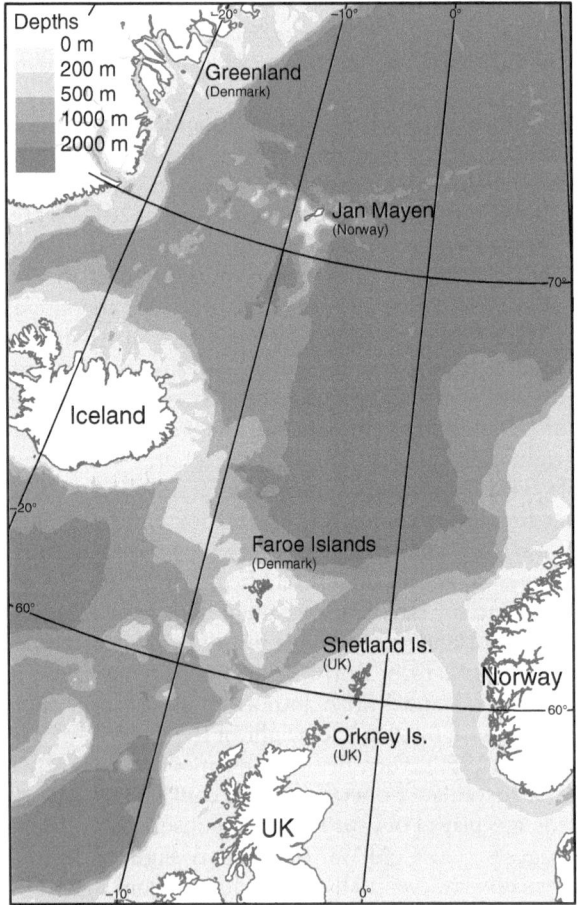

Figure 5.1 Two vital Cold War passages in the Atlantic: the Straits of Gibraltar and the Faroe Shetland Channel. Maps by Hans van der Maarel, Red Geographics

surveillance facility at the Royal Air Force station of Brawdy (Wales) brought an end to the quarrel.

Surveillance drove scientific and military strategic agendas throughout the Cold War and was a decisive factor in shaping relations between Western allies.[3] US military agencies strived for integrated systems of monitoring that would ensure global coverage. This ambition, however, was not accepted by all US allies and was definitely different from what British military defense and scientific circles sought to achieve through reconnaissance. Britain's "hidden hand" had been responsible for surveillance operations abroad for more than a century and represented a pillar of its now-crumbling empire. During World War II, the pursuit of signal intelligence (SIGINT) had extended British intercept networks across the world, and the British chiefs of staff subsequently advocated coordinating surveillance operations with US agencies, especially in strategically vital study areas such as the Soviet atomic weapons program.[4] British planners understood surveillance as ensuring coverage of strategically sensitive spots rather than panoptic viewing. US planners had a much grander vision—that of encompassing everything from the seafloor to space, in a reconnaissance network. Employing hydrophones, radar, spy satellites, seismographs, and various electronic surveillance (ELINT) techniques to intercept "enemy" communication networks, they aimed to place the whole of the earth under surveillance.

The Cold War has often been described rather simplistically as a bipolar conflict between two homogenous alliances, whose solidarity was maintained by mutual enmity toward the "other" bloc. However, the Western alliance was in reality much more fragmented.[5] At least up until the mid-1960s American and British military plans were based on diametrically opposed visions, and, as noted by commentators at the time, "unlike the United States, where military planning is predicted mainly on assessment of enemy *capabilities,* British planning gives somewhat greater weight to enemy *intentions.*"[6] Leading military officers in the United Kingdom feared mainly airborne threats, such as intermediate-range ballistic missiles (IRBMs) carrying hydrogen bombs, and believed that sea surveillance should be limited to a minimum. In 1962 British vice-admiral Varyl C. Begg aptly argued that "the facts of life are that to destroy this country overnight there would be no point in the enemy coming by sea!"[7] That said, since control of communications across the Atlantic was going to play a key role in any future conflict, both the US and British navies sought to invest in electronic equipment securing sea communications and detection of enemy vessels. However views in the United States on how to use it differed dramatically. The advent of submarines carrying nuclear missiles led the US Navy to set up arrays of hydrophones far away from its country's coasts so as to offer advanced warning of imminent Soviet attacks to North America. Conversely, the British Admiralty had no intention of investing substantially in a detection system that offered no protection to Britain.

Despite the lack of enthusiasm for comprehensive marine surveillance, the British Admiralty eventually agreed with its foremost ally about the need for oceanographic surveys enabling the emplacement of fixed surveillance technologies on the seabed.[8] The chapter thus traces the history of Anglo-American ocean surveillance projects in Gibraltar during the early 1960s and studies of the "northern flank"—specifically of the Greenland–Iceland–UK gap—during the later 1960s and early 1970s. It concludes with an assessment of how these military-scientific relationships unfolded.

Competing Philosophies on Ocean Surveillance

When World War II ended, British and US naval commanders had very different ambitions with regards to ocean surveillance. In 1948 the UK Cabinet Defence Committee stressed that Anti-Submarine Warfare (ASW) had to become the Royal Navy's chief responsibility. This assessment was based on the accepted belief that Britain would not be in an economic position to fight a war before 1957, and that if war did come, extensive support from across the Atlantic would be needed in naval operations.[9] Renunciation of a more proactive role in naval defense globally, in favor of one which the British economy could sustain, was thus inevitable. The Royal Navy's role in patrolling the Eastern Atlantic on behalf of its allies had only been secured after Churchill delivered an impassioned speech beseeching President Truman to "make room for Britain to play her historical role 'upon that Western sea whose floor is white with the bones of Englishmen.'"[10] However, the Yangtze incident (1949), the Korean War (1950–3), and Malayan Emergency (1948–60) demonstrated that the Royal Navy still retained some influence in international sea operations. And when, in 1949, NATO assigned to the Royal Navy responsibility for monitoring the Eastern Atlantic, the new task provided a lifeline to the costly force. Although in 1957 Defence Secretary Duncan Sandys still described its role in the future as "somewhat uncertain" and his White Paper also argued for a reduction of future investments in Britain's naval force, the Royal Navy's responsibility in the Eastern Atlantic enabled the funding of novel ASW techniques and surveillance systems.[11]

By contrast, the US Navy could lavish funds on very expensive pieces of ASW equipment and design multiple far-reaching projects to deploy surveillance virtually everywhere. While the Royal Navy had seen deep funding cuts, the reality for the US Navy was the opposite. During the early Cold War, the US Navy expanded rapidly overseas establishing bases throughout the world, often in direct competition with existing Royal Navy facilities. One notable case was the US Naval Station at Rota (Spain) established in 1953, just over 100 kilometers northwest of Gibraltar. Moreover, the US Navy could also count on a constant flow of new oceanographic knowledge and the production of new technologies for sea surveillance, through a well-oiled mechanism, which secured regular disbursements from the US Office of Naval Research to a number of military and civilian research agencies and private companies.[12]

The difference between what the Americans and the British could afford fully emerged when plans for new sonar equipment were being outlined. By the early 1950s both the US and UK navies had independent projects for the development of a passive sonar system.[13] However, Corsair, a project jointly undertaken by the British Admiralty's Research Laboratory and its Underwater Detection Establishment, failed and was cancelled in 1957.[14] Its US equivalent, project Jezebel, succeeded thanks to the involvement of an industrial partner, the Bell Laboratories, which made the system more effective through the use of target identification equipment. Sonar came in two varieties—active and passive. Active systems worked in much the same way as radar, emitting pulses (in this case sound waves), which bounced off targets in the water. Any returning sound was registered by hydrophones (sort of powerful water-resistant microphones) on the ship allowing the tracking of underwater objects. Active systems had been used extensively in World War II; however, the emission of the pulse also gave away the position of the emitting vessel.[15] Passive systems omitted the pulse and instead relied on listening for sounds in the oceans (in much the same way a doctor uses a stethoscope). This is exactly what the US Navy opted for, thus making the surveillance system silent and impossible to detect—a powerful weapon in itself. From the mid-1950s the passive Sound Surveillance System (SOSUS) was deployed on the eastern Atlantic seaboard and there were plans to deploy it in the Pacific as well.[16] Conversely, as a consequence of the Corsair failure, British attitudes to maritime surveillance changed considerably, leading to placing greater emphasis on developing an active sonar system instead.

This divergence emerged at a crucial point in time for Anglo-American relations. The period from the International Geophysical Year (1957–8) to the Limited Test Ban Treaty (1963) was typified by US–UK military-scientific mutual interdependence in oceanographic research exactly because of the growing need for reliable ocean surveillance.[17] Variations in currents, temperature, salinity, and pressure (depth) caused phenomena known as *thermoclines*, which altered the characteristics of sound waves travelling through sea water. Thermoclines caused the path of sound waves to be bent rather than remaining linear. As a result, submarines could hide in plain sight, invisible to sonar. Naval commanders viewed thermoclines alternatively as a defensive challenge and an offensive advantage, depending on whether they were trying to hide or seek. Nevertheless the pursuit of surveillance though passive and active sonar-instigated collaboration between naval commanders and oceanographers, since marine scientists had already investigated waves, currents, and temperature variations in the Atlantic and other oceans in the previous decade. In fact, they had been responsible for transforming oceanography into a "big science" by using expensive vessels and instruments for this purpose.[18]

The implementation of surveillance systems depended upon the study of oceanographic characteristics and the setting up of surveys enabling marine scientists to chart them. As will be shown, it was principally the Strait of Gibraltar that focused the attention of naval officers and oceanographers alike in Britain and the United States. This led them to join forces, producing oceanographic knowledge and competing in elaborating new surveillance schemes and systems.

Sea Lines of Communication: Surveillance of the Strait of Gibraltar

In the mid-twentieth century Gibraltar was not a typical colonial possession; it was, in fact, a large, although not modern, overseas naval base.[19] The "Rock," as it was known, had been in British hands since 1704, and the Royal Navy had permanently kept a contingent there. After World War II, the naval base continued to play a role as a key communication and transit center for the management of British interests and colonies abroad.[20] Vessels travelling to British colonies used the facilities at Gibraltar to refuel and make temporary repairs. The base gave significant geostrategic and geopolitical capital and provided justification for

Britain's continued involvement in the Mediterranean, since its location allowed the monitoring of all shipping moving to and from the Atlantic.[21] British and American planners agreed that the sea areas adjacent to Gibraltar ought to be surveyed; however, they actually had very different ambitions on how they should be monitored.

A number of Cold War developments, including some close to Gibraltar, prompted the US Navy to challenge Britain's unilateral control of the Strait and patrol it more. The 1956 Suez crisis heightened tensions in North Africa, while France and Spain were forced to withdraw from their African colonies shortly afterward.[22] These withdrawals were the first major steps toward decolonization and with Western European forces retreating, the Soviet Union positioned itself as the bastion of international freedom and anticolonialism in these regions.[23] The supply of arms and machinery to the North African states resulted in an increased number of Soviet freighters operating in the Mediterranean and close to Gibraltar.[24] The Soviet presence represented a threat to American and British hegemony in the Mediterranean and was a major catalyst for designing new surveillance plans.[25]

Surveillance at Gibraltar was primarily a concern for the US Navy ASW director, Rear-Admiral Lawson P. Ramage, but the Royal Navy's flag officer (senior British officer) at Gibraltar was also preoccupied about the protection of communication lines near the strait. Gibraltar would be the last line of detection (and defense, if necessary) before Soviet submarines leaving their bases in the Black Sea could reach the Atlantic.[26] In 1960 the Royal Navy's director of undersurface warfare, George Symonds, reported on the Strait of Gibraltar's strategic role and recalled the threats derived from the recent construction of a Soviet base in the Mediterranean.[27] If Symonds agreed with Ramage on the nature of the Soviet threat, he disagreed on the exact shape and form of surveillance system to be deployed.

At the heart of the dispute was a different assessment of cost-effectiveness, efficiency, and adaptability of competing systems. The American proposal, in line with the ongoing development of the SOSUS network, was to establish a chain of fixed passive sonar installations facing the Atlantic side of Gibraltar.[28] Symonds resisted this proposition and argued instead for an active sonar system. He believed active sonar to be the best surveillance technology to be deployed at Gibraltar because the knowledge and expertise that Project Jezebel had given to the US Navy was not held within the British Admiralty or the Royal Navy. Furthermore, the Admiralty ceased research into fixed passive sonar arrays following the Corsair failure.[29] Finally, Symonds believed it necessary to deploy a system capable, if necessary, to seek and destroy nuclear submarines, especially since these would represent a direct threat to British territories.[30] Thus, he suggested countering the US proposal with a British plan to produce helicopters fitted with deep-dangled sonar on a long cable to unite surveillance and offensive weaponry capable of destroying submarines.[31] Since 1958 the Admiralty had been developing the Wasp helicopter specifically for this role.

In order to verify the efficiency of the proposed ASW system, the Admiralty arranged a survey of the sea area adjacent to Gibraltar. In the summer of 1960 the Admiralty Underwater Weapons Establishment (AUWE), the Admiralty Research Laboratory and the National Institute of Oceanography (NIO) jointly undertook the exercise with the assistance of NIO's research vessel *Discovery II*.[32]

The survey allowed for a better understanding of acoustic propagation in the strait's waters and testing signal strength at different depths so as to ascertain if the deep-dangled sonar would succeed.

US naval officers now sought to convince their allies of the urgent need of surveying the strait's waters, thus juxtaposing the Admiralty's project with one bringing together NATO partners. Joint Canadian, US, and British oceanographic research had already taken place in Gibraltar in 1959.[33] And in that year NATO took responsibility for sponsoring collaborative oceanographic projects in light of ballistic missile-carrying submarines, which entered service in both the Soviet (1956) and US navies (1959). This compelled the alliance to secure a better understanding of key sea passages such as the Gibraltar and the Turkish straits, and the Norwegian Sea. Through its Science Committee, NATO established a Sub-Committee on Oceanographic Research, responsible for organizing the surveys.[34]

By the end of 1960 both the US and British naval authorities were looking at NATO as the space where decisions about Gibraltar would be taken, but while the US Navy was lobbying for joint oceanographic exploration, the British Admiralty hoped to convince NATO allies to pay for the sonar to be made in Britain. The final report of the Admiralty survey showed that the acoustic properties of the seabed required a more detailed study and the lack of knowledge of Spanish territorial waters prevented the deployment of the new British device. And handling and controlling the equipment in the strait's "strong and variable tidal conditions" was going to be challenging.[35] Despite these limitations, Symonds now advocated the adoption of the deep-dangled sonar, arguing that it was going to be cheaper and more effective than the US-made sonar array. Moreover, as the strait would be an important choke point in a future war, he argued, a purely defensive surveillance system would not work. If Gibraltar was now to become an "unofficial" NATO base, the Admiralty felt that all their allies should be compelled to contribute to its surveillance.[36] Symonds's summary thus launched the ambitious British project and argued for NATO assistance: "we cannot ignore the defence of the Straits of Gibraltar, but on the other hand we cannot 'go it alone' and NATO countries must assist."[37]

However, the 1960 UK Defence White Paper further restricted the budget for military research and gave the Admiralty limited funds for science projects. Now a discussion developed between the Admiralty's director of plans, Peter Ashmore, the director of general weapons, Michael Le Fanu, and Symonds about the best use of limited funds.[38] All three agreed that the construction of a surveillance system at Gibraltar had implications for wider NATO strategy and should not be solely paid for by the Royal Navy.[39] But since both the Canadian and the US navies had shown no interest in the British system, the deep-dangled sonar scheme floundered in Admiralty bureaucracy. In 1961 even the under-secretary of state, Nigel Abercrombie, failed to find sufficient funds for full-scale trials.[40]

The Civil Lord of the Admiralty, Ian Orr-Ewing, who had been brought in by Macmillan to curb spending in the Royal Navy, now took responsibility for assessing the feasibility of the British project and requested that the Third Sea Lord, Peter Reid, prepare a report. Reid's document was strongly worded. SOSUS was described as being fantastically expensive and ineffective: "The reason for our basic disagreement with the American concept of a static [anti-submarine] Defence of the Straits in peace time is that it is a purely global war requirement which, by government policy, is at the bottom of our list of priorities."[41] Static defenses were of "no use either in peace or war unless there is a force of [anti-submarine] vessels...to classify and follow up the contacts which the static defences report."[42] Reid convinced Orr-Ewing that the British active sonar represented the way forward, but when the Civil Lord of the Admiralty enquired again about NATO's European allies taking on some of the expense and research for the project, he found out that none was willing to do so.[43] The project thus stalled and, meanwhile, it became apparent that placing the Strait of Gibraltar under surveillance had become even more urgent.

In 1963 the French forces abandoned the Tunisian city of Bizerte, which hosted a French naval base and the facilities of the Allied Forces Mediterranean (AFMED).[44] As a result, convoy routes in the southern Mediterranean which NATO forces patrolled and planned to use in time of conflict were now exposed to a greater risk. The NATO commander in chief of Allied Forces Mediterranean, Vice Admiral A. B. Cole, claimed that the offensive capability to defend the eastern approaches to the Strait of Gibraltar was compromised. The Gibraltar naval base now had to project naval power in a 180-degree south-facing sweep. The Soviet threat had also increased due to submarine activities (potentially involving nuclear submarines) and "hydrographic/ELINT" ships in the Mediterranean.[45] In that year one NATO report speculated that Soviet ships gathered both scientific data and intelligence, but in fact, little was known about the true purpose of these vessels. The memo underlined the effects of decolonization, the rise of Soviet naval power, and the reduction of NATO military geographic deployment. Inevitably Gibraltar was to become the main NATO naval and military base and focus for offensive antisubmarine operations. The question remained however as to whether and how it should patrolled.[46]

Cole, a New Zealander who had been in the Royal Navy since the end of World War II,[47] vigorously argued for effective surveillance of the Strait of Gibraltar in order to "detect, identify and destroy transiting enemy submarines." He also emphasized the benefits to be derived from surveillance of merchant shipping transiting the straits. This wider conception of the purpose of surveillance lent support to the British Admiralty's plans for a deep-dangled sonar emphasizing its role in peacetime surveillance. In a clear reflection of the Cuban missile crisis of the previous year, the report suggested that the surveillance system would provide "useful intelligence information in peace or in times of tension."[48] Cole also stated that although NATO had already spent considerable sums on improving the military strength of the naval base, there was no progress in securing "comprehensive and effective surveillance system in the Gibraltar Straits."[49] Attached to the memo was a report detailing operational requirements and suggesting that the British system was a cheaper alternative to be adopted in any case before a fully fledged fixed installation could be completed.

Cole's appeal yielded little. The only item agreed upon at NATO level was to continue surveying the strait, a project that found some NATO countries such as France, Belgium, Norway, and Italy very sympathetic and, conversely, was reluctantly carried forward by British oceanographers. By contrast, from 1963 the US Navy intensified its efforts and sought now to "investigate by analysis and/or model study the effect of currents on transducer orientation," while the US

Office of Naval Research in London collected new data on subsurface currents in the strait in order to improve the understanding of acoustic properties and to aid the design a more effective passive system[50] Although there was cooperation on oceanographic work between the British and US Navies, researchers had little to share in terms of what kind of surveillance Gibraltar really needed. British plans for "on-the-spot" detection and response with ASW helicopters did not persuade the Americans, who—by contrast—looked at Gibraltar only as a knot in a vast network of SOSUS arrays, extending from the seas surrounding Alaska to the Indian Ocean. These competing visions overlapped with competing strategies to inform diplomatic relations with countries neighboring Gibraltar and in particular with Spain. Actually, it would be Britain's changing relations with Franco's regime that sanctioned the final defeat of the British surveillance system and the adoption of an American one.

The Role of Spain

Decisions on surveillance at Gibraltar did not depend solely on the availability of oceanographic data or the cost-effectiveness of new sonar technologies. Relations with Spain were important as well. In the 1960s Franco's Spain was still isolated internationally and, although a member of the United Nations from 1955, it had no representation in NATO or any other pan-European organization. The US administration had offered Spain a way out of isolation, mainly because of Spain's strategic positioning between Atlantic, Mediterranean, and North Africa, following the signing of the 1953 Pact of Madrid. One important consequence of the agreement was the offer of military aid in exchange for placing US naval and air bases on Spanish territory. As a consequence of that, as we have seen earlier, the US Navy could establish a station in Rota. When in the 1960s the NATO project for oceanographic surveys covering the Strait of Gibraltar and the adjacent Alboran Sea developed, French and Belgian oceanographers invited their Spanish colleagues to join in. Consequently the Spanish vessel *Xauen* of Madrid's Oceanographic Institute could take part in the scientific exercise, while the NIO's *Discovery II,* which was meant to lead the expeditions, ended up playing a more peripheral role.[51]

Since the British Admiralty had failed to convince the US Navy or a NATO ally to adopt the deep-dangled sonar and Britain's role in NATO oceanographic efforts was being marginalized, one Royal Navy official now turned to the old Spanish enemy looking for support. Presumably this officer believed that the US Navy would be more sympathetic toward the British sonar system if the two European countries controlling the strait agreed on the need for adopting it. So, in 1963 the British commander in Gibraltar, Rear Admiral Errol Sinclair, arranged a private meeting with the Spanish minister of marine, Pedro Nieto Antúnez, in order to restore cordial relations, especially as these had been strained for many years.

During the meeting, the issue of securing a new surveillance system for Gibraltar was discussed together with other pressing issues including the presence of Soviet vessels in the Gibraltar Bay area and the possibility of joint scientific and naval exercises.[52] The decision to hold talks with a non-NATO ally on surveillance issues shows that the local British commander perceived coordination in the strait area as urgent. This demonstrated that in order to favor the adoption of the new deep-dangled sonar, Sinclair was now even prepared to talk with representatives of a country, Spain, which had been unfriendly up until then. He even disclosed restricted information to Antúnez, eventually reporting to his superiors that he had just presented his personal views, "coloured by his task at Gibraltar."[53]

Sinclair and Antúnez *did* discuss sensitive issues, for instance, how Soviet merchant vessels feigned machinery damage while secretly offering military equipment to Moroccan forces in their anticolonial struggle against Spain. Sinclair explained to Antúnez that Soviet trawlers covertly gathered intelligence and scientific data. And he hinted at the Anglo-American surveillance plans, hoping that he could win the Spaniard's support for the new British surveillance scheme.[54] While analyzing Spanish plans to adopt a fixed active surveillance installation in Ceuta (on the southern shore of the Strait), he attempted to "sell" the British ASW Helicopter methodology, suggesting that the scheme was more in line with the financial capabilities of Spain. Helicopters were "invaluable as weapons carriers" too.[55] Antúnez replied that as far as he was concerned there were no difficulties in going ahead with adopting a joint surveillance scheme. "You may quote me on this as you wish," he even added.[56] Notwithstanding Sinclair's security breach, the Foreign Office's Permanent Under-Secretary showed enthusiasm for his initiative, which finally seemed to cast some positive light on the future of the deep-dangled sonar. "Apart from the defence aspect, co-operation between the respective naval authorities may help to remove the political difficulties between Britain and Spain in relation to Gibraltar," he argued.[57]

But in 1965 the Royal Navy unsuccessfully attempted to establish an Anglo-Spanish agreement on cooperation in oceanography.[58] In the same year, Spain requested that Gibraltar be returned to the Iberian country at the UN's Anti-Colonisation Committee. A Red Book published by the Spanish Foreign Ministry now openly divulged that Britain had no interest in the Gibraltarians and planned to keep the Rock mainly because it needed a military base from which to track Soviet rockets and submarines. The recent secret conversation between Sinclair and Antúnez clearly proved it.[59]

Now the new Labour government in London shifted firmly away from cooperating with Franco. In 1966 Spain's government retaliated by placing restrictions on British aircraft crossing Spanish airspace when flying toward Gibraltar. These restrictions later extended to motor vehicles crossing the border by land. The British government wished to prove the Spaniards wrong in their allegation about keeping Gibraltar only because of its surveillance implications and called for a referendum, to demonstrate that the Gibraltarians wished to continue being UK citizens. The day before the referendum NATO conducted naval exercises launched from Gibraltar, which, unsurprisingly, excluded Spanish naval forces.[60] As tensions escalated, Franco's government chose to close the border in 1969, and it remained closed until well after his death in 1975. By then, however, the situation with regards to the surveillance of the Strait of Gibraltar had rapidly evolved in favor of the US proposal. Since Britain had failed to find countries— within and outside NATO—sympathetic to its surveillance project, this was mothballed. Conversely by 1974 US plans for establishing a worldwide sea surveillance network including a fixed installation on the Atlantic side of Gibraltar had reached completion, highlighting that in the end the British Admiralty had to agree with the US plans for fixed installations since an alternative scheme no longer existed.[61] Even more worryingly for the Admiralty, Gibraltar was not going to be the only case when British surveillance plans were cast aside in favor of more ambitious and expensive US surveillance projects.

Global Seas, Global Surveillance: The GIUK Gap

Oceanographic surveys helped to address the problem of how to improve detection in the presence (or absence) of currents and other oceanic conditions in Gibraltar. In the North-East Atlantic, the confluence of cold Arctic water and warm Atlantic water in the Southern Norwegian Sea had been a subject of enquiry for oceanographers from the early twentieth century and much more needed to be understood about this convergence, especially if this strategically vital chokepoint granting the Soviets access to the Northern Atlantic could be efficiently monitored by Western naval forces.[62] Unsurprisingly, the US–UK dispute on competing surveillance schemes that typified Gibraltar continued in the cooler Atlantic waters off the Norwegian coast. This time, however, British officers were especially worried about the somewhat forceful way in which their US colleagues sponsored NATO oceanographic surveys in line with their ambitions for ensuring global coverage in surveillance operations.

In 1960 the NATO Sub-Committee on Oceanographic Research planned more surveys focusing this time on the Faroe-Shetland Channel. The committee's chairman, the Norwegian oceanographer Håkon Mosby, took responsibility for leading the NATO expeditions in these strategically vital sea areas in order to produce novel data on salinity, temperature, and currents.[63] The NATO study was openly geared toward assisting in the development of the SOSUS network as plans to extend it to the Northern European coastal line existed. There was no opposition in Britain to these plans and, actually, the NIO was among the leading oceanographic institutions in the endeavor. During the expeditions, the NIO's vessel *Discovery II* carried the Swallow Float, a new device developed by oceanographer John Swallow in an effort to take current measurements at predetermined depths.[64] The study produced novel data on thermoclines and allowed for the production of synoptic charts illustrating how water temperature changed in the sea channel (and thus affected sonar detection).[65]

By sponsoring the NATO expedition, the US Navy wished to generate consensus on its plans for sea surveillance and address—with the assistance of new oceanographic data—the limitations of the SOSUS technology. However NATO's oceanographic research actually had a very disruptive effect on other oceanic surveys that had no surveillance ambitions, something that worried British government officials. In 1960 the International Council for the Exploration of the Seas (ICES) executed the Overflow expedition, aiming to chart oceanic characteristics of the Faroe–Shetland channel, the very same sea area now chosen by NATO.[66] ICES was primarily interested in fishery and executed, in contrast with NATO, both physical and biological data collections (including plankton sampling).[67] Crucially the United States was not an ICES member, while the Soviet Union was represented in the council.

The planning of the NATO survey took the collaboration between some European oceanographers who had been part of ICES to an end—at least for the next 10 years—and brought them into closer collaboration with US colleagues. Those such as the NIO's director George Deacon and Mosby, who had manifested some dissatisfaction with the ICES expedition, now took the opportunity to promote NATO's plans.[68] After taking part in the Overflow expedition, NIO's *Discovery II* ceased to further participate to the ICES program. Likewise, Mosby and the Norwegians in the *Helland-Hansen* (that had also taken part in Overflow) now agreed to more surveys, which had surveillance ambitions. This was in line with the stance of the Norwegian government that surveillance in the North Sea needed to be improved. Other experts—involved in the ICES exercise and interested primarily in fishery studies—did not welcome NATO's intervention in oceanographic affairs. The "invasion" of NATO oceanographers into a delicate environment for fisheries research caused outrage among scientists that rumbled on in Britain for the next five years.[69] In particular, J. B. Tait of the Torry Marine Laboratory in Aberdeen (Scotland) continued to be part of the NATO survey up until 1962, but after that he became a staunch critic of Deacon.

That said, NIO's contribution to the Faroe-Shetland channel also waned somewhat despite the initial enthusiasm shown by Deacon for taking it to completion. The chief reason for the diminished interest in carrying it forward was the criticism put forward at UK cabinet level for the ways in which NATO work had disrupted traditional fishery research. When the Working Group on Oceanography of the UK Cabinet Committee on International Co-operation met on September 3, 1962, a representative of the Ministry of Agriculture and Fisheries argued that "[…] NATO studies […] were *likely to give offence* to the USSR and jeopardise Russian collaboration in the field of fishery conservation." Deacon replied at the meeting that "he saw the NATO project as supplementing fisheries research in the area, not as duplication."[70] Although no official decision was taken at the cabinet level to address the resentment of fishery scientists, the participation to NATO surveys of British oceanographic groups was affected by the altercation.

Meanwhile, the Americans capitalized on the NATO studies and swiftly moved from the stage of oceanographic research to that of setting up new sonar installations. The Cuban missile crisis of October 1962 revealed that monitoring Soviet submarines from the Eastern Atlantic seaboard was not sufficient. If tracking was to be effective throughout the North Atlantic, then SOSUS ought to be extended.[71] The tracking of Soviet merchant vessels carrying nuclear missiles and launching equipment to Cuba had revealed that another threat existed aside from submarines. The SOSUS system deployed off the Florida coast was effective but could not ensure coverage of other Atlantic sea areas. In particular. the sonar array ceased to offer clear detection when climatic conditions changed around the mid-Atlantic ridge.[72] These changeable conditions, therefore, dictated the maximum range of detection daily. In the months following the crisis, it became apparent to US Navy officials that a second line of subsurface sonar stations in the North East Atlantic (see Figure 5.1b) would have to be given greater priority and the installation program accelerated.

When the NATO surveys ended, a wealth of new oceanographic data was made available to Western navies' officers. Those of the US Navy now agreed to discuss with the British Admiralty about the possibility of improving the efficiency of SOSUS in the Atlantic. The conversation was obviously facilitated by the conclusion of the dispute in Gibraltar and the American officers knew that since the Admiralty no longer had the intention to sponsor its sonar system in the southern European strait, it could be persuaded to adopt the American one further north. In June 1967 a small team from the Admiralty Underwater Weapons Establishment (AUWE) was invited to the United States.[73] During the trip, the Admiralty was invited to observe a SOSUS facility and the US Navy officials even proposed that innocent sounding cover names could be adopted so as to make it easier to believe that the facility's "activities are concerned entirely with Oceanography."[74] The AUWE team now realized that the offer was being made because the central Atlantic ridge seriously affected detection performance. Setting up a SOSUS station in Britain could save the US Navy up to $19 million by enabling them to relay and route the signals differently.[75] Weighing the pros and cons of UK participation, the AUWE officers reported that the scheme was beneficial to Britain's defense and foreign policy objectives.[76]

After so much bickering with the Americans, the new project for a SOSUS station in Britain restored a "special" relationship on sea surveillance affairs. Yet, in casting a positive light on the collaborative deal, the Admiralty team's report now suggested that the British sonar scheme proposed in previous years was not *alternative* to the SOSUS but actually *complementary.* "Collaboration in SOSUS would be concrete evidence of sincerity in collective ASW defence," it argued, and "the large UK investment in ASW frigates, helicopters, submarines and aircraft could be put to effective use."[77] In fact, the Admiralty document indicated that it was *exactly* because of SOSUS that the British investment in sea surveillance could be made cost effective: "it would significantly improve the UK military control [. . .] without incurring large overseas expenditure from forces abroad."[78]

A major problem, however, was "the effort involved in 'Anglicisation' of US equipment."[79] In any case, the project was given the go ahead; even though "it is obviously of some importance to the Anglo/US alliance that we assist each other whenever possible but this does not presuppose UK acceptance of large bills unless we are receiving full value for them."[80] Project Backscratch (the cover name for the construction of a UK SOSUS Station) was meant to include a British contribution of £5.6 million (mainly to contribute to the cost of undertaking surveys and providing shore facilities). Arrays, cables, and hydrophones were all provided by the US Navy.[81] The Royal Navy had to be careful since it was about to close overseas bases to fund its existing projects and the endorsement of a costly Anglo-American scheme seriously impacted on the management of ongoing programs abroad.[82]

In 1970 the SOSUS project was revisited as the new Conservative prime minister Edward Heath entered office. The original commitment of £5.6 million had by then been significantly revised. Although there was still to be a full intelligence-sharing agreement, the Ministry of Defence was now required to provide the building to house the shore station to be built at the Royal Air Force facility of Brawdy in Wales. The site was chosen because it allowed the US planes to land and resupply the base. It also suited British ASW plans. The new minister for defence, Peter Carrington, presented these plans in the cabinet meeting of December 28, 1970. Britain would pay only £1.5 million, a small fraction of the US funding of SOSUS (£80 million).[83]

The facility at RAF Brawdy came into service in 1974 and remained operational until 1995, when it was decommissioned. The facility was not welcomed by antinuclear campaigners, who believed it hosted a nuclear missile command center, and it was even picketed during the 1980s by the women of Greenham Common.[84] In reality, it was staffed by US Navy officials operating the SOSUS detection equipment and had no obvious military installation. However, much like with Gibraltar, its significance was not in fixed installations at sea monitored from Brawdy, but rather in the integration of signals from any remote place in the world that stations like that in Brawdy picked up and transmitted to other nodes in the SOSUS network. Brawdy, as with many other similar bases around the world, demonstrated that American plans for global sea surveillance had proven successful and exactly in the country where they had been initially firmly opposed.

Conclusions

Speaking at the 1982 conference on the implementation of the United Nations Convention on the Law of the Sea (UNCLOS), the then director of the Institute of Oceanographical Studies (as the NIO had by then become), Anthony Laughton, reflected on oceanographic research of the 1960s:

> Most oceanographic research has a complex mixture of motives and funding, sometimes related to the long-term needs of defence, sometimes to the needs of scientific curiosity, sometimes to the assessment of potential resource exploitation, sometimes commissioned by government or industry to achieve defined objectives related to the utilization of the oceans.[85]

In separating the various motives for funding oceanographic research, Laughton did not mention the use of oceanographic science as a means for surveillance and intelligence-gathering, or the significant investment in technologies securing surveillance of the sea. Military support for oceanography, which had led to the rapid expansion of the discipline during the 1950s and in particular the 1960s, was premised on the fear that the sea had hidden threats deriving especially from the movement of enemy submarines. This fear continued to blur the boundaries between military operations and oceanographic work even when the systems of surveillance discussed in this chapter were finally operational. In 1968 the USS Pueblo was "captured" by communist North Korea, while carrying an assortment of ELINT and SIGINT equipment when the vessel was supposedly carrying out oceanographic survey work.[86] For Western navies, the rapid expansion of a Soviet oceanographic research fleet chimed with continued fears about the activities of Soviet trawlers operating off the coasts of Scotland and Iceland and in the Mediterranean.[87]

This chapter has explained how oceanographers and naval officers jointly pursued sea surveillance and looked in particular at how this happened in the context of the "special relationship" between the US and UK navies and the NATO alliance. I have shown that very different plans existed to enforce surveillance of the sea in NATO's two leading countries. The United States saw this problem as a global one. Its fiscal supremacy enabled costly projects to be implemented throughout the world's oceans. Of the three military surveillance technologies adopted in the last century—sonar, radar, and satellite—the first remains the most secret and hidden. The case of the United Kingdom is quite different. Here the development of ocean surveillance has to be seen in the light of an increasingly restricted world presence, limited budgets, and a specific geographical location. For the Royal Navy, the threat was from Soviet submarines launching a conventional attack against its home bases. So the Royal Navy saw that standing alongside its American counterparts was important, but not carte blanche to squander precious resources of its own.

I have also shown that despite the disagreement, in both countries (and NATO more generally) naval officers in charge for surveillance programs understood the importance that novel knowledge on the characteristics of the ocean had for the pursuit of sea surveillance. I have therefore argued that oceanographers were mobilized to survey those sea areas that needed routine reconnaissance as lack of knowledge of sea transit's physical characteristics could jeopardize these patrolling operations. Throughout the period British and American oceanographers worked jointly on these areas, both civilians and military scientists collaborated. From 1963 the US Office of Naval Research gathered all known data existing in Europe on the chokepoints at Gibraltar and the Faroe-Shetland channel and making use of the data processing facilities available in the United States started to complete the panoptical study of oceanic waters surrounding the continents in the Northern Hemisphere. However, the offer of collaboration in oceanographic enterprises came also in an attempt to dissuade the British Admiralty from adopting its own surveillance scheme.

By looking at both the case of Gibraltar and the Greenland-Iceland-UK Gap, I have shown how the British position changed considerably over time, also because of lobbying from the US Navy and financial difficulties. The early proposal for the active sonar system in Gibraltar found little support among NATO allies and was quietly dropped. It was reconsidered in light of an improbable alliance with Spain and eventually abandoned. After that, the US Navy gave the Admiralty the opportunity to prevent criticism at home by finding a different use for the sonar made in Britain that never materialized in Gibraltar. By proposing to set up a SOSUS station in Wales as part of a major scheme to better reconnoiter the North Sea, the US Navy effectively provided the Admiralty with an opportunity for utilizing the otherwise unusable British surveillance gizmos.

That said, the differences that typified British and US views on what surveillance meant and what one could aim to achieve with it continued. The problem was not, of course, just about sea and its invisible threats, but more generally about the changing role that Britain and the United States were about to play on the geopolitical chessboard. Approaches to surveillance in Britain were consistent with the geopolitical ambitions of a declining colonial power that had lost most of its possessions abroad and wished therefore to exclusively monitor what threatened its national security in selected parts of the world; incidentally those strategic bases retained after the empire had collapsed. Conversely, the United States sought to reaffirm its role as Cold War *hegemon* by investing significantly in technologies that would ensure a global vision on existing threats (at sea and elsewhere). The investment was consistent with a geopolitical stance that recognized that territorial occupation was a far less effective means to ensuring worldwide domination than securing the surveillance of distant places. The new American empire, in contrast with the old British one, was managed through the placing of invisible "electronic ears" underground, on land, at sea, and in space all around the globe.

Notes

1. Admiral J. Peter L. Reid, Report for the Civil Lord of the Admiralty Ian Orr-Ewing, August 17, 1961, Secret, in ADM 1/29275, The National Archive, Kew, London [TNA from now on].

2. In particular papers of the British Admiralty (ADM) at the TNA; those of the Archivio General the la Administración in Alcalá de Henares (Spanish national archive, AGA from now on); and those of the Ministerio de Asunto Exteriores in Madrid (Ministry of Foreign Affairs Archive, MAE).

3. Kristine Macrakis, "Technophilic Hubris and Espionage Styles during the Cold War," *Isis* 101:2 (2010): 378–385.

4. Richard Aldrich, "British Intelligence and the Anglo-American 'Special Relationship' during the Cold War," *Review of International Studies* 24:3 (1998): 331–351, on 331.

5. See Jeffrey A. Engel, *Cold War at 30,000 Feet: The Anglo-American Fight for Aviation Supremacy* (Cambridge, MA: Harvard University Press, 2007), 17–52; Alan Dobson, *Anglo-American Relations in the Twentieth Century: Of Friendship, Conflict and the Rise and Decline of Superpowers* (London: Routledge, 1995).

6. W. P. Snyder, *The Politics of British Defence Policy* (Athens, OH: Ohio University Press, 1964), 29.

7. Vice Chief of the Naval Staff to Flag Officer (Senior Royal Navy Commander), Scotland, August 15, 1952, ADM 1/28093, TNA.

8. Three other nations: Italy, Turkey, and Norway host US equipment for sea surveillance on their bases but the level of involvement and cooperation is as yet unclear.

9. Eric Grove, *Vanguard to Trident: British Naval Policy since World War II* (Annapolis: Naval Institute Press, 1987), 39.

10. Sean M. Maloney, *Securing Command of the Sea: NATO Naval Planning 1948–1954* (Annapolis: Naval Institute Press, 1995), 135.

11. Grove, *Vanguard to Trident*, 211. See also: Michael Dockrill, *British Defence since 1945* (Oxford: Wiley-Blackwell, 1988), 68.

12. On ONR funding to US oceanographic institutions see: Gary E. Weir, *An Ocean in Common. American Naval Officers, Scientists, and the Ocean Environment* (College Station: Texas A&M University Press, 2001) and Jacob D. Hamblin, *Oceanographers and The Cold War. Disciples of Marine Science* (Seattle, WA: University of Washington Press, 2005).

13. Stephen Twigge, E. Hampshire, and G. Macklin, *British Intelligence Secrets, Spies and Sources* (London: TNA, 2009), 157.

14. Following the Portland Spy Scandal, this establishment was consolidated into the larger, Admiralty Underwater Weapons Establishment (AUWE), which combined the Underwater Detection Establishment, Torpedo Experimental Establishment, Underwater Launching Establishment, in 1961. See Thomas Wright, "Aircraft Carriers and Submarines: Naval R&D in Britain in the Mid-Cold War," in Robert Bud and Philip Gummett, *Cold War Hot Science: Applied Research in Britain's Defence Laboratories 1945–1990* (London: Harwood/Science Museum, 1999), 153–154.

15. J. K. Petersen, *Understanding Surveillance Technologies: Spy Devices, Privacy, History & Applications* (New York: Auerbach Publications, 2007), 231.

16. Jeffrey T. Richelson and Desmond Ball, *The Ties That Bind: Intelligence Cooperation between the UKUSA Countries* (London: Allen and Unwin, 1985), 200.

17. Hamblin, *Oceanographers and The Cold War*, xvii–xix. See also: Simone Turchetti, "Sword, Shield and Buoys: A History of the NATO Sub-Committee on Oceanographic Research 1959–1973," *Centaurus* 54:3 (2012): 205–231. On the test ban treaty see: Vojtech Mastny, "The 1963 Nuclear Test Ban Treaty:

A Missed Opportunity for Détente?" *Journal of Cold War Studies* 10:1 (2008): 3–25; Glenn T. Seaborg, *Kennedy, Khrushchev and the Test Ban* (Berkeley: University of California Press, 1981).

18. Hamblin, *Oceanographers and The Cold War,* 264–5

19. Peter Gold, *A Stone in Spain's Shoe: The Search for a Solution to the Problem of Gibraltar* (Liverpool: Liverpool University Press, 1994), 3–4; Kent E. Calder, *Embattled Garrisons: Comparative Base Politics and American Globalism* (Princeton: Princeton University Press, 2010), 11.

20. Jeremy Black, *The British Seaborne Empire* (London: Yale University Press, 2004), 242 and 356; Tillman Nechtman, "'...for it was founded upon a Rock': Gibraltar and the Purposes of Empire in the Mid-Nineteenth Century," *The Journal of Imperial and Commonwealth History* 39:5 (2011): 749–770.

21. Donald C. Watt, *Succeeding John Bull: America in Britain's Place 1900–1975* (Cambridge: Cambridge University Press, 1984), 12; Melvyn P. Leffler, *A Preponderance of Power: National Security, the Truman Administrations, and the Cold War* (Stanford: Stanford University Press, 1992), 123, 171, and 273.

22. Egya N. Sangmuah, "Eisenhower and Containment in North Africa, 1956–60," *Middle East Journal* 44:1 (1990): 76–91; Matthew Connelly, "Rethinking the Cold War and Decolonization: the Grand Strategy of the Algerian War for Independence," *International Journal of the Middle East* 33:2 (2001): 221–245. See also: Peter Hahn, *The United States, Great Britain, Egypt, 1945–1956: Strategy and Diplomacy in the Early Cold War* (Chapel Hill: University of North Carolina Press, 1991); Tony Chafer, *The End of Empire in French West Africa: France's Successful Decolonization?* (Oxford: Berg, 2002).

23. John L. Gaddis, *Strategies of Containment: A Critical Appraisal of American National Security Policy during the Cold War* (Oxford: Oxford University Press, 2005), 179–180.

24. Record of a meeting between Flag Officer, Gibraltar, and Spanish Minister of Marine, April 2, 1963, FO 371/169501, TNA.

25. Ennio Di Nolfo, "The Cold War and the transformation of the Mediterranean," in M. P. Leffler and O. A. Westad, *The Cambridge History of the Cold War* (Cambridge: Cambridge University Press, 2010), Vol. 2, 238–257.

26. Harriet Critchley, "Polar Deployment of Soviet Submarines," *International Journal* 39:4 (1984), 836–837.

27. Note by George Symonds, Director of Undersurface Warfare, December 21, 1960, ADM 1/29275, TNA.

28. There are no official government documents that have been declassified which confirm its existence. However, several works claim that a SOSUS line west of Gibraltar exists: John Craven, *The Silent War: The Cold War Battle beneath the Sea* (New York: Touchstone, 2002), 93; Peter Huchthausen and Alexandre Sheldon-Duplaix, *Hide and Seek: The Untold Story of Cold War Naval Espionage* (London: John Wiley, 2009), 113; W. Craig Reed, *Red November: Inside the Secret US-Soviet Submarine War* (London: Harper Collins, 2010), 253–254; Sherry Sontag and Christopher Drew, *Blind Man's Bluff: The Untold Story of American Submarine Espionage* (London: Harper Collins, 1998), 40–1, 68, 122; Robert E. Harkavy, *Bases Abroad: The Global Foreign Military Presence* (Stockholm: SIPRI, 1989), 193–194; William E. Burrows, *Deep Black: Space Espionage and National Security* (New York: Random House, 1988), 177–181; Jeffrey T. Richelson and Desmond Ball, *The Ties That Bind,* 198–202.

29. The Admiralty, however, continued to promote the study of passive sonar systems. It simply halted work on arrays of hydrophones. The demise of Corsair was divulged to nonscientific staff in the Admiralty only in 1968.

30. "Any form of detection device needs a weapon to back it up." Note by George Symonds, Directory of Undersurface Warfare, December 21, 1960, ADM 1/29275, TNA.

31. The system was known as ASDIC. Straits of Gibraltar Anti-Submarine Group, Minutes of Meeting held in Washington, February 5, 1960, ADM 1/29275, TNA.

32. Flag Officer, Gibraltar, to Commander-in-Chief, Mediterranean, July 11, 1960, ADM 1/29275, TNA.

33. In 1959 the CANUKUS group (made up of the three main NATO powers: Canada, the United Kingdom, and the United States) commissioned a study of surveillance measures in Gibraltar. Minutes of Informal Meeting in Washington, February 5, 1960, ADM 1/29275, TNA.

34. Turchetti, "Sword, Shield and Buoys," 214.

35. Captain Superintendent, AUWE to George Symonds, November 23, 1960, ADM 1/29275, TNA.

36. See Memo by A. G. Draper, Military Branch II, and Admiralty for T.A.K. Elliott, Foreign Office, March 16, 1964, ADM 1/29275, TNA; and Statement by Sir B. Burrows at the North Atlantic Council on March 29, 1968 [available at www.gib-action.com/docs/nato1968.htm].

37. Note by George Symonds, Director of Undersurface Warfare, December 21, 1960, ADM 1/29275, TNA.

38. Correspondence by Director of Plans, January 3, 1961; Director of General Weapons, February 3, 1961; Director of Undersurface Warfare, March 20, 1961 in ADM 1/29275, TNA.

39. Note, Director of Undersurface Warfare, March 20, 1961, ADM 1/29275, TNA.

40. Note by Nigel Abercrombie, USS, May 26, 1961, ADM 1/29275, TNA.

41. Admiral J. Peter L. Reid, Report for the Civil Lord of the Admiralty Ian Orr-Ewing, August 17, 1961, in ADM 1/29275, TNA.

42. Ibidem.

43. Note from Department of National Defence Royal Canadian Navy communicated through the Senior Naval Liaison Officer (British Defence Liaison Staff), March 23, 1962, ADM 1/29275, TNA.

44. Memo by Vice Admiral A. B. Cole, Chief of Allied Staff—NATO Headquarters, Allied Forces Mediterranean, Malta, April 23, 1963, ADM 1/29275, TNA.

45. ELINT or "electronic signals intelligence" refers to the monitoring of signals not used for communications. On ELINT trawlers see Harkavy, *Bases Abroad*, 208–10.

46. Memo from NATO Headquarters, Allied Forces Mediterranean, Malta, April 23, 1963 (sgn. Vice Admiral, A. B. Cole), ADM 1/29275, TNA.

47. In the "Royal New Zealand Navy," at: http://nzetc.victoria.ac.nz/tm/scholarly/tei-WH2Navy-c29.html Cole had served during World War II as a commander of a antisubmarine destroyer, during the 1950s he rose within the Admiralty severing as assistant chief of naval staff in 1959–62 before taking up a NATO role.

48. Memo from NATO Headquarters, Allied Forces Mediterranean, Malta, April 23, 1963 (sgn. Vice Admiral, A. B. Cole), ADM 1/29275, TNA.

49. Ibidem.

50. Ibidem.

51. Letter from Amirante Jefe del E. M. de la Armada to Sub-secetario del Ministerio Asuntos Exterores, on "Investigaciones Oceanográficas en el Estrecho de Gibrlatar en colaboración con Francia en su aspecto military," Madrid, May 12, 1958, MAE, Leg. R. 5334, Exp. 20.

52. Record of a meeting between Flag Officer, Gibraltar, and Minister of Marine, Spain, April 2, 1963, FO 371/169501, TNA. Foreign Office's earlier stances are discussed in the file *US and Spanish Attitudes to Gibraltar 1952*, FO 371/102020, TNA.

53. Record of a meeting between Flag Officer, Gibraltar, and Minister of Marine, Spain, April 2, 1963, FO 371/169501, TNA.

54. Record of a meeting between Flag Officer, Gibraltar, and Minister of Marine, Spain, April 2, 1963, FO 371/169501, TNA.

55. Ibidem.

56. Ibidem.

57. Note by of N. J. A. Cheetham, FO Permanent Undersecretary, April 2, 1963, FO 371/169501, TNA.

58. See, CO 926/2061, TNA.

59. Fernando María Castiella [Spanish Minister of Foreign Affairs], "Borrador de la introducción del Libro Rojo", April 1965, Leg. 69 Rel. 1 n15/19: 57, AGA.

60. Fernando María Castiella to Spanish Minister of Navy, San Sebastián, August 26, 1967, Leg 70. Rel 1. 15/19. 2, Carpeta 2. 213/220, AGA.

61. R. E. Harkavy, *Strategic Basing and the Great Powers, 1200–2000* (London: Routledge, 2007), 127–129.

62. On this see Gunnar Ellingsen, "Varme havstrømmer og kald krig: 'Bergensstrømmåleren' og vitenskapen om havstrømmer fra 1870-årene til 1960-åreme [Warm Ocean Currents and the Cold War: the Bergen Current Meter and the Science of Ocean Currents from the 1870s to the 1960s]," PhD dissertation, Universitetet I Bergen, 2012.

63. Turchetti, "Sword, Shield and Buoys," 215.

64. On Swallow see: W. John Gould, "From Swallow floats to Argo—the Development of Neutrally Buoyant Floats," *Deep-Sea Research,* 52:3 (2005): 529–543 [available at: www.argo.ucsd.edu/Gould_Float_history.pdf, accessed October 9, 2013]. See also John Swallow, interview by Margret Deacon, 1994 [also available at http://www.argo .ucsd.edu/Gould_Float_history.pdf]; Herny Charnock, "John Crossley Swallow," *Biographical Memoirs of the Fellows of the Royal Society* 43 (1997): 505–519.

65. W. John Gould, "From Swallow floats to Argo," 530.

66. James Brian [J.B.] Tait (ed.), *The Iceland-Faroe Ridge International "Overflow" Expedition, May-June, 1960: An Investigation of Cold, Deep Water Overspill into the North-Eastern Atlantic Ocean,* ICES Report 157 [available at: http://ocean .ices.dk/Project/OV60/RPV157.pdf, last accessed October 9, 2013].

67. Helen M. Rozwadowski, *The Sea Knows no Boundaries: A Century of Marine Science under ICES* (University of Washington Press, 2004), 133–134.

68. Hamblin, *Oceanographers and the Cold War,* 233.

69. Ibidem, 232–234.

70. Minutes of Meeting, September 3, 1962, Working Group on Oceanography, CAB 124/2170, TNA.

71. Ryurik A. Ketov, "The Cuban Missile Crisis as Seen Through a Periscope," *The Journal of Strategic Studies* 28:2 (2005): 217–231; Geoff Till, "Holding the Bridge in Troubled Times: The Cold War and the Navies of Europe," *The Journal of Strategic Studies* 28:2 (2005): 309–337.

72. Norman Friedman, *Seapower as Strategy: Navies and National Interests* (Annapolis: Naval Institute Press, 2001), 58.

73. Head of Home NAVAL, Note on UK Participation in SOSUS, December 28, 1967, DEFE 69/8, TNA.

74. Visit of Team to naval facility, Lewis, Delaware, and Norfolk SOSUS evaluation centre, June 1967, DEFE 69/8, TNA.

75. Oceanographic Research Committee Position Paper, UK Participation in the SOSUS System, draft by DUSW (N), undated, DEFE 69/8, TNA.

76. Head of Home NAVAL, Note on UK Participation in SOSUS, December 28, 1967, DEFE 69/8, TNA.

77. ORC Position Paper, UK Participation in the SOSUS System, draft by DUSW (N), undated, DEFE 69/8, TNA.

78. Ibidem.

79. AUWE Working Party Report, "Research and Development Implications" (Annex 8), September 1967, in UK Participation in the SOSUS System, DEFE 69/8, TNA.

80. AUWE Working Party Report, "Political Considerations" (Annex 4), September 1967, in UK Participation in the SOSUS System, DEFE 69/8, TNA.

81. AUWE Working Party Report, 'Financial Considerations' (Annex 3), September 1967, in UK Participation in the SOSUS System, DEFE 69/8, TNA.

82. Ibidem.

83. Minutes of the Minister for Defence, Lord Carrington, for the Prime Minister, December 28, 1970, PREM 15/291, TNA.

84. Greenham Common (Berkshire, England) was the site and name of a women's protest camp, established in 1982, outside the local RAF base. The women campaigned against nuclear weapons being sited at the base.

85. Anthony S. Laughton, "The future of oceanographic research in the light of the UN convention," in E. D. Brown, and R. R. Churchill. (eds.), *The UN Convention on the Law of the Sea: Impact and Implementation* (Honolulu: University of Hawaii, 1987), 388.

86. On the episode see Mitchell B. Lerner, *The Pueblo incident. A Spy Ship and the Failure of American Foreign Policy* (Lawrence: University Press of Kansas, 2002); F. Baldwin, "Patrolling the Empire: Reflections on the USS Pueblo," *Bulletin of Concerned Asian Scholars* (Summer 1972): 54–75.

87. Instructions to ships "Operation CAIRNDUFF," Shadowing of Russian ELINT Trawlers, 1962, ADM 1/28093, TNA.

Chapter 6

Scientists and Sea Ice under Surveillance in the Early Cold War

Peder Roberts

On February 24, 1958, 85 individuals from the United States, Canada, Europe, Japan, and the Soviet Union gathered in the small town of Easton, Maryland, for a conference on Arctic sea ice.[1] Over four days they discussed a range of issues including the characteristics and physical composition of sea ice, its distribution and drift, and issues related to navigation and ice forecasting. Organized by the Earth Sciences Division of the United States National Academy of Sciences (NAS), at the behest of the Office of Naval Research (ONR), the conference was billed as an opportunity for the "productive exchange of facts and ideas."[2] The event reflected the rapid growth of interest in a scientific field with profound implications for military strategy in North America and commerce in the Soviet Union—and the desire for the United States to pick the brains of counterparts from around the world in a field where others almost certainly led, especially in terms of practical experience.

My claim in this chapter is that for the United States and its allies, sea ice forecasting and associated research agendas can best be understood when viewed through a dual prism of surveillance and engagement: the drive to know the earth through observation and calculation and to know the enemy and its capabilities through contact with its scientists. Military planners in the Cold War West recognized understanding the geophysical environment as the first step toward controlling it.[3] For the United States, sea ice became an issue of state significance really only after 1945, with the need to send large naval convoys to far northern locations in order to construct and supply military facilities. This need was in turn created by the "polar strategy" of concentrating American capacity for nuclear attacks on the Soviet Union in bases at the northern edge of the Western Hemisphere.[4] By contrast, the Northern Sea Route (NSR) from the Barents Sea to the Bering Straits had attracted significant investment from the Soviet government in the 1930s, leading to a higher level of understanding of sea ice and its related problems. Soviet researchers—most notably Nikolai Nikolaevich Zubov—were translated and cited frequently by American researchers, especially as Americans came to recognize that improved sea ice forecasting relied upon understanding the physics of sea ice in addition to more (and better) observations of ice conditions and associated meteorological and hydrological phenomena.[5]

The Easton conference took place at a time when scientific exchange between the United States and the Soviet Union was once again becoming possible, permitting a flow of information that had been impossible prior to 1953 and placing individual researchers in direct contact. Large-scale events such as the International Geophysical Year (IGY, 1957–58) demonstrated that the open performance of science, in addition to the open dissemination of its results, was a source of prestige.[6] Furthermore, the exchange of scientists was becoming formalized through an official arrangement between the NAS and the Academy of Sciences of the USSR (AoS). There was considerable interest in illuminating the hidden world of Soviet science (an endeavor that scientists in the West could also use to win support for their own research). These considerations undoubtedly helped to secure official American backing for Soviet participation at the Easton conference.[7]

Sea Ice and Its Forecasting to 1947

Advancing in the winter and receding in the summer, sea ice has long been an important factor in navigation within Arctic and Antarctic waters, and within particular areas such as the Gulf of Saint Lawrence and the Baltic Sea. Anticipating when sea ice forms and breaks up is essential to planning when shipping can arrive at northern destinations. Until recently, sea ice made the Arctic coastlines of Eurasia and North America sufficiently difficult to navigate that the first complete traverses were completed comparatively recently and over periods of several years, from 1878–80 and 1903–06, respectively.[8] As Figure 6.1 demonstrates, during winter months a considerable part of the north was icebound.

Knowledge of the characteristics of sea ice and of its annual cycles has always been important to communities living on ice-affected coasts, given its relevance to hunting, fishing, and transportation. Navigation made sea ice an issue for European vessels sailing in and out of what is now northeast Canada, and around Greenland, especially connected to fur trade shipments. As exploration of the polar regions became an end in itself during the nineteenth century, knowledge of the extent of Arctic sea ice improved through studies of its composition and behavior in addition to observations of its distribution.[9] From the early twentieth century the military-strategic and commercial value of sea ice research became more important. While trade along Russia's Arctic coast was growing in the late nineteenth century, state-sponsored investigation of the NSR began in earnest after the disastrous Battle of Tsushima in 1905, in which Russian generals felt they might have offered stronger resistance against the Japanese Navy had reinforcements been able to arrive more quickly.[10]

More than 60 years later, the best study of the early history of the NSR and its investigation remains *The Northern Sea Route: Soviet Exploitation of the North East Passage*, originally written as a doctoral dissertation by Terence Armstrong (1920–1996). Armstrong (whose name will reappear later in this chapter as a participant in sea ice research during the 1950s) was trained as a Russian language expert before his skills were harnessed to monitoring Soviet Arctic activity at the Scott Polar Research Institute (SPRI) in Cambridge from 1947, which at the time functioned as an information center for the British state in addition to aiding polar research and exploration.[11] His ability to locate and interpret Russian sources proved as valuable as the work he did translating the few reports

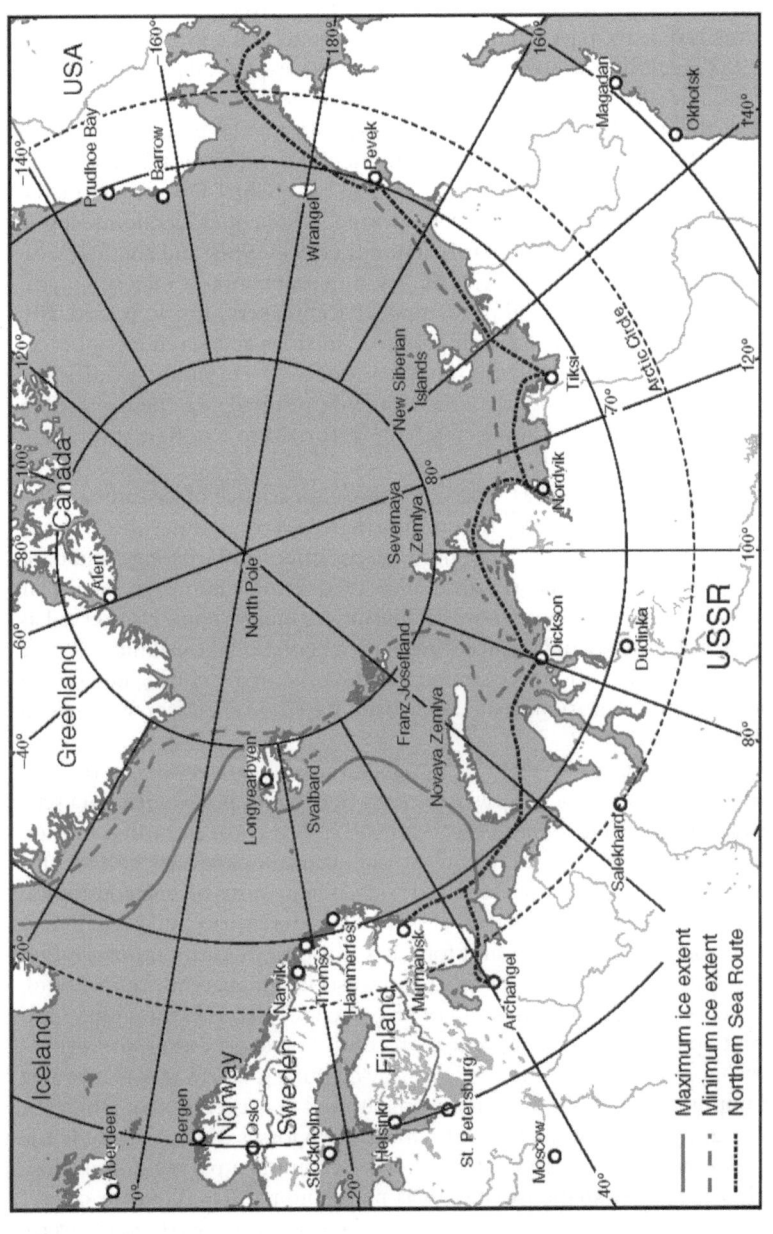

Figure 6.1 Average maximum and minimum sea ice extent on the Northern Sea Route, 1930s. Map by Hans van der Maarel, Red Geographics

of Arctic activity to emerge from the post-1947 Soviet Union, and provided the foundation for his disciplinary transition to geography.[12]

The NSR was important to Russian planning from the late nineteenth century, as partial traverses to the Ob', Yenisei, and Lena rivers connected internal Siberian river traffic to markets in Russia and beyond. This process continued after the 1917 revolution as Soviet planners sought to exploit all possible domestic resources. Already in the late 1920s, icebreaker-led expeditions conducted meteorological, hydrological, and sea ice observations along the NSR, with the goal of better understanding the environmental conditions and improving the route's function as a shipping lane.[13] In 1932 the Chief Directorate of the Northern Sea Route (*Glavsevmorput'*) was formed, a scientific-economic-logistical empire led by the charismatic Otto Schmidt (1891–1956) and charged with overseeing almost all aspects of the NSR.[14] Using a network of meteorological stations and reconnaissance flights to support icebreakers, traffic passed 100 ships in 1936, but during the following season 26 ships and seven of eight icebreakers were forced to overwinter in the Arctic due to severe ice and insufficient reconnaissance. For Armstrong, the events of 1937 resulted in a "new phase" in which the sprawling "state within a state" was returned to its original function of running a transport route.[15]

What Armstrong described as "the very important science of ice forecasting" emerged from this effort to survey Russia's northern waters. Annual observations of breakup and freeze-up dates from shore-based offices and existing shipboard data collection were complemented after 1937 by drifting stations on the ice far toward the North Pole, stations whose establishment and resupply also served as a basis for effective propaganda concerning Soviet operational capabilities.[16] The establishment of the North Pole-1 station in 1937, the first to be located on sea ice (although Antarctic stations had earlier been based on shelf ice), was indeed a "powerful political act."[17]

Long-term forecasting began to be developed in the early 1930s, with forecasts initially issued four times yearly. Although they built upon similar data sets, the individuals responsible for each forecast constructed them according to their individual beliefs in which factor was most important (a particular current, for instance). Short-term forecasting followed, dependent more on meteorological and ice observations (though hydrological data mattered too), and leading to regular ice reconnaissance flights and forecasts made from local outposts rather than from a central bureau.[18] Both types of forecasts were apparently quite useful: Armstrong reported a 1940 claim of up to 75 percent "correctness" on long-term forecasts and up to 85 percent "average correctness" on short-term forecasts.[19]

While economic activity along the NSR slowed during World War II, the area remained important strategically. Armstrong argued consistently that while the NSR was a significant drain on Soviet finances, such losses did not preclude the possibility that Moscow envisaged it *eventually* becoming a paying proposition, particularly when its strategic function was paramount.[20] The German cruiser *Komet* made a rapid (just over 21 days) and at the time secret west–east traverse of the NSR in 1940 with the aid of the Soviet authorities.[21] After Germany invaded the Soviet Union on June 22, 1941, German submarines and raiders attacked Soviet convoys, especially in the western part of the route.[22] The NSR proved largely useless for sending naval reinforcements to either extremity of the

USSR, though at least one large cruiser sailed west to east prior to the planned Soviet invasion of Japan.[23]

After 1945, the flow of information out of the Soviet Union concerning the NSR dried up almost completely. The new enemy became ever more shadowy, and the state of Russian knowledge only guessed at—just at the moment when the Arctic began to attain importance to North American military planners.

Sea Ice and Security in North America

In the years following 1945, the North American Arctic became important both to military planners, to whom it was a frontier separating the superpower enemies, and to civilian administrators of a space where increased economic development could be anticipated.[24] The Swedish glaciologist Hans Ahlmann, who had documented a recent trend toward warmer northern climates, briefed the Pentagon in 1947 on the strategic implications of changing Arctic ice conditions—a specific topic that would continue to attract attention among both earth scientists and military planners even before the modern era of anthropogenic climate change.[25] Canada and the United States cooperated to form the Arctic Institute of North America in 1945, a nongovernmental body that nevertheless was envisaged by individuals in both states as a means to acquire knowledge useful to statecraft. Bodies such as the United States Navy Hydrographic Office and the Canadian government's Geographical Branch (established 1947) soon began to work on sea ice problems, anticipating the need for better understanding of conditions in a space that was already attracting more activity than before the war.

The Hydrographic Office's 1946 "ice atlas" of northern waters was not an example of forecasting, but a fixed estimate of the probability of meeting with ice at a given point at a given time of year,[26] which in any case had already been found "inaccurate in various areas."[27] The Geographical Branch organized a Canadian Ice Distribution Survey from 1951 on much the same principles, viewing lack of data as the chief issue—despite growing evidence that the Arctic was warming, meaning that past observations could not necessarily be taken as accurate guides to ice extent in the future (an issue that Zubov had already discussed in print).[28] Ice observations, initially gleaned from libraries and archives, were entered on to cards and used to ground research papers on the distribution of ice both within the Canadian mainland and on its maritime periphery. The project was expanded in 1955 to include first-hand observations from the field, with a focus on the eastern Canadian Arctic and the Gulf of Saint Lawrence.[29]

For the United States Navy, sea ice presented a challenge when the Arctic became both a theater for potential conflict, and—more importantly—a transit lane for the construction and supply of defense installations. The initial catalyst for the Navy's active interest was the need to construct the Thule base in northwest Greenland.[30] In 1950 there was also discussion about basing bombers on fast sea ice,[31] an idea that was theoretically feasible but probably never seriously considered.[32] As Nikolaj Petersen has argued, Thule became important due to a shift in United States planning from a "perimeter" strategy in which nuclear bombers would depart from territory nearer the Soviet Union—such as the United Kingdom or the Middle East—to a "polar" strategy involving bases on North American soil, as the range of the new generation of bombers increased (see Figure 6.2).[33]

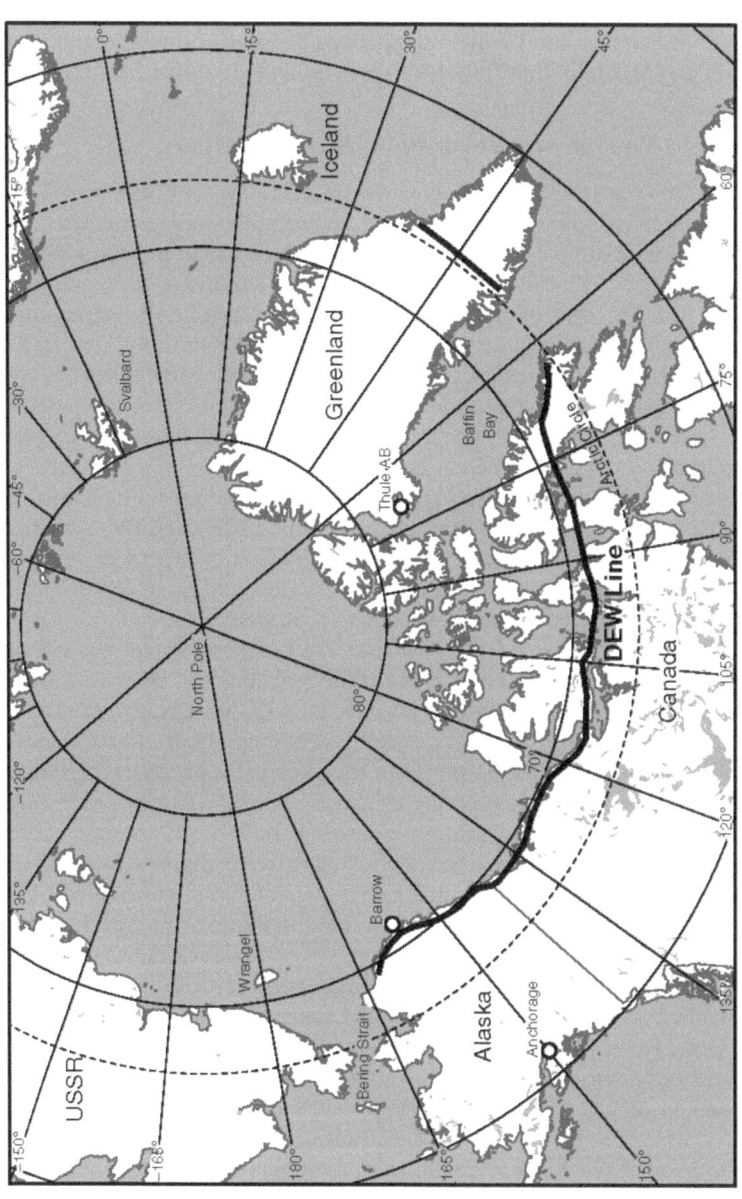

Figure 6.2 Northern North America and Greenland, showing the DEW Line and the Thule Air Base. Map by Hans van der Maarel, Red Geographics

Given the short window for delivering supplies by sea, and the huge quantities of material involved, knowledge of ice conditions was essential. A popular account of Operation Bluejay, the codename for the project to construct the Thule base, noted that when the ice broke up later than usual in summer 1951, "nearly 50 ships" were waiting to commence an operation "with all the apparatus of a wartime landing, without the shooting."[34] A less propagandistic account from much later recalled how lack of ice information meant the Navy "squandered weeks of valuable time," leading directly to increased investment in both ice observation and forecasting.[35] By 1953 the Hydrographic Office was issuing short-term forecasts that leaned upon theoretical insights from abroad (including Russia and Austria)[36] while incorporating improved observational technologies.[37]

The other Canadian body dealing with sea ice was the Defence Research Board (DRB), also formed in 1947.[38] By 1949 the DRB was liaising with the Royal Canadian Air Force to organize ice reconnaissance flights between Hudson Bay and Baffin Bay,[39] which helped it to provide ice reports for the state's Joint Intelligence Bureau, "an all-source intelligence clearing-house" that reported to the Canadian government in Ottawa.[40] While part of a wider network of British intelligence-gathering, the bureau established close ties with the Central Intelligence Agency (CIA) before it did so with other British Dominions such as Australia, reflecting the primary importance of the North Atlantic relationship above Commonwealth ties.[41] The bureau's first leader, Graham Rowley, was an Arctic expert whose focus on domestic geography rather than foreign intelligence-gathering (the bureau's central mission) doomed his tenure.[42] Ice data collected in this context was intended mostly to support aviation—including landing on frozen waterways—but its relationship to forecasting was to aid its development rather than refine its practice.[43]

Interest in forecasting picked up within the DRB in 1952 when the United States Weather Bureau and the United States Military Sea Transportation Service began seeking cooperation, and when the Canadian Weather Bureau furnished its American counterparts with measurements of sea ice thickness from far northern coastal communities, helping to provide clues on how close the winter ice was to melting.[44] An early attempt to provide short-term forecasts for ice conditions of Labrador involved superimposing a weather map over an ice map and inferring ice movement from predicted meteorological changes.[45] Even then, while Rowley, now the DRB's Arctic expert, declared in 1953 to the United States naval attaché in Ottawa that Canada was "very interested in ice forecasting and 'hope[d] to take a more active part in it next year," an internal department memorandum written by his co-worker Moira Dunbar just nine days later admitted that "the whole question of ice forecasting is very new in Canada and the situation somewhat confused."[46]

As in Canada, authorities in the United States relied heavily upon reconnaissance flights and field observations to assess ice conditions. For the most part, ice reports taken from the air were part of a wider attempt to acquire environmental knowledge, with observers concerned primarily with meteorological data collection. Perhaps the most notable example was Operation Ptarmigan, a series of reconnaissance flights along a triangular route with its apex at the geographic North Pole, which included observations of sea ice in addition to their primary mission of collecting meteorological data.[47] Such activities provided the basis for analytic work, providing records throughout time (seasons) as well as space, leading to theory that could underpin forecasting. In Armstrong's judgment, American instructions assumed "less highly trained observers" than their Soviet counterparts, quite likely a reflection of the greater importance sea ice had historically held there as a subject in its own right.[48]

In addition to ice atlases and probability studies, the Hydrographic Office recognized that reconnaissance was an important component of forecasting, but that it was also important to understand the principles behind ice movement—a subject already dealt with at length in Soviet literature. Indeed, when the US Navy began synoptic ice forecasting in 1952, much of the theoretical foundation was borrowed from Soviet work.[49] As the veteran of American sea ice research W. F. Weeks has noted, the United States Strategic Air Command (SAC) as well as the North East Air Command (NEAC) had a significant interest in Arctic ice forecasting during these years in order to determine if and when aircraft could use sea ice as a landing strip.[50] This particular requirement helped stimulate research on the ice physics side of sea ice studies. From 1954 the Air Force Cambridge Research Center joined the Hydrographic Office and the Snow, Ice, and Permafrost Research Establishment (SIPRE), a branch of the Army Corps of Engineers, in investigating a range of questions related to sea ice.[51] The result was a growing focus on sea ice in itself, with consequences for forecasting (even if this was not always the primary goal), sometimes drawing upon a deep set of data from a single location to fully understand the interplay between temperature gradients (above and below the ice sheet), wind gradients, and solar radiation on the properties of the ice itself.[52] While the early burst of American research enabled improved short-range forecasting (up to several months), more complex methods were needed to tackle larger-scale phenomena. One such was ice potential climatology, the capacity for a given body of seawater to form stable ice at a specific location, which could in turn be mapped—hence the analogy to climate maps.[53]

The advance of radar installations northward—most notably to the Distant Early Warning (DEW) Line, a series of radar outposts designed to detect incoming Soviet nuclear bomber aircraft that became operational in 1957[54]—only reinforced the need for the United States to be able to operate ships effectively in ice-infested waters and potentially land supply aircraft in the winter months to accelerate the construction timetable (see Figure 6.2).[55] During the summer of 1955, no fewer than 969 ice forecasts were delivered in connection with DEW Line construction efforts.[56] A similar imperative spurred the Air Force to sponsor research on sea ice conditions in the Baffin Bay/Davis Strait region, a project that would ultimately require detailed, large-scale investigation of how sea ice was formed in addition to "long-range aerial surveys."[57] When the Navy later became interested in using submarines for resupply missions, Arctic drifting stations attained some importance, though much of the data they produced remained classified and had only limited impact in developing sea ice studies.[58]

Russian workers had taken the lead in studies of Arctic sea ice forecasting during the first decades of the twentieth century, a direct consequence of the importance the state attached to the NSR, and by 1954 North American researchers still lagged behind in terms of both the infrastructure for collecting observations and the capacity to convert data into forecasts.[59] Terence Armstrong, trained as a linguist rather than a natural scientist and thus well aware of the limitations posed by lack of specialized background, noted that in contrast to the quite recent past, "ice physics can no longer be importantly advanced by the gifted amateur studying the natural history of sea ice."[60] The resources that the United States devoted to sea ice research were not at the same level as funding for other areas of science during this period, but the increase in support was consistent with the development of sea ice forecasting as both a research program and a military instrument. In 1955 the Hydrographer of the Navy prefaced a report on ice potential climatology (linking the stability and thickness of ice to environmental conditions) by noting that in order to improve "the techniques of ice forecasting, it has been necessary to formulate theories concerning the large-scale movements of ice in the Arctic Basin."[61] Pursuing that task involved increasing the level of material assistance while also mastering and developing theoretical methods—and keeping abreast of developments behind the Iron Curtain.

Learning from Others

In the years immediately following 1945 Soviet science retreated to the "isolationist nationalism" that Nikolai Krementsov argues was characteristic of the 1930s—but in this case, even more so.[62] This was true of almost all fields,[63] but particularly of Soviet activity in the Arctic, which had earlier been such a powerful source of national propaganda. The high visibility of Soviet work in the 1930s followed by the lack of subsequent information fostered a sense of uneasy fear. Glimpses such as the lavish jubilee celebrations of the Soviet Academy of Sciences in the summer of 1945, which afforded Hans Ahlmann a rare look at the impressive Arctic Research Institute in Leningrad, prompted fear as much as admiration.[64]

Terence Armstrong, who followed Soviet developments as closely as anyone from his office in Cambridge, noted in 1952 that "virtually no information on the working of the Northern Sea Route" had become available since 1946.[65] The situation only began to change after 1954, with Stalin's death in 1953 followed by the public reappearance of the Arctic ice drift stations that had been such powerful symbols of Soviet Arctic strength during the 1930s.[66] In the meantime, Armstrong was reduced to joining scattered dots, such as his conclusion that a number of airmen and scientists who worked together in the 1930s had probably continued to do so during the intervening years, based on their collective reappearance in the 1954 floating station—which in turn hinted at considerable investment and presumably progress.[67] Brian Roberts, who oversaw the British Foreign Office's desk for the polar regions, told his superiors in 1955 with some certainty that "the magnitude of [Soviet Arctic] operations exceeds that of all other countries put together."[68]

From 1951 both Roberts and Armstrong worked simultaneously for the Scott Polar Research Institute and the British government—Roberts for the Foreign Office, Armstrong for the Admiralty. Armstrong combined surveillance of Soviet work with an increasing personal interest in sea ice issues. These interests came together in 1953 through a contract for his services between the DRB and SPRI, which placed him aboard the Canadian icebreaker HMCS *Labrador* on its inaugural traverse of the North-West Passage (a passage smoothed by recent US investment in ice forecasting and reconnaissance),[69] and enabled him to compile ice atlases in addition to reports on the state of Soviet research.[70] With his near-perfect command of Russian and access to Soviet publications through SPRI, Armstrong was able to sketch an outline of the directions in work, despite the Stalin-era cloak that covered much Soviet science.[71] As Armstrong put it, "although some of the principles are known, it is not certain which are used, nor how and with what success they have been applied."[72] He concluded in 1954 that lack of sufficient observational data had held back earlier Soviet forecasting, a drawback he was certain had since been addressed.[73]

In July 1955 SPRI organized a reciprocal exchange that saw Armstrong and Roberts visit a number of polar-related institutions in the Soviet Union during the summer of 1956, with a specific hope of obtaining information on sea ice forecasting. Their report of the visit was unclassified (though a prefatory note advised against it being kept on open library shelves, as its comments on certain individuals were frank).[74] Using his command of Russian and thorough knowledge of the literature, Armstrong was able to detect when his interlocutors were lying about the existence of certain publications—such as ice yearbooks—and gain a more nuanced understanding of what kind of information remained classified.

While diligent charting of past records could increase their utility for probability forecasts in the future, a "cardinal need" remained for greater knowledge of physical processes, to which the US Navy Electronics Laboratory could contribute—as would access to Soviet literature, given that Armstrong felt American work had reached a level in 1954 equal only to Soviet work in 1939.[75] One of the DRB's sea ice specialists, Moira Dunbar, learned Russian for this reason.[76] The few articles by Soviet researchers that had appeared in English became important resources, especially those such as Zubov and Arkady Georgievich Kolesnikov who used equations to analyze ice formation and behavior.[77] Zubov's 1945 book *Arctic Ice* has aptly been described as "an innovative synthesis of widely scattered work from both the Russian and the Western literature ... there are insights on every page."[78] This work was translated in parts into English during the 1950s and referenced frequently in American papers, leading the US Air Force Cambridge Research Center to eventually pay for a complete translation in 1963.[79]

The most visible symbol of Soviet strength in Arctic research was their drifting ice stations. Like their American counterparts, the first of which (T-3) was established in 1952, the Soviet drift stations could be imagined as military sites without too much difficulty, even if that function remained theoretical.[80] The British Foreign Office suspected that knowledge gained from these stations (especially in meteorology) would enhance the Soviets' understanding of the strategically important Northern Sea Route while aiding aviation in high latitudes, and perhaps helping to move the Soviet radar defence network northward[81]—a fear apparently confirmed by a former US Air Force member who participated in reconnaissance flights from Thule in early 1959.[82] US-operated stations eventually became sites for research in underwater acoustics (relevant for detecting submarine-launched missile systems), as did Soviet drifting stations.[83]

At the same time, the drifting stations were national symbols that bore the values of their sponsors (and the values attributed by rivals), located within an emerging dynamic of Cold War competition that extended from the space race to the Olympic games. Weeks has recalled the impact a published map of Soviet explorations from ice stations had on American bureaucrats in the early 1950s, ultimately driving a matching response from the United States and Canada.[84]

The Soviet media eagerly linked American stations with an aggressive militarization of the Arctic, similar to their depiction of American bases in Greenland, while characterizing the development of the Northern Sea Route as an aid to economic development and nation-building.[85] The publicity given to Soviet ice stations by the Soviet media, thought the British Foreign Office, might also have been part of a broader campaign to legitimize Soviet authority over the sector it had unilaterally declared in 1926, even if international law suggested additional territorial claims based on occupation of sea ice stations were unlikely to succeed.[86]

The drifting stations were only the tip of the iceberg when it came to the world of Soviet Arctic activity, most of which remained hidden. The United States was determined to improve its understanding of sea ice, leading to the 1958 formation of Sea Ice Central (a joint Canadian-US forecasting body using personnel trained at the US Navy Hydrographic Office but with Canadian aircraft and field stations), and a major facility upgrade to the Army's Cold Regions Research and Engineering Laboratory.[87] One means of achieving this goal was to bring leading researchers to the United States, a task to which I now turn.

The National Academy of Sciences Sea Ice Conference

In September 1956 the ONR approached the NAS's earth sciences branch to organize an event that would aid "in furthering studies that would lead to more effective sea-ice forecasting…The Navy mission of forecasting sea-ice conditions needs the stimulus of better basic knowledge of sea-ice formation and growth, drift, and decay and disintegration; analytical techniques in forecasting would also benefit from an exchange of data and views."[88] As the number of workers in the field remained "rather small," the ONR felt a single gathering would provide an "inventory" of current work and future directions.[89] Up to 12 Europeans would be invited. Plans developed slowly through 1957, including one crucial development: the potential participation of Soviet scientists.

The precise reason why Soviet scientists were initially excluded is unclear, but it is highly likely their participation became conceivable thanks to contemporaneous events that fostered East-West scientific connections. Planning for the sea ice conference took place at the same time as planning for the International Geophysical Year (IGY) and—like other activities from the Antarctic to the Mediterranean Sea and beyond—to some extent rode upon its coattails. The IGY's overt focus on surveillance of the earth and surveillance of scientists, most visible in the massive Antarctic program and its formalized series of mutual observers and inspections, was less a denationalization of science as a recognition that science could simultaneously be a source of strategically relevant data and a vehicle for articulating foreign policy objectives.[90] The infrastructure of World Data Centers at which information was deposited for access by other states made concrete a rhetorical devotion to international cooperation that nevertheless served the interests of the superpowers. It also permitted them to establish additional stations in strategically sensitive locations that might otherwise have attracted controversy, including on floating ice.

The sea ice conference similarly appealed for international exchange in a field where the United States knew that its knowledge was imperfect, the conference an opportunity to become acquainted with theoretical advances in addition to expanding the geographic scope of data from which conclusions might be drawn. The IGY also provided a discursive framework in which research that had such strategic implications could be presented as value-neutral contributions to basic science—something that accorded with prevailing Soviet rhetoric. This relationship between basic and applied research was articulated neatly in the three-page press release that accompanied the conference's opening. Making no mention of the military, this document instead stressed the economic benefits of improved sea ice forecasting, linking hopes for significant advances to drift stations, exchange of information with Soviet researchers, and the quest to know the earth in terms of "mathematical formulae" linked to the IGY.[91] Of course, practical importance could mean military importance: the topics and participants were chosen in consultation with both the ONR and the US Navy, and the planning committee included men such as Gordon G. Lill and Louis Quam who played key roles in the geophysics–military interface.

Around this time the NAS had also initiated discussions with the Soviet AoS on a series of formal exchanges within specific branches of science and technology.[92] The interacademy nature of the arrangement ensured a level of state control—or at least, a guarantee that activity would conform to state expectations. Fields of interacademy exchange were negotiated, participants vetted, and the reports submitted to the NAS on their return examined in terms of whether they had resisted Soviet brainwashing as well as the technical information they had gleaned. While the conference did not come under the aegis of this agreement, NAS president Detlev Bronk and his Soviet counterpart Alexander Nikolaevich Nesmeyanov had established a framework within which international exchange became feasible, if by no means routine.

Bronk, a biophysicist who served as president of Johns Hopkins University and later Rockefeller University while also presiding over the NAS, was a key figure in the postwar links between science and government in the United States.[93] This included liaising with the Operations Coordinating Board, a body established by the Eisenhower administration in 1953 to link national security with psychological warfare and strategy,[94] which tasked Bronk with delivering reassuring statements on matters from radiation to Soviet space achievements.[95] Indeed, on February 26, 1958, while the NAS sea ice conference was underway, the OCB suggested that Bronk take a key role in executing President Eisenhower's publicly announced plans for "a full-scale cooperative program of science for peace."[96] The fact that the board had been asked to implement this lofty goal is a reminder—like many others from the era of the IGY—that international science invariably possessed a strategic dimension.

The existing relationship between Bronk and Nesmeyanov, and the NAS's imprimatur on the conference, provided an opening through which invitations could be sent once the US Navy had been convinced that no classified American information would be disclosed.[97] As other chapters in this volume emphasize, the flow of information at international scientific meetings was rarely if ever unidirectional.[98] The Navy's stance was a reminder that in addition to the propaganda value of participating in international scientific cooperation (which the USSR had embraced during the IGY), the Soviet scientists would be eager to learn useful things from their US counterparts, necessitating limits on what could be shared. Even after the NAS had invited the AoS to nominate participants, the fact Easton was part of the wide swathes of American territory normally off-limits to Soviet citizens had to be addressed, through the East-West Contacts branch of the State Department.[99] Thurston argued that Soviet presence was essential because its scientists

> are believed to have special competence and experience extending over a longer span of time than other national groups…The utilitarian objective of sea ice research is to produce reliable forecasts of sea ice conditions for as long a range as possible. Many organizations gather data but only three are known to venture forecasts: the Japanese, the American-Canadian, and the Soviet groups. In this field the Soviet organizations are thought by some well-informed specialists to have the greatest experience and possibly to have developed advanced procedures…It is well worth admitting them to the closed area of Easton, MD.[100]

Soviet participation was only confirmed on February 5, 1958—less than three weeks before the conference was scheduled to begin—and it was only then that the four delegates were nominated. Bronk suggested a list of nine researchers (not including Zubov, who may have been presumed to have already retired).[101] One of those suggested (Kolesnikov) joined the delegation, along with Pavel Afanas'evich Gordienko, a veteran of the drift stations and a practical rather than theoretical man. Telegrams between Bronk, Nesmeyanov, and the glaciologist Grigory Aleksandrovich Avsyuk (leader of the Soviet delegation) revealed a chaotic rush to obtain exit visas and finalize travel plans.

The conference proceeded smoothly and cordially under seven sections (running consecutively rather than parallel): distribution and character of sea ice; sea ice observing and reporting techniques; physics and mechanics of sea ice; sea ice formation, growth, and disintegration; drift and deformation of sea ice; sea ice prediction techniques; and sea ice operations. Soviet researchers presented in the first five of these areas, although their state clearly possessed particular expertise in the final two (reflected in the questions directed to Gordienko in the discussions following the session on prediction). Whether much of the information presented had significant novelty is unclear—Soviet participants frequently answered questions with general references to ongoing or as yet incomplete work—but the opportunity to interact with researchers whose work had only previously been known through translated texts was valuable, an opinion confirmed by both the Hydrographer and the chief of the ONR.[102] Approval for Soviet participants to visit unclassified government facilities following the conference arrived just before the conference began.[103] Despite the NAS attempting to construct a detailed program for their Soviet visitors (including visits to national monuments in Washington DC and opportunities for shopping), the short time frame made precise planning difficult as institutions such as SIPRE in Illinois got wind of the Soviets' presence and sought to invite them for a visit.[104]

If the event succeeded as its hosts had hoped, how was it viewed from the Soviet side? Archival evidence is thus far lacking, but it is possible to say something about how the conference was used as a propaganda instrument within the Soviet public sphere. Gordienko's impressions of the conference were published in the magazine *Priroda* [Nature] after his return, providing an official account of how the event's importance was mediated in the Soviet Union.[105] Upon arrival in New York the delegation was taken to Easton—which

Gordienko found "quaint."[106] Gordienko was sure also to record that the Soviet papers had provoked "animated discussions...because of the scope and acuteness of formulation of the subject matter and because of the high theoretical level of the papers"[107] (a favorable impression that was apparently not shared by all his Western listeners).[108] Gordienko also noted the importance of Russian work in observational practice and theories (especially that of Zubov) to Western researchers, reinforcing a sense that this had been an opportunity to demonstrate Soviet leadership.[109] A similar concern for emphasizing Soviet superiority probably drove Gordienko's claim that while he had heard some interesting and novel papers, the United States was recognizably behind other countries "in the study of physical properties of ice."[110] He was more impressed by the Smithsonian Museum of Natural History of America than by a visit to George Washington's childhood home of Mount Vernon, which was described in two dry and opinionless sentences.[111] When Gordienko claimed that the 12-day visit had laid the basis for "developing cultural and scientific communication" between the superpowers, his readers could be reassured that such communication had confirmed rather than challenged their own strength.[112]

Conclusion

Sea ice forecasting continued to be an important research field after the Easton conference. Weeks has noted that understanding ice dynamics became important for detecting submarines (a fact noted also noted in the USSR), especially after it became possible to base nuclear-powered and -armed submarines under Arctic ice, helping to shift the focus of investigations away from using sea ice for engineering toward understanding how objects could hide (or be detected) beneath it.[113] Moreover, the economic and social development of Arctic North America continued apace, and the Alaskan North Slope became particularly important once oil was discovered in large quantities in 1968, marking a return to relevance of sea ice in engineering terms.

The events described in this chapter demonstrate that sea ice forecasting, like many other geophysical disciplines that attained relevance to military planners during the early Cold War, involved its practitioners and patrons in the dual process of knowing the enemy and knowing the earth. Observations from field stations and reconnaissance flights were prerequisites for successful forecasting, but that enterprise also required theoretical competence that at least for the first postwar years, was suspected to be more highly developed in the Soviet Union. Scientific exchange and interaction was a means of acquiring knowledge with military-strategic value, and even small, focused events like the Easton conference were possible only within a larger political framework that regulated East-West contacts.

Another point might be made concerning the nature of knowledge-gathering concerning sea ice. Building upon the work of Thomas Hughes, Kristie Macrakis has described a "technophilia" within United States intelligence-gathering services during the Cold War, a preference for instrumental over human espionage—the opposite of the agent-centered approach pursued by the Eastern Bloc.[114] I would certainly not classify events such as Armstrong's visit to the Soviet Union or the Easton conference as espionage, but they did obtain intelligence that provided a more detailed picture of Soviet scientific achievements. Just as Macrakis has pointed to the inability of satellite images provide accompanying context for data (most notably by illuminating political intentions), knowing the location of Soviet sea ice stations and the rough outlines of their administrative structures could not reveal the level of technical advancement in sea ice forecasting. As the documentation surrounding its planning make clear, the Easton conference was more than just a venue for scientific interaction: it was a forum for assessing the state of a strategically sensitive field of knowledge.

Sea ice research continued to be strategically important from the 1960s onward, but—as Sebastian Grevsmühl demonstrates later in this volume—satellite imagery emerged as the dominant source of data. The quest to place the entire earth under surveillance produced a number of developments within the geosciences, expected and otherwise, among them a far greater capacity to map sea ice. Military applications such as identifying the potential surfacing spots of submarines justified the pursuit of research that is today important not only for the ships that navigate the Northern Sea Route in ever greater numbers, or for the geoscientists for whom sea ice constitutes an important signal. Sea ice extent has become an emblematic element of climate change science, its forecasting and measurement a matter with ramifications far beyond the immediately practical.

Notes

1. The proceedings of the conference were published as National Academy of Sciences, *Arctic Sea Ice: Proceedings of the Conference Conducted by the Division of Earth Sciences and Supported by the Office of Naval Research* (Washington, DC: National Academy of Sciences-National Research Council Publication 598, 1958).

2. National Academy of Sciences, *Arctic Sea Ice*, preface (unpaginated). Much of the organizational correspondence for the event is held in the National Academy of Sciences Archives (hereafter NASA) Division of Earth Sciences: Conference on Arctic Sea Ice, February 1958.

3. See, most notably, Ronald E. Doel, "Constituting the Postwar Earth Sciences: The Military's Influence on the Environmental Sciences in the USA after 1945," *Social Studies of Science* 33 (5) 2003: 635–666.

4. See, for instance, Nikolaj Petersen, "SAC at Thule: Greenland in the US Polar Strategy," *Journal of Cold War Studies* 13 (2) 2011: 90–115.

5. On Zubov's importance to American scholars, see for instance Sverker Sörlin and Julia Lajus, "An Ice-Free Arctic Sea? The Science of Sea Ice and its Interests," in Miyase Christensen, Annika E. Nilsson, and Nina Wormbs (eds), *Media and the Politics of Arctic Climate Change When the Ice Breaks* (New York: Palgrave Macmillan, 2013), 70–92.

6. Much has been written on the International Geophysical Year. Notable works include Walter Sullivan, *Assault on the Unknown: The International Geophysical Year* (New York: McGraw Hill, 1961); Jacob Darwin Hamblin, *Oceanographers and the Cold War: Disciples of Marine Science* (Seattle: University of Washington Press, 2005); Roger D. Launius, James Rodger Fleming, and David H. DeVorkin, eds, *Globalizing Polar Science: Reconsidering the International Polar and Geophysical Years* (New York: Palgrave Macmillan, 2008); and Christy Collis and Klaus Dodds, "Assault on the Unknown: The Historical and Political Geographies of the International Geophysical Year (1957–8)," *Journal of Historical Geography* 34 (4) 2008: 555–573.

7. John Krige, "Atoms for Peace, Scientific Internationalism, and Scientific Intelligence," in Krige and Barth (eds), *Global Power Knowledge: Science and*

   Technology in International Affairs, 161–181. On the use of Soviet achievements
   as leverage for even non-Americans to press for more US investment in science,
   see Peder Roberts, "Intelligence and Internationalism: The Cold War Career of
   Anton Bruun," *Centaurus* 55 (3) 2013: 243–263.

8. There are numerous works on the history of exploration in these two waterways,
   especially the North-West Passage. See for instance Terence Armstrong, *The
   Northern Sea Route* (Cambridge: Cambridge University Press, 1952); Pierre Berton,
   *The Arctic Grail: The Quest for the North West Passage and the North Pole, 1818–
   1909* (New York: Viking, 1988); Glyn Williams, *Voyages of Delusion: The North West
   Passage in the Age of Reason* (New Haven, CT: Yale University Press, 2003).

9. W. F. Weeks (with W. D. Hibler), *On Sea Ice* (Fairbanks: University of Alaska
   Press, 2010), 11–16.

10. Armstrong, *Northern Sea Route*, 15.

11. Lawson Brigham, "Armstrong, Terence," in Mark Nuttall (ed.), *Encyclopedia of the
   Arctic: A-F* (New York: Routledge, 2005), 154–155. On this period in SPRI's history, see Peder Roberts, *The European Antarctic: Science and Strategy in Scandinavia
   and the British Empire* (New York: Palgrave Macmillan, 2011), Chapter 7.

12. On the isolation of Soviet science during the period 1947–1953 and the reasons
   for the "thaw" following the death of Stalin, see Mark B. Adams, "Networks in
   Action: The Khrushchev Era, the Cold War, and the Transformation of Soviet
   Science," in G. E. Allen and R. M. MacLeod (eds), *Science, History and Social
   Activism: A Tribute to Everett Mendelsohn* (Dordrecht and Boston: Kluwer, 2001),
   255–276; and Konstantin Ivanov, "Science after Stalin: Forging a New Image of
   Soviet Science," *Science in Context* 15 (2) 2002: 317–338.

13. Armstrong, *Northern Sea Route*, 35.

14. On the history of this organization, see also John McCannon, *Red Arctic: Polar
   Exploration and the Myth of the North in the Soviet Union, 1932–1939* (New York:
   Oxford University Press, 1998).

15. Terence Armstrong, *The Russians in the North: Aspects of Soviet Exploration and
   Exploitation of the Far North, 1937–57* (London: Methuen, 1960), 91–92.

16. Armstrong, *Northern Sea Route*, 89; McCannon, *Red Arctic*, Althoff, *Drift Station:
   Arctic Outposts of Superpower Science* (Washington, DC: Potomac Books, 2007).

17. Althoff, *Drift Sstation*, 39.

18. Armstrong, *Northern Sea Route*, 96–97.

19. Ibidem, 98.

20. Armstrong, *Russians in the North*, 97.

21. Armstrong, "The Voyage of the *Komet* Along the Northern Sea Route, 1940,"
   Polar Record 5 (37–38) 1949: 291.

22. On the strategic value of the NSR during World War II, see Constantine Krypton,
   The Northern Sea Route and the Economy of the Soviet North (London: Methuen,
   1956), 166–170.

23. Armstrong, *Russians in the North*, 102, citing Henry A. Wallace (with Andrew
   J. Steiger), *Soviet Asia Mission: 12,000 Air Miles Through the New Siberia and
   China* (New York: Reynal & Hitchcock, 1946).

24. On Canada, see J. Keith Fraser, "Activities of the Geographical Branch in
   Northern Canada, 1947–1957," *Arctic* 10 (4) 1957: 246–250.

25. Ronald E. Doel, "Why Value History?" *Eos, Transactions of the American Geophysical
   Union* 83 (47) 2002: 544–45. See also Sverker Sörlin, "Narratives and Counter-
   Narratives of Climate Change: North Atlantic Glaciology and Meteorology, ca
   1930–1955," *Journal of Historical Geography*, 35 (2) 2009: 237–255. For a graphic
   account of the potential implications of a warming Arctic for ice conditions, see
   Graham Rowley, Defence Research Board memorandum dated May 19, 1952, in

Libraries and Archives Canada (hereafter LAC) file number DRBS 135–760–267–7 collection Defence Research Board—Arctic Research Sea Ice. On the history of climate change science, see, most notably, Spencer Weart, *The Discovery of Global Warming* (Cambridge, MA: Harvard University Press, 2003).

26. United States Navy Hydrographic Office, *Ice Atlas of the Northern Hemisphere* (Washington, DC: United States Navy Hydrographic Office, 1946).

27. Canadian Ice Distribution Survey Project Presentation (file number C-2-2-9), produced by Department of Mines and Technical Surveys, Geographical Branch, 1951. A copy is held in LAC DRBS 135–760–267–7.

28. N. N. Zubov trans. U.S. Navy Oceanographic Office and American Meteorological Society, *Arctic Ice* (Washington, DC: United States Navy Hydrographic Office, 1963). The work was originally published in Russian in 1943, and translated in bits and pieces over the following years until a full translation appeared in 1963.

29. F. A. Cook, "Ice Studies of the Canadian Geographical Branch," *Polar Record* 10 (65) May 1960: 123–125.

30. See, for instance, D. G. Lindsay, "Sea Ice in the Canadian Archipelago," MSc thesis submitted to the Department of Geography, McGill University, 1968, 8.

31. Dansk Udenrigspolitisk Institut, *Grønland under den kolde krig: dansk og amerikansk sikkerhedspolitik 1945–68* (Copenhagen: Dansk Udenrigspolitiks Institut, 1997), 114.

32. W. F. Weeks, personal communication.

33. Petersen, "SAC at Thule," 90–91.

34. Charles Michael Daugherty, *City under the Ice: The Story of Camp Century* (New York: Macmillan, 1963), 36; 37.

35. Charles C. Bates, Thomas F. Gaskell, and Robert B. Rice, *Geophysics in the Affairs of Man: A Personalized History of Exploration Geophysics and Its Allied Sciences of Seismology and Oceanography* (Oxford and New York: Pergamon, 1982), 144.

36. Owen S. Lee and Lloyd S. Simpson, *A Practical Method of Predicting Sea Ice Formation and Growth* (Washington, DC: United States Navy Hydrographic Office, 1954), Technical Report TR-4.

37. Bates, Gaskell, and Rice, *Geophysics in the Affairs of Man,* 144.

38. On the history of this institution, see Jonathan Turner, "Politics and Defence Research in the Cold War," *Scientia Canadensis* 35 (1–2) 2012: 39–63; and Turner, "The Defence Research Board of Canada, 1947–1974," PhD diss., University of Toronto, 2013.

39. See the correspondence from March to October 1949 in LAC DRBS 135–760–267–7.

40. Huw Dylan, "The Joint Intelligence Bureau: (Not So) Secret Intelligence for the Post-War World," *Intelligence and National Security* 27 (1) 2012: 37.

41. Kurt F. Jensen, *Cautious Beginnings: Canadian Foreign Intelligence 1939–51* (Vancouver: University of British Columbia Press, 2009), 153–154.

42. Jensen, *Cautious Beginnings,* 144.

43. Canadian Ice Distribution Survey Project Presentation (file number C-2-2-9). LAC DRBS 135–760–267–7.

44. F. H. Gardner (United States Naval Attaché) to A. Ironside (Defence Research Board), March 27, 1953. LAC DRBS 135–760–267–7.

45. Rowley, memorandum to Bowen (Joint Intelligence Board), October 18, 1952. LAC DRBS 135–760–267–7.

46. Rowley to Gardner, April 18, 1953; Dunbar, untitled memorandum, April 27, 1953. Both held in LAC DRBS 135–760–267–7.

47. On the Ptarmigan flights, see Aubrey O. Cookman, "Top of the World Weather Run," *Popular Mechanics* (November 1948): 97–101; 262; 264. See also Edward L.

Corton, *The Ice Budget of the Arctic Pack and its Application to Ice Forecasting* (Washington, DC: United States Navy Hydrographic Office, September 1954).

48. Terence Armstrong, "Sea Ice Studies," *Arctic* 7 (3–4) 1954: 203.
49. Ibidem, 203.
50. Weeks, *On Sea Ice,* 21.
51. See, for instance, W. F. Weeks and O. S. Lee, "Observations on the Physical Properties of Sea-Ice at Hopedale, Labrador," *Arctic* 11 (3) 1958: 134–55.
52. J. H. Brown and E. E. Howick, "Physical Measurements of Sea Ice," Navy Electronics Laboratory Research and Development Report 825, February 11, 1958: 16.
53. Edward L. Corton, *Climatology of the Ice Potential as Applied to the Beaufort Sea and Adjacent Waters* (Washington, DC: United States Navy Hydrograhic Office Technical Report TR-30, 1955), 1.
54. For a map of the shipping lanes required to supply the DEW Line, see Roy J. Fletcher, "Military Radar Defence Lines of Northern North America: An Historical Geography," *Polar Record* 26 (159) 1990: 265–276. On the DEW Line more generally see P. Whitney Lackenbauer, Matthew J. Farish, and Jennifer Arthur-Lackenbauer, *The Distant Early Warning (DEW) Line: A Bibliography and Documentary Research List* (Calgary: Arctic Institute of North America, 2005).
55. Weeks, *On Sea Ice,* 21; Edmund A. Wright, *CRREL's First 25 Years: 1961–1986* (Hanover, NH: United States Army Corps of Engineers, 1986), 16. The term "ice-infested" refers to any waters in which ice is present.
56. Bates, Gaskell, and Rice, *Geophysics in the Affairs of Man,* 145.
57. Henry S. Kaminski, *Distribution of Ice in Davis Strait and Baffin Bay* (Washington, DC: United States Navy Hydrographic Office, 1955), v.
58. Weeks, personal communication.
59. Armstrong, "Sea Ice Studies."
60. Ibidem, 202.
61. J. B. Cochran, preface to Corton, *Climatology of Ice Potential,* iii.
62. N. L. Krementsov, *Stalinist Science* (Princeton: Princeton University Press, 1997), 130.
63. For a case from oceanography, see Roberts, "Intelligence and Internationalism."
64. On Ahlmann's visit to Moscow, see Robert Marc Friedman, "Background to the Establishment of Norsk Polarinstitutt: Postwar Scientific and Political Agendas," The Northern Space: The International Research Network on the History of Polar Science, Working Paper 2 (1995); Roberts, *The European Antarctic,* 114–115; and Julia Lajus and Sverker Sörlin, "Melting the Glacial Curtain: The Politics of Scandinavian-Soviet Networks in the Geophysical Field Sciences Between Two Polar Years, 1932/33–1957/58." *Journal of Historical Geography* 44 2014: 44–59.
65. Armstrong, *Northern Sea Route,* 50.
66. Armstrong, *Russians in the North,* 104.
67. Ibidem, 105.
68. Brian Roberts, draft information paper circulated within the Foreign Office, March 12, 1955. The National Archives of the United Kingdom (hereafter TNA) FO 371/114000. While there was a break of 12 years (1938–1950) between the drift stations NP-1 and NP-2, high-altitude aerial surveys were conducted during this time, as detailed in B. T. Sokolova, ed., *High Latitude Air Expeditions "North" [Sever]* (St. Petersburg: Arctic and Antarctic Research Institute, 2000). Available online at http://elib.rshu.ru/files/img-213115508.pdf.
69. Bates, Gaskell, and Rice, *Geophysics in the Affairs of Man,* 145.
70. See for instance T. Armstrong, *Sea Ice North of the USSR* (London: Admiralty Hydrographic Department, 1958). The value of the contract was declared in early 1955 to be between C$1,000–3,000 (Roberts, draft information paper, March 12,

1955. TNA FO 371/114000.) The subject was discussed frequently at the Scott Polar Research Institute's Committee of Management during the years 1953–55. Thomas Manning Archives, Scott Polar Research Institute, Committee of Management Papers (uncatalogued).

71. See, for instance, Terence Armstrong "Soviet Work on Sea Ice Forecasting," *Polar Record* 7 (49) 1955: 302–311.

72. Ibidem, 309.

73. Ibidem, 310.

74. An abridged copy of this report is held at the Scripps Institution of Oceanography Archives (hereafter SIO), Robert S. Dietz Papers, MC 28 26/4.

75. Armstrong, "Sea Ice Studies," 204.

76. Campbell Thomas, "Moira Dunbar: Woman Scientist the Navy Refused to Take Aboard," *The Guardian*, January 12, 2000. Available online at http://www.theguardian.com/news/2000/jan/12/guardianobituaries1

77. For a good example, see Elliott B. Callaway, *An Analysis of Environmental Factors Affecting Ice Growth* (Washington, DC: United States Navy Hydrographic Office, September 1954).

78. Weeks, *On Sea Ice*, 18.

79. Zubov, *Arctic Ice*.

80. Weeks, personal communication.

81. Earl (George) Jellicoe, Foreign Office minute on Soviet Arctic activities, August 9, 1954. TNA FO 371/111750.

82. Petersen, "SAC at Thule," 109–110. Weeks (personal communication) notes that while it was possible to establish radar installations on drifting stations, their range (and thus military value) would have been very limited.

83. Weeks, *On Sea Ice*, 23; personal communication. On underwater acoustics, see Robinson in this volume.

84. Weeks, *On Sea Ice*, 22.

85. See for instance Terence Armstrong to Earl Jellicoe, June 14, 1955. TNA FO 371/116226.

86. See the correspondence in TNA FO 371/116764.

87. For a brief overview, see "Naval Affairs," *Naval Review* 46 (3) July 1958: 370–371; "Canadian Sea Ice Forecasting Service," *Polar Record* 9 (63) September 1959: 581–582.

88. William J. Thurston (NAS Division of Earth Sciences) to S. D. Cornell (NAS Executive Officer), 24 September 1956. NASA Division of Earth Sciences: Conference on Arctic Sea Ice, February 1958.

89. Thurston (NAS Division of Earth Sciences) to S. D. Cornell (NAS Executive Officer), September 24, 1956. NASA Division of Earth Sciences: Conference on Arctic Sea Ice, February 1958.

90. Allan Needell, *Science, Cold War, and the American State: Lloyd V. Berkner and the Balance of Professional Ideals* (Amsterdam: Harwood, 2000), 317.

91. "Scientists Seek Methods for Predicting Arctic Ice," press release from National Academy of Sciences and National Research Council, release date February 24, 1958. NASA Division of Earth Sciences: Conference on Arctic Sea Ice, February 1958.

92. Details of this program are held in NASA Division of Earth Sciences: Conference on Arctic Sea Ice, February 1958.

93. On Bronk, see Frank Brink Jr, *Detlev Wulf Bronk 1897–1975: A Biographical Memoir* (Washington, DC: National Academy of Sciences, 1978).

94. See Shawn J. Parry-Giles, *The Rhetorical Presidency, Propaganda, and the Cold War: 1945 – 1955* (Westport, CT: Praeger, 2002), 134.

95. See, for instance, the resumes of the OCB luncheon meetings held on March 11, 1958 and OCB luncheon meeting May 28, 1958. Accessed through http://www.foia.cia.gov/search-results?search_api_views_fulltext=bronk&field _collection=.

96. Resume of OCB luncheon meeting held on February 26, 1958. Accessed at http://www.foia.cia.gov/sites/default/files/document_conversions/5829 /CIA-RDP80B01676R002700050044-2.pdf.

97. Chief Naval Operations to Chief Naval Research Chief of Naval Operations, September 20, 1957. NASA Division of Earth Sciences: Conference on Arctic Sea Ice, February 1958.

98. See, for instance, Turchetti, in this volume.

99. Bronk would continue to pressure the State Department to ease restrictions on visits from Soviet scientists after the conference. See for instance George Bogdan Kistiakowsky, *A Scientist at the White House: The Private Diary of President Eisenhower's Special Assistant for Science and Technology* (Cambridge MA: Harvard University Press, 1976), 254.

100. Thurston to Frederick T. Merrill (Director East-West Contacts Staff, State Department), January 23, 1958. NASA Division of Earth Sciences: Conference on Arctic Sea Ice, February 1958.

101. Bronk to Nesmeyanov, August 30, 1957. NASA Division of Earth Sciences: Conference on Arctic Sea Ice, February 1958.

102. H. C. Daniel (United States Navy Hydrographer) to Bronk, March 12, 1958; Rawson Bennett to Bronk, March 29, 1958. Both held in NASA Division of Earth Sciences: Conference on Arctic Sea Ice, February 1958.

103. Thurston to Frederick T. Merrill, March 13, 1958. NASA Division of Earth Sciences: Conference on Arctic Sea Ice, February 1958.

104. Thurston to Merrill, February 14, 1958. NASA Division of Earth Sciences: Conference on Arctic Sea Ice, February 1958.

105. Gordienko, "Arctic Sea Ice Research," *Priroda* 9 1958: 68–71. Translated from the original Russian by Judith I. B. Danner. P. A. Gordienko Collection, Rauner Library, Dartmouth College, MSS-161 folder 4. My sincere thanks to Sarah Hartwell for tracking down the provenance of the translation held at the Library.

106. Gordienko, "Arctic Sea Ice Research," 2.

107. Ibidem.

108. Weeks, personal communication.

109. Gordienko, "Arctic Sea Ice Research," 3.

110. Ibidem, 4.

111. Ibidem, 8.

112. Ibidem, 8.

113. Weeks, personal communication. On the development of Arctic submarine capacity in the United States and its relationship with sea ice research from the late 1950s, see William M. Leary, *Under Ice: Waldo Lyon and the Development of the Arctic Submarine* (College Station: Texas A&M University Press, 1999). On antisubmarine warfare, see Robinson's chapter in this volume.

114. Kristie Macrakis, "Technophilic Hubris and Espionage Styles during the Cold War," *Isis* 101 (2) 2010: 378–385.

Section IV

Surveillance Technologies

Chapter 7

Space Technology and the Rise of the US Surveillance State

Roger D. Launius

Electronic surveillance entered the mainstream during the Cold War, especially with satellites, as technologies were pursued by all combatants to gain an advantage in ensuring victory in this rivalry.[1] U.S. president Lyndon B. Johnson did not overestimate the importance of this technology in 1967 when he said that the United States probably spent between $35 and $40 billion on it, but "[i]f nothing else had come of it except the knowledge we've gained from space photography, it would be worth 10 times what the whole program has cost."[2]

Both Cold War rivals raced to develop the satellite reconnaissance technology that would give them the edge in understanding what the other side was doing. The Americans launched the first generation of spy satellites in 1960, and the Soviet Union followed a few years later. Accordingly, the CORONA and successor satellites, as well as a range of signals intelligence, early warning, ASAT, and missile launch detection spacecraft have dominated the national security interests since the beginning of the space age. Additionally, this development created an experience whereby everyone in Western Civilization began to live life as a target during the Cold War since everyone could be on the receiving end of an attack.[3] As electronic spying became the norm, it has portended important ramifications for both personal privacy and national security. The National Security Agency (NSA) had the beginnings of this capability and developed greater capability over time, but other intelligence organizations played key roles as well.[4]

But the story really begins earlier, as nation-states have pursued surveillance technologies that would give them advantage on the battlefield. Balloons were the first overhead reconnaissance vehicles to be deployed and date to the Napoleonic wars when the French used them extensively to track troop movements, both enemy and others, and to spot for artillery. Both the Union and the Confederate armies used balloons for the same purpose in the American Civil War of the 1860s and both the French and the German armies used them in the Franco-Prussian War. Accordingly, overhead reconnaissance had a long history prior to the invention of the airplane.[5]

Reconnaissance was certainly an early understood use of powered aircraft immediately from the Wrights brothers' first flight to the present. Indicative of this view was the comment of American army lieutenant Frederick E. Humpheries in 1910:

> From a military standpoint, the first and probably the greatest use will be found in reconnaissance. A flyer carrying two men can rise in the air out of range of the enemy, and passing over his head out of effective target range, can make a complete reconnaissance and return, bringing more valuable information than could possibly be secured by a reconnaissance in force. This method would endanger the lives of two men; the other would detach several thousand men for a length of time and endanger the lives of all.[6]

When the war began, therefore, the air doctrine of all combatants involved using the airplane for observation and message carrying, not really for combat.

Although used earlier in the Balkan wars of the first decade of the twentieth century, during World War I the airplane emerged as the dominant force in providing reconnaissance for both sides. As the British Expeditionary Force (BEF) retreated from German invaders in France in 1914, some two-dozen Royal Air Force observation airplanes watched from above and warned of impending attack.

Most aircraft were not even armed during the first months of the war. The pilots of the various nations, moreover, had been prewar flying enthusiasts and many knew each other from European air meets. It was a little like "old home week" as the opposing pilots hailed each other in the air. This did not last long, however, for by the fall of 1914 the airplane had proven its worth as a reconnaissance vehicle. Field commanders recognized that it was important to keep opposing observation planes from accomplishing their missions, because of the massive ground attacks that were sure to head for a weak sector discovered from the air. Antiaircraft fire quickly became coordinated and deadly, but opposing pilots also began taking weapons up with them to take occasional pot shots at enemy planes. Within a very short period, machine guns had been mounted on aircraft, either for the observer to fire or fixed so that the pilot could aim the plane like a weapon. The machine gun, as had been the case on the ground, became the master of the aerial battlefield. For the rest of the war the legendary "Knights of the Air" dueled for control of the skies over the European battlefields, and later used to employ the aircraft as a ground attack weapon. As both fronts settled into trench warfare, aircraft came to define their missions even more succinctly, a significant part of which involved observation by flying over enemy lines and bringing reports back about enemy positions, the locations of munitions, and movements of troops and materiél.[7] By the end of World War I a special niche for overhead reconnaissance had been established and it would remain critical to military operations worldwide thereafter.

Overhead reconnaissance in World War II was a key method for obtaining intelligence about the enemy and its activities for all sides in the war and all theaters. Photographs provided concrete evidence on enemy locations, movements, and resupply efforts. The combatants also assessed battle damage and used such data to make decisions about new efforts. This reconnaissance played a central role in the Allied victory, and its activities on the Axis side may well have extended the war by enabling the enemy to fight more effectively for a longer period. The statement of General Werner Von Fritsch of the German High Command in 1938 proved prophetic: "The military organisation with the best aerial reconnaissance will win the next war."[8]

The Cold War era, conducted as it was not by direct combat between the United States and the Soviet Union, raised the significance of overhead reconnaissance even higher than in earlier eras. Keeping tabs on what the other side was doing proved essential to each side's strategic place in the Cold War rivalry.[9] The emphasis on reconnaissance emerged from four basic forces, according to political scientist Glenn Hastedt:

> The first was the problem of strategic surprise as symbolized by Pearl Harbor. This was the event that, in the minds of many, national security policy had to make sure was not repeated. The second force was the solution of greater centralization and cooperation at the national level among bureaucracies involved in foreign diplomatic, military, and economic policy. Pearl Harbor occurred in spite of warning; intelligence was present but it was not recognized or acted upon. The inherent validity of this solution was reinforced by the wartime experience of ad hoc military centralization that came about out of the need to cooperate with the British. To bring this about, the 1947 National Security Act created the Central Intelligence Agency (CIA), the National Security Council (NSC), and unified the military services under a Secretary of Defense in a national defense establishment that would soon become the Department of Defense (DoD). The third force was the de facto establishment of an intelligence community that was to work together to prevent another Pearl Harbor. Along with the newly created CIA the other founding members were the Bureau of Intelligence and Research (INR) at the State Department, and U.S. Army, Navy, and Air Force intelligence. The final force that exerted great influence on the origins of American national security policy was the advent of the Cold War. It presented the United States—and national security policy—with a clearly identifiable enemy in the Soviet Union and then a strategy—containment—around which policy makers could unite.[10]

The quest for overhead reconnaissance proved an especially difficult problem in international law since both the United States and the Soviet Union desired to preserve the status quo rather than destabilize the global situation.

To pave the way for strategic reconnaissance, the Eisenhower administration convened the top secret Technological Capabilities Panel (TCP) in March 1954 to assess Cold War prospects in meeting a threat from the Soviet Union. As discussed by national security policy analyst Peter L. Hays,

> The TCP completed a secret two-volume report and briefed the National Security Council (NSC) in February 1955. The report strongly recommended rapid development of U.S. technical intelligence-gathering capabilities and supporting policies for overflight. The TCP process and report were critical drivers behind development of America's first high-tech intelligence collection platforms: the Lockheed U-2 aircraft and the weapons system (WS)-117L reconnaissance satellite program that eventually led to successful operation of the Corona system in August 1960.[11]

The TCP report persuaded the Eisenhower administration to pull out all stops in developing this reconnaissance capability. As a result, the National Security Council issued in 1955 a secret document, NSC 5520, which emphasized development of overhead reconnaissance capabilities.[12]

Some of the methods of gathering intelligence were inelegant and somewhat bizarre. For example, on December 27, 1955, the Eisenhower administration approved the development of GENETRIX, a project involving balloons equipped with cameras capable of flying up to 90,000 feet in altitude. The cameras, developed by the Boston University Physical Research Laboratory, produced optical reconnaissance camera systems. GENETRIX balloons, launched either from ship or land, traveled across the area of the Soviet Union between January 22 and February 24, 1956. Over 500 balloons were launched altogether, but only 46 were recovered and of those only 34 brought back any usable imagery. The 13,813 usable photos covered 1,116,449 square miles of the Soviet Union and China but because the flight path could not be controlled much of the photography was of limited intelligence value. The balloons also made easy targets for the Soviet air force and very quickly the GENETRIX project became anything but secret. Protests from the Soviets about them created international tensions not seen since the Berlin blockade of 1948–1949, but in this case the United States was the offending nation rather than the Soviet Union. In February 1956, the Red Army displayed for the international press recovered GENETRIX cameras, equipment, and photographs, in the process condemning the United States for spying on a sovereign nation. The violation of Soviet airspace was obvious and in the face of massive world criticism on February 6, 1956, Eisenhower terminated this particular program.[13]

Oddly, this fiasco (there is really no other term for it; what did the proposers think would happen?) did not dissuade the DoD from undertaking later balloon flights. A second series, designated the WS-461L program, began in July 1958. Predicated on the knowledge that during six weeks each summer there was an east-to-west jet stream over the Soviet Union at an altitude of approximately 110,000 feet, they believed they could deploy balloons to that altitude with effective cameras they could reconnoiter the Soviet Union. Eisenhower reluctantly approved this scheme in June 1958, although he should not have done so. The DoD deployed three balloons from the aircraft carrier *U.S.S. Windham Bay* cruising in the Bering Sea, but because of poor calculations and inaccurate knowledge of currents all three came down in Poland. This ignited a second firestorm of controversy about overhead reconnaissance, prompting a red-faced Eisenhower to declare an end to all balloon reconnaissance flights.[14]

But what about alternative clandestine endeavors to gather overhead reconnaissance data? Surely there must be more effective means of determining the actions of a potential enemy. As explained by historian Coy Cross,

[a]t first, conventional American aircraft, usually bombers converted for reconnaissance, gathered signals intelligence along the Soviet borders. Occasionally, they would take advantage of gaps in radar coverage to overfly and photograph Soviet cities. But the Soviets became more aggressive in defending their borders. The cost for continuing intelligence-gathering operations against the Soviet Union with contemporary aircraft was becoming too high.[15]

Like the balloon flights, in this case Americans and Soviets violated each other's airspace and incited heated diplomatic exchanges. In some cases there were also shootdowns.[16]

One example will suffice. In December 1956 Eisenhower approved a military overflight of Vladivostok and vicinity, opposite Japan, with three high-altitude RB–57D aircraft, converted bombers with lengthened wingspans to allow higher altitude flights. These three aircraft entered Soviet airspace on December 11,

1956, overflying the home base of the Soviet Pacific fleet and collecting some useful data on its composition, anchorage, and disposition. Again, the Soviet Union protested vehemently about this violation of its airspace. The formal protest on December 15 confirmed that the Soviets had excellent radar tracking capabilities, even though it did not have as yet interceptors with the speed, range, and altitude to engage the American reconnaissance aircraft. The Soviet Union, however, aggressively worked on developing this capability and promised to respond to American incursions in the future, threatening that "the United States of America will have to bear the full responsibility for the consequences of such violations."[17] These actions took place periodically, and the Soviet shoot-down of Korean Airlines flight 007 on September 1, 1983, a Boeing 747 with 269 passengers onboard, brought home the serious nature of these incursions.[18]

Because of this, officials at the U.S. Air Force's Air Development Command, Richard Leghorn and John Seaberg, proposed development of a specially designed reconnaissance plane with a powerful turbojet engine and long, high-efficiency wings. They believed such an airplane could fly at 65,000–70,000 feet, where it could elude Soviet fighters and hopefully be impervious to attack from the ground, at least until about 1960. As John Carter, a Lockheed strategist, suggested: "In order that this special aircraft can have a reasonably long and useful life, it is obvious that its development must be greatly accelerated beyond that considered normal." In July 1953 the Air Force issued solicitations for proposals for building such a new high-altitude reconnaissance aircraft.[19]

Nonetheless, they persisted and by the mid-1950s aerial reconnaissance technology had advanced to the state that the Eisenhower administration believed that it could see Soviet capabilities and likely military intentions more clearly than ever before. U-2 flights over the Soviet Union, begun during the summer of 1956, kept the U.S. government more well-informed than ever before of Russian defense developments.[20] The aircraft flew at more than 70,000 feet above the Earth's surface, prompting US intelligence officials to conclude that "the Soviets would not be provoked into quickly developing a capability to track and shoot down objects at that extreme altitude." Indeed, the Soviet Union did identify U-2s immediately upon flight in 1956, but lacked the technology to "shoot them down." The Soviet Union did not admit their inability, however, so as "not to admit their vulnerability to American reconnaissance technology."[21] In the process, the CIA confirmed that Soviet military might did not yet match American capabilities.[22]

The advent of the U-2 spy plane proved a boon for overhead reconnaissance. The first flights over the Soviet Union were quickly followed by observation of the Eastern Mediterranean, where photography of French and British warships during the 1956 Suez Crisis proved useful in resolving the stalemate. The CIA instigated overhead flights over Syria, Iraq, Saudi Arabia, Lebanon, and Yemen in the latter 1950s, as well as over China—both the People's Republic of China and Nationalist China—and other hot spots around the globe. Throughout the latter 1950s, the U-2 proved its mettle in both the Soviet Union and beyond.[23] The famous incident on May 1, 1960, when Francis Gary Powers had his U-2 shot down over the Soviet Union—and he was captured in the process—indicated that even this technological marvel was not invulnerable to Soviet missile attack. Again, US leaders—especially Eisenhower—suffered a strategic public relations blow from this embarrassing episode. There had to be a better way.[24]

Open Skies, Satellite Overflight, and
Strategic Reconnaissance

That better way, at least for the Cold War rivals, came in the form of the new realm of space operations. The establishment of "freedom of space," of the rights of overflight from Earth orbit, was critical to this development. Few today appreciate the desperate nature of the Cold War rivalry with the Soviet Union and the potential for any misstep to instigate nuclear confrontation. The rivals nearly stepped over the line during the Cuban missile crisis of 1962 and also came close on other occasions. The national security space regime established in the 1950s made possible a less tense set of relations than would have been the case otherwise, because of a little greater ability to separate fact from fiction in the opposite's intentions, but it was certainly tense enough even with those space capabilities. As historian R. Cargill Hall has concluded, this regime was "predicated on a maritime analog. In maritime law, the vessels of all nations possess the right to ply the high seas while adhering to the treaties and customs that detail the terms of navigation and accepted rules of the road."[25] Collectively these principles offered some of the building blocks of an effective national security strategy.

The centerpiece of this national security space strategy rested on "freedom of space," sometimes referred to as the "open skies" doctrine. Eisenhower sought to establish the right of international overflight with satellites, making possible the free use of reconnaissance spacecraft in future years. From the perspective of the Eisenhower administration, which was committed to the development of an orbital reconnaissance capability as a national defense initiative, an international agreement to ban satellites from overflying national borders without the individual nation's permission was unacceptable.

In a critical document, "Meeting the Threat of Surprise Attack," issued on February 14, 1955, US defense officials raised the question of international law governing territorial waters and airspace, in which individual nations controlled those regions as if they were their own soil. That international custom allowed nations to board and confiscate vessels within territorial waters near their coastlines and to force down aircraft flying in their territorial airspace. This has resulted in shootdowns on occasion. But in the 1950s space as a territory had not yet been defined and US leaders argued that it should be recognized as beyond the normal confines of territorial limits. An opposite position, however, argued for the extension of territorial limits into space above a nation into infinity.[26]

"Freedom of space" became an extremely significant issue for those concerned with orbiting satellites, because the imposition of territorial prerogatives outside the atmosphere could legally restrict any nation from orbiting satellites without the permission of nations that might be overflown. Since the United States was in a position to capitalize on "freedom of space" it favored an open position. Many other nations had little interest in establishing a free access policy that allowed the United States to orbit reconnaissance satellites overhead. Eisenhower had tried to obtain a "freedom of space" decision on July 21, 1955, when he proposed it at a US/USSR summit in Geneva, Switzerland. Soviet premier Nikita Khrushchev rejected the proposal, however, saying that it was an obvious American attempt to "accumulate target information." Eisenhower later admitted, "We knew the Soviets wouldn't accept it, but we took a look and thought it was a good move."[27] The Americans thereafter worked quietly to establish the precedent.

Soviet *Sputniks* 1 and 2, launched in the fall of 1957, overflew international boundaries without provoking a single diplomatic protest. They effectively established this important overflight precedent. On October 8, 1957, deputy secretary of defense Donald Quarles told the president: "the Russians have…done us a good turn, unintentionally, in establishing the concept of freedom of international space."[28] Eisenhower immediately grasped this as a means of pressing ahead with the launching of a reconnaissance satellite. The precedent held for *Explorer 1* and *Vanguard 1*, and by the end of 1958 the tenuous principle of "freedom of space" had been established. With the Soviet Union in the lead, the nations of the world established the US-backed precedent for free use of space.[29]

While some have emphasized Eisenhower's prescience in establishing this precedent of "freedom of space," it was essentially serendipity from the circumstances of 1957.[30] It was an important serendipity, without question, and as is so often the case in history the unintended consequences of actions turn out to be as important as the intended ones. Indeed, political scientist Peter Hays has concluded that American scientific satellite efforts masked the more important objective of establishing "freedom of action in space, using the benign IGY program as a 'stalking horse' to establish the precedent of space over flight and legitimize eventual operation of military reconnaissance satellites."[31] The story of the establishment of "freedom of space" is a critical case of an unintended consequence of momentous importance for the rest of the Cold War.

Throughout the rest of the Eisenhower administration it reaffirmed the free-access-to-space position already established in precedent and declared that space would not be used for warlike purposes. At the same time that it asserted the necessity of reconnaissance satellites and other military support activities that could be aided by satellites, such as communications and weather, Eisenhower's subordinates went a step further to insist that they were actually peaceful activities that assisted in strategic deterrence and therefore averted war. This was a critical space policy decision as it provided for open use of space and fashioned a virtual "inspection system" to forewarn of surprise attack through the use of reconnaissance satellites.

Indeed, an irony too great to ignore is that both of the superpowers locked in Cold War struggle for more than a generation cooperated to ensure satellite reconnaissance remained inviolate despite everything else that divided them. The Kremlin, in addition to seeing the value of this technology in relation to the United States, also found it critical in understanding what the Chinese were doing on their long border to the southeast.[32] As then-Air Force Lieutenant Colonel Larry K. Grundhauser commented in *Aerospace Power Journal* in 1998, "over time the two superpowers established a 'practice of the parties' as the legal basis for legitimizing the use of satellites for reconnaissance—an unspoken and unrecorded 'gentleman's agreement' that respected the immunity of each other's reconnaissance satellites."[33]

"Freedom of space," established as a practical reality by *Sputnik*, received official sanction through a variety of actions. For example, the United Nations General Assembly officially recognized "freedom of space" in 1961 as a part of a joint resolution.[34] It also gained formal status in the "Treaty on Principles Governing the Activities of States in the Exploration and Use of Outer Space" in 1967. This treaty declared that space, "including the moon and other celestial bodies, shall be free for exploration and use by all States without discrimination of any kind, on a basis of equality."[35] This has remained the effective law of space since that time and no one has suggested that this right of overflight be overturned.

The Overhead Reconnaissance Harvest of CORONA

Despite the establishment of the right of over flight, the United States's development of a viable satellite reconnaissance program proved a major challenge. Under development in the latter 1950s, Project CORONA was a major successful reconnaissance satellite program of the United States. Contracted to Lockheed, Glenn L. Martin Co., and RCA under the codename "Pied Piper" in 1955, by July 1956 the development plan for the covert CORONA spacecraft was approved.[36] This effort featured an Atlas booster with a spacecraft stabilized in orbit on three axes for high pointing accuracy of still cameras using film weighing thousands of pounds. At the same time it pursued television capabilities in a satellite later named *Samos* (see Figure 7.1).[37]

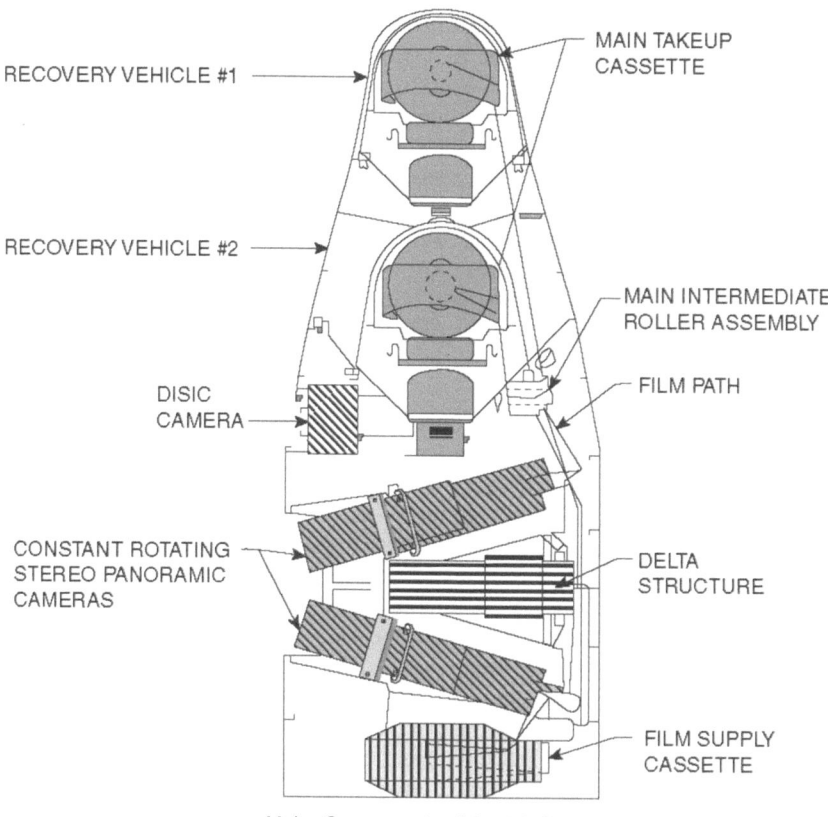

Major Components of the J-3 System

Figure 7.1　Interior layout of the Corona reconnaissance satellite
Source: US National Reconnaissance Office.

In the aftermath of the *Sputnik* crisis of 1957–1958, the DoD raised the priority of the reconnaissance satellite program and increased funding. Reassigned to the CIA with Richard Bissell Jr. leading the effort, the CORONA film reconnaissance approach raced to an early deployment. An Itek camera, built for the satellite, featured a 12-inch focal length lens in a camera mounted on a three-axis stabilized satellite. It would take 70-degree wide photographic swaths with a resolution of 12 meters (40 feet) from an orbit with a perigee of about 190 kilometers (120 miles). As the camera's acetate film was exposed, it would be fed into a return capsule at the top of the spacecraft. After a few orbits, a small solid-propellant recovery rocket could decelerate a recovery capsule into a reentry trajectory, a parachute would deploy, and the reentry vehicle would be snatched mid-air by a C-119 recovery plane (as showed in Figure 7.2).[38]

CORONA progressed at a frantic pace in the later 1950s, covering its activities with the ruse of the codename "Project Discoverer," a test program to develop new technologies required for the study of the space environment, including biomedical experiments that had to be recovered from space. The reality was that this was all about satellite reconnaissance of the Soviet Union. The first such test of this capability came on January 21, 1959, with the attempted launch of a Thor-Agena booster combination that failed on the launch pad. Additional tests had their problems as well, and it was not until Discoverer 13, launched on August 18, 1960, that the CORONA system reached its orbit and then correctly

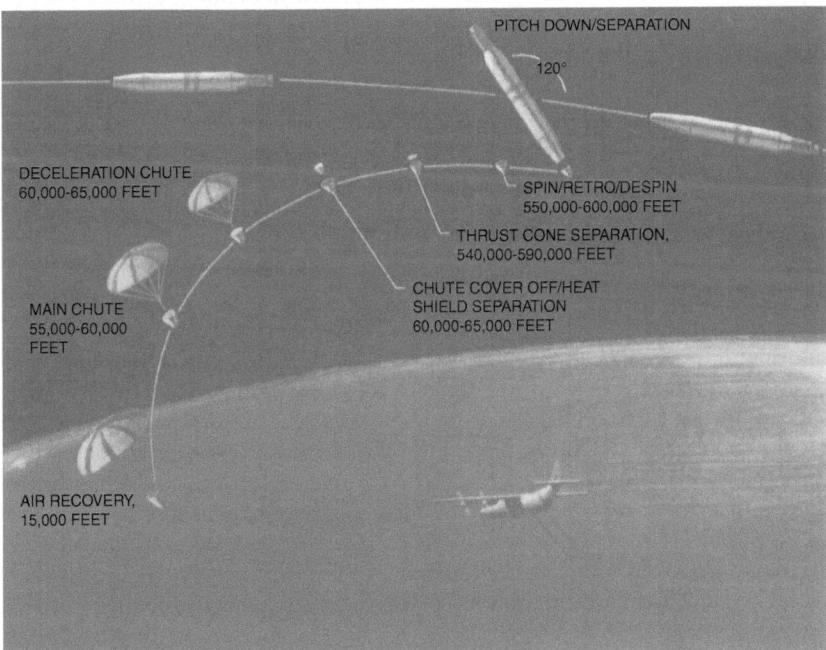

Figure 7.2 This illustration demonstrates the manner in which the Corona reconnaissance satellite profile took place through the 1960s

Source: US National Reconnaissance Office.

returned its reentry vehicle containing photographs of the ICBM base at Plesetsk and the bomber base at Mys Schmitda in the Soviet Union. After this flight, CORONA became an operational mission and functioned through 1973 when it was succeeded by later generations of reconnaissance satellites.

In assessing the 13-year-long CORONA program, one can only call attention to the treasure trove of imagery that enabled much more intelligence analysis than ever before in the Cold War. Through six versions of progressively more sophisticated satellites, named KH-1 through KH-4B (KH stood for "Key Hole"), CORONA had 144 satellites launched, of which 102 returned usable imagery. An official history of CORONA concluded:

> In the context of its operational utility, exploitation of technology, and enhancement of the nation's fund of intelligence information, Corona had to be rated an outstanding success. Originally considered an interim system and assumed to have, at best, three of four years of operational utility, Corona remained the sole source of overflight intelligence for the United States for nearly five years, and was a primary source of basic information used to shape national defense policy for 12 years. Although designed as a search system, at the end Corona was providing better detail and resolution than several of the surveillance systems earlier touted to supplement it.[39]

Throughout the 1960s the system provided critical data about Soviet military capabilities, among other things confirming that there was no missile gap as alleged in the 1960 presidential election with the United State trailing the Soviet Union in capability, offering early intelligence on the deployment of Soviet missiles to Cuba in 1962, and adding to understanding about the various conflicts in the Middle East, Asia, Africa, the Sino-Soviet border, and Central Europe. An appropriate conclusion is: "In its years of service, Corona had identified and accurately located all operational Soviet ballistic missile sites. More need not be said."[40]

Beyond CORONA

Since the retirement of the CORONA program in 1973, the US National Reconnaissance Office (NRO) has undertaken a succession of observation programs that continue to the present (see Figure 7.3).

These remain the most highly secret efforts of the United States and any information in the public domain about them must be carefully weighed to ascertain its veracity. Legitimate information, along with cover stories and what appears for all the world to be so much nonsense, has comingled to create a morass of beliefs about this subject. For example, disinformation about a secret military human spaceplane has circulated throughout popular culture. It was the subject of a story arc in the television series, *The West Wing*, and it has been featured in articles in respected periodicals despite it being a best a fringe belief.[41]

In addition to reconnaissance satellites, by the early 1960s the US military, with the Air Force leading the way, developed a comprehensive range of space capabilities that would enhance the deployment of military might around the globe. Most of the Cold War era efforts were designed to support strategic nuclear operations, such as the infrared sensors on the Midas as well as Defense Support Program (DSP) satellites that provided early warning of ICBM launches (see Figure 7.4).[42] This led naturally into a desire to develop early warning for

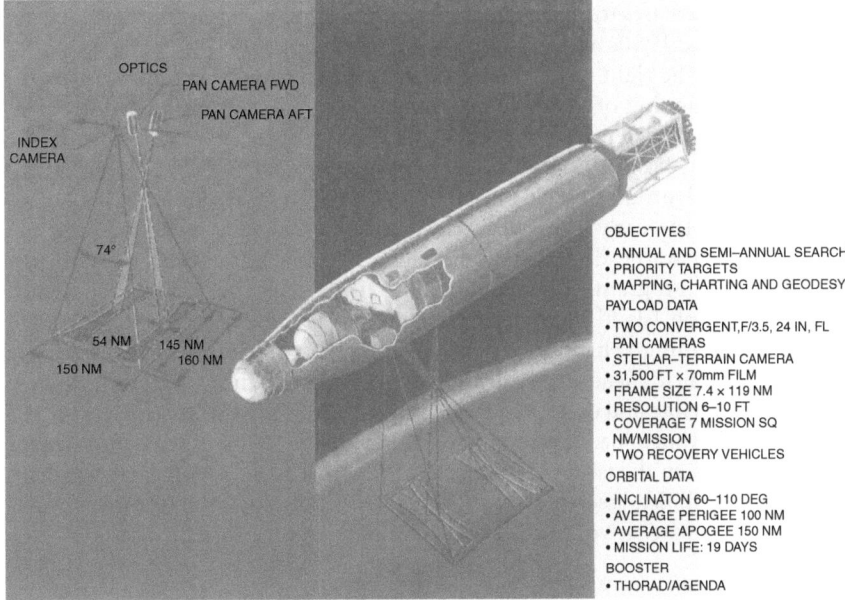

OPTICS
PAN CAMERA FWD
PAN CAMERA AFT
INDEX
CAMERA

74°

54 NM   145 NM
150 NM   160 NM

OBJECTIVES
• ANNUAL AND SEMI–ANNUAL SEARCH
• PRIORITY TARGETS
• MAPPING, CHARTING AND GEODESY
PAYLOAD DATA
• TWO CONVERGENT,F/3.5, 24 IN, FL
  PAN CAMERAS
• STELLAR–TERRAIN CAMERA
• 31,500 FT x 70mm FILM
• FRAME SIZE 7.4 x 119 NM
• RESOLUTION 6–10 FT
• COVERAGE 7 MISSION SQ
  NM/MISSION
• TWO RECOVERY VEHICLES
ORBITAL DATA
• INCLINATON 60–110 DEG
• AVERAGE PERIGEE 100 NM
• AVERAGE APOGEE 150 NM
• MISSION LIFE: 19 DAYS
BOOSTER
• THORAD/AGENDA

Figure 7.3 This illustration from the National Reconnaissance Organization (NRO), charged with managing satellite reconnaissance, depicts the ingredients in the reconnaissance system put into place with the Corona program of the 1960s

Source: US National Reconnaissance Office.

Figure 7.4 An artist representation of an Air Force Defense Support Program satellite

and countermeasures to deal nuclear threats, the domain of the Strategic Defense Initiative of the 1980s and thereafter.

Of course, the right to defend against attack explicitly emerged as a prerogative at the beginning of the space age. No one has seriously questioned the right of any nation to defend itself from attack. The manner in which that may be done, however, has been open to reinterpretation over the years. The United States pursued ground-based antisatellites (ASAT) capabilities on two occasions, during the early 1960s with a modified Nike Zeus missile that could launch nuclear warheads to destroy satellites in a low-Earth orbit. Second, the DoD pursued Program 437 near the same time, deploying nuclear Thor missiles at Johnston Island.[43] Another possibility emerged when an F-15-launched Miniature Homing Vehicle, tested on September 13, 1985, launched a two-stage kinetic kill vehicle that successfully homed in using an infrared targeting system on a target satellite and destroyed it on impact.[44] Even so, ASATs have not proven effective over time. Space policy analyst Dwayne A. Day has referred to them as "blunt arrows" in the larger arsenal of defensive space assets with a modest demonstrated capability, asserting that "the United States does not need to pursue a more active, provocative, or expensive ASAT development than what it already

Figure 7.5   A Delta II rocket lifts off from Space Launch Complex-2 at Patrick Air Force Base, Florida, on April 28, 2006

Source: US Air Force.

has. The threat does not justify it, and rarely has."[45] Other related efforts over the years, including missile defense initiatives that achieved both some success and political notoriety, have drawn similar pointed criticism and stalwart defense.[46]

By the time that the Cold War came to an end in 1989, national security space capabilities had matured significantly in their use to enhance terrestrial war fighting. The experience of the First Gulf War in 1991 demonstrated this in fundamental ways, but there had been earlier hints, such as the overtaxing of communications satellites in Grenada in 1983, about American reliance on space technology for military purposes. Retreading space capabilities designed to support Cold War strategic and nuclear operations became a major task in the 1990s as they now supported decidedly nonnuclear battlefield operations, including long-range precision guided munitions tracked and controlled from space. Couple this with the profound changes coming in the twenty-first century with unmanned aerial vehicles (UAV) piloted via space-based communications systems from remote locations far beyond the theater of deployment, and the nature of surveillance has profoundly changed. Both reconnaissance and weapons platforms have been deployed increasingly and used to gather intelligence and undertake strikes against potential enemies (see Figure 7.5). This raises profound questions about the nature of these assets, the propriety of their use, and their overall effectiveness in the quest for American (in)security.[47]

Defining and Dealing with the "Watchers"

When it comes to the rise of the modern surveillance state, however, James Bond was all so very twentieth century. Intelligence and spying is largely focused on electronic data collection in the early twenty-first century, and the United States for good and ill leads the world, much of its capacity being space-based. The greatest leaps toward this capability since the first reconnaissance satellites of the 1960s came when Admiral John Poindexter, President Ronald Reagan's national security advisor, expressed worry over how more effective intelligence might have prevented the 1983 terrorist attack on the military barracks in Beirut that led to the deaths of 241 Marines. The problem in Poindexter's mind was not the failure to collect information but in the inadequacy of the collation and analysis of data in real time from diverse sources. He dedicated his efforts thereafter toward building a system, along with the technology, which could gather and sift terabytes of data from all sources for signs of a potential enemy's activity. He called it "Total Information Awareness," and its success in Poindexter's mind ensured that traditional rights of personal privacy ensconced in law in the United States had to be curtailed. As a matter of record, he cared not a whit about the privacy of non-US citizens in this arena.[48]

The system that Poindexter masterminded essentially played the "six degrees of separation" game in analyzing seemingly unrelated data. This idea—if someone is one step away from each person they know then everyone is at most six steps away from any other person on Earth—suggests that there may be connections that may be analyzed connecting a known terrorist to others who were unknown but were planning acts of terrorism. To make this connection required the sweeping up of massive amounts of electronic data and then analyzing it using sophisticated technologies. As journalist Shane Harris commented, for intelligence analysts "to find signals in the noise one had to collect information from far and wide."[49] The National Security Agency (NSA) had the beginnings of this capability in the 1980s and developed greater technology over time, but other intelligence organizations also played key roles as well.

But the Total Information Awareness program that Poindexter envisioned tread ruthlessly over US laws in place to ensure personal privacy. Telephone calls for Americans, for instance, are viewed as sacrosanct and may not be listened in on legally except by explicit court order. But to find the few signals in all of the noise the NSA demanded the overturning of this long-standing right so it could collect this data without explicit warrants. The result was legislation by Congress allowing warrantless wiretapping in the aftermath of the 9/11 attacks, and a host of other actions systematically overturning civil liberties. Even when parts of these efforts were rolled back after being exposed, some became covert activities and continued thereafter.[50]

Where does this leave the United States? At some level, the technology seems to have outstripped political thinking. The US government has demonstrated that it can do quite a lot to collect data and engage in real-time analysis, in the process enhancing the surveillance state beyond anyone's expectation when the space age began. But there is less certainty about the quality of what is found through this process of data mining. There has been considerable disappointment, even from such individuals as Poindexter, in the risk this process subjects personal privacy to in the modern era. In the end many have advanced this new approach to national security. As Shane Harris commented: "John Poindexter envisioned a world. Mike Hayden made it a reality. Mike McConnell enshrined it in law. And Barack Obama inherited it. In broad strokes, that's how we got where we are now."[51]

The Question of Space Weaponization: Sanctuary, Stars Wars, or Something Else?

This discussion leads naturally to a central policy debate relative to national security space in the last 20 years: the weaponization of space as a means of preserving American primacy in the arena for all manner of space-based applications. For more than 50 years, the world has engaged in activity in outer space for military, scientific, and commercial purposes, but without placing weapons there or engaging in serious efforts to target objects in space. Working effectively during the Cold War, since then the space arena has witnessed the entry of many more actors and a much broader array of vested interests than during the Cold War, resulting in a variety of positions regarding future space activities. For example, humans have been in space more or less continuously since 1961 and since November 2000 have been permanently in place on the International Space Station, a peaceful, cooperative venture of 16 nations that represents at more than $100 billion, the largest nonmilitary cooperative effort in world history. At the same time, almost 700 spacecraft are operating in continuous Earth orbit, each serving a range of scientific, military, civilian, and commercial uses. And the hegemonic status of the United States and the Soviet Union (Russia) has been demolished in the last 20 years. Over 60 new launches take place every year, and at least 40 nations had satellites in orbit in 2013.[52]

In this increasingly chaotic environment with so many actors, the United States remains the dominant player and wants to ensure that it does so indefinitely, hence the desire to protect space assets. As one policy analyst put it: "Given the U.S. reliance on its space systems for national security, would the United States (as some have argued) face a future 'space Pearl Harbor' if it did not first acquire the means to protect its space systems from deliberate harm?"[53] The answer to ensuring US hegemony in space rests in no small part with the protection of the nation's satellites and other space-based capabilities while denying that same capability to potential adversaries. There may be a range ways in which that might be accomplished, but one of the most important is the placement of systems in space to protect against attack. Depending on how one interprets these assets, it may represent the weaponization of space, thereby overturning a 50-something-year-old decision not to do so.

Debate over this issue has been marked by two extreme positions, neither of which are representative of the majority of those debating the subject. The first is the "sanctuary" concept, which asserts that space "should not be used for military purposes," as Malcolm Mowthorp has written:

> The intrinsic value space provides for national security is that satellites can be used to examine within the boundaries of states, since there is no prohibited over flight for satellites as there is for aircraft. This enables arms limitation treaties to be verified by satellites in space serving as a national technical means of treaty verification. Early warning satellites serve to strengthen strategic stability since they provide surveillance of missile launches which increases the survivability of retaliatory strategic forces. The sanctuary school sees the importance with which space systems provides these functions that space must be kept free from weapons, and antisatellite weapons must be prohibited, since they would threaten the space systems providing these capabilities.[54]

Sanctuary advocates have argued that space weaponization by the United States would ensure an arms race in space in which all would ultimately lose. They have opposed it on moral grounds, but more importantly because of long-standing predispositions in favor of arms control, conflict resolution, and global collective stability. Any move beyond limited national security operations such as satellite reconnaissance, arms control verification, early warning, and communications represents for them a "slippery slope" to an arms race in space. As Lt. Col. Bruce M. DeBlois wrote more than a decade ago in a thoughtful essay in *Airpower Journal*, "Unlike the strategy for nuclear weapons, there exists no obvious strategy for employing space weapons that will enhance global stability. If the precedent of evading destabilizing situations is to continue—and that is compatible with a long history of US foreign policy—one ought to avoid space-based weapons."[55] Noting the long-standing successful policy put into place by Eisenhower in the 1950s, opponents of space weaponization have seen little positive in trying to alter this national security space environment.

The most radical conception on the other side, "star wars," is also a caricature. It essentially seeks to ensure American hegemonic status in space. It is a retreading of the "high ground" argument but one carried to its logical conclusion through weaponizing space and using the region as an American "lake" while denying others its use for military purposes. This is a position not unlike the long-standing policy of the United States toward the Western Hemisphere first enunciated in the Monroe Doctrine and reaffirmed in numerous policy statements since 1822, opposing European involvement in the region. The Commission to Assess United States National Security Space Management and Organization in 2001 concluded: "We know that every medium—air, land and sea—has seen conflict. Reality indicates that space will be no different. Given this virtual certainty, the United States must develop the means both to deter and to defend against hostile acts in and from space."[56] Everett C. Dolman of the Center for Advanced Airpower Studies at the USAF's Maxwell Air Force Base, Alabama, certainly the most eloquent advocate of the necessity of taking proactive measures to ensure American hegemony in space, has stated:

> No nation relies on space more than the United States—none is even close—and its reliance grows daily. A widespread loss of space capabilities would prove disastrous for American military security and civilian welfare. America's economy would collapse, bringing the rest of the world down with it. Its military would be obliged to hunker down in a defensive crouch while it prepared to withdraw from dozens of then-untenable foreign deployments. To prevent such disasters from occurring, the United States military—in particular the United States Air Force—is charged with protecting space capabilities from harm and ensuring reliable space operations for the foreseeable future.[57]

Space power theorists such as Dolman and others see no option but to place weapons in space to ensure the survivability of American space assets in any future conflict.

Advocates of space weaponization note that new capabilities, broader uses, and greater efficiencies have made the US military far more dependent on space systems than even since the 1991 Persian Gulf war, to the extent that their loss might mean the difference between victory and defeat in a major war. US Air Force general Lance Lord spoke for many on this side when he wrote in a 2005 article thus: "Space Superiority is the future of warfare. We cannot win a war without controlling the high ground, and the high ground is space." He argued that at every turn in history an opponent always sought to prohibit the "high ground" and such an opponent must challenge the United States in space at some time, perhaps not far into the future.[58] The recent "illumination" of an American satellite by a Chinese system suggests that Lord may well be right and that a major challenge may loom just around the corner.[59]

Recent developments suggest that the United States may seek to overturn the common law of a ban on weapons in space. On December 13, 2001, for example, President George W. Bush announced that the United States was withdrawing from the 1972 Anti-Ballistic Missile (ABM) Treaty, and it officially did so in 2003. Abrogation of this treaty removed the only legal prohibition against the United States developing a space-based ABM system. The Bush administration also committed to deploying a missile defense system that could include a space-based element. This was highly controversial. Even the conservative-leaning Cato Institute analysts concluded: "The current threat to U.S. satellites does not warrant the near-term weaponization of space." Instead, Cato analysts recommended making greater use of commercial resources and redundant or distributed systems. Commercial space should drive US space policy. It "should strive to foster an environment that allows commercial space activity to grow and flourish rather than create a new area for costly military competition."[60] Also, lest anyone conclude that this is an entirely partisan issue, since 1995 the United States has been blocking a movement at the United Nations for an official prohibition of weapons in space despite its widespread support in other quarters, and both Republican and Democratic administrations have championed the cause.[61]

The 2006 US space policy provided further evidence of this change in the policy arena. It drew sharp criticism from a wide range of observers for opening the Pandora's box of weapons in space and the belligerence of their use against American rivals. Bronwen Maddox, writing in the *London Times* on October 19, 2006, began by asserting that space was "no longer the final frontier but the 51st state of the United States. The new National Space Policy that President Bush has signed is comically proprietary in tone about the US's right to control access to the rest of the solar system." He noted that "[t]he eye-catching declaration is that the US asserts the right to deny access to space to anyone 'hostile to US interests,' although it gives no basis for that right. It also rejects arms control talks that would limit future US actions in space."[62] Former vice president Al Gore even weighed in on it, declaring that this new space policy

> has the potential, down the road, to create the [same] kind of fuzzy thinking and chaos in our efforts to exploit the space resource as the fuzzy thinking and chaos the Iraq policy has created in Iraq. It is a very serious mistake, in my opinion. We in the United States of America may claim that we alone can determine who goes into space and who doesn't, what it's used for and what it's not used for, and we may claim it effectively as our own dominion to the exclusion, when we wish to exclude others, of all others. That's hubristic.[63]

Michael Krepon and Michael Katz-Hymen of the Henry L. Stimson Center remarked of this situation: "The central dilemma of US space policy—the essential and vulnerable nature of satellites used for national and economic security—is highlighted by recent developments. There is no exit from this dilemma. The more we seek to protect our satellites by the use of force in space, the more vulnerable our satellites will become if our own practices are emulated by others."[64]

In reality, there was little new in the 2006 US space policy. As a former NASA Jet Propulson Laboratory (JPL) project manager put it: "What is new is that world opinion, energized by other unilateral statements and actions of this [Bush] Administration, sees this statement as a realization of what people in the more belligerent parts of America's space enterprise have wanted all along; namely an ability to control space and deny it to others."[65] Regardless, the outcry from around the world has been strong and sustained. Persistent space critic Robert L. Park remarked: "The first goal of the 1996 policy was to: 'Enhance knowledge of the Earth, the solar system and the universe.' Now the first goal is to: 'further U.S. national security, homeland security, and foreign policy objectives.'"[66]

Despite recent developments, most of the space weaponization debate has confined itself to the middle part of the policy spectrum even as it has been both strident and sometimes uncharitable. Of course, it represents a fascinating subject for future study in the history of space policy, one that could occupy several researchers for a considerable period just sorting out the various perspectives. The simplistic "either/or" discussion of popular media fails to unpack the nuances of the debate and tends to obscure the truly important differences. In so doing, one must always distinguish between the militarization of space—force enhancement through communications, navigational, early warning, intelligence, and other types of satellites—and the deployment of weapons in space. This dichotomy tends to polarize the discussion in ways that misdirect it from the central issue: devising the best approach toward ensuring national and global security in space.

So what are the priorities for national security space and issues for the development of space policy? It makes sense to recognize that the place the United States is in the early twenty-first century is the best place to be from the standpoint of national security space issues and therefore a status quo for the United States is not a bad future. The United States has pursued a three-point program relative to space security issues, and this appears both prudent and in retrospect quite prescient. First, the United States has ensured that peer competitors did not step beyond the space technological capabilities that this nation possessed through a range of hard and soft power efforts, treaties and arms control measures, and other initiatives. Second, the United States has long made clear that it would take harsh action should a competitor alter the national security regime in space. A long history of declaratory statements condemning actions viewed as belligerent in space and warning of appropriate repercussions has helped to create the current favorable situation for the United States. A continuation of those methodologies is appropriate and completely expected by the other nations of the globe. Third, the United States has pursued on the whole a reasonable program of research and development (R&D) to ensure that any rival capabilities can be destroyed if necessary. This has taken the form of ASAT and ballistic missile defense projects, directed energy weapons development, targeting of ground infrastructure, and other objectives.

Weapons in space, therefore, might not be the only way, or even the best way, to protect American surveillance capabilities. In the last few years, the United States has aggressively pursued redundancy and hardening of potential space targets. Efforts to build small, inexpensive, easily replaced space assets have also offered an alternative. If a satellite were to be destroyed by a foe, another replacement could immediately be placed in space. Ground-based ASATs, both kinetic energy and other types, are reasonable investments in future security, despite the technological stretch required. So are efforts to target from the ground rival space ground stations and other support systems. At the same time, if the United States has become overdependent on space assets for achieving its national security objectives then perhaps the DoD should also take action to reduce that dependence. There are a range of possibilities for delivering the force enhancements possible through space-based resources. For example, some communications or other capabilities could be offered via high-altitude balloons or UAVs. That does not resolve the vulnerabilities, but less dependence would obviate some of the concerns present among those charged with ensuring US capability to conduct military operations.

Conclusion

The development of the technologies necessary for understanding the unique situation of the Cold War bipolar confrontation between the 1940s and 1980s created the conditions whereby Americans pursued the surveillance state. Once established, whether needed or not in the post–Cold War era, these capabilities have been maintained and enhanced and broadened and twisted in so many different directions as to make them virtually unrecognizable to those who first fashioned the national security framework of the 1950s. It has led the national leaders of the United States to embrace these technological fixes for dealing with security affairs; indeed the answer to virtually any question relative to intelligence-gathering is to compound and advance technological answers to questions. Have we gone too far? Throughout this process whatever the United States did was within the power of its citizenry. The nation, no doubt, has allowed an erosion of civil liberties in the name of greater security, a path began early on and advanced over the last half of the twentieth century. Perhaps Dwight D. Eisenhower said it best in his "Farewell Address" of January 17, 1961, presented just as the 34th president departed the White House. It is remembered today chiefly for his warning about the potency of the military-industrial complex, which he said had the "potential for the disastrous rise of misplaced power." What has been mostly forgotten is Eisenhower's equally strong warning about the "danger that public policy could itself become the captive of a scientific-technological elite." He cautioned that this scientific-technological elite was closely tied to the power of the military-industrial complex, indeed the technological revolution made possible by this elite largely fueled the sweeping changes in the industrial/military posture during and after World War II.[67] It was this scientific-technological elite that created the surveillance state now present in America. We find it necessary at present, with the erosion of civil liberties and personal privacy, to ask "who's watching the watchers?" The answer to that question may well shape fundamentally the trajectory of the surveillance state in the future.

Notes

1. Four important recent books on the satellite reconnaissance program have been published: Dwayne A. Day, John M. Logsdon, and Brian Latell, eds., *Eye in the Sky: The Story of the Corona Spy Satellite* (Washington, DC: Smithsonian Institution Press, 1998); Robert A. McDonald, *Corona Between the Sun and the Earth: The First NRO Reconnaissance Eye in Space* (Bethesda, MD: ASPRS Publications, 1997); Curtis Peebles, *The Corona Project: America's First Spy Satellites* (Annapolis, MD: Naval Institute Press, 1997); Jeffrey T. Richelson, *America's space sentinels: DSP satellites and national security* (Lawrence: University of Kansas Press, 1999). See also, William E. Burrows, *Deep Black: Space Espionage and National Security* (New York: Random House, 1987); Jeffrey T. Richelson, *America's Secret Eyes in Space: The U.S. Keyhole Spy Satellite Program* (New York: Harper and Row, 1990).

2. Quoted in *The NRO at the Crossroads: Report of the National Commission for the Review of the National Reconnaissance Office* (Washington, DC: National Reconnaissance Office, November 1, 2000), Appendix E, 120.

3. See Roger D. Launius, "American Memory, Culture Wars, and the Challenge of Presenting Science and Technology in a National Museum," *The Public Historian* 29 (2007): 13–30.

4. R. Cargill Hall, "From Concept to National Policy: Strategic Reconnaissance in the Cold War," *Prologue: The Journal of the National Archives* 28 (1996): 106–23; R. Cargill Hall, "The NRO in the 21st Century: Ensuring Global Information

Supremacy." *Quest: The History of Spaceflight Quarterly* 11:3 (2004): 4–11; David H. Onkst, "Check and Counter-Check: The CIA's and NRO's Response to Soviet Anti-Satellite Systems, 1962–1971," *Journal of the British Interplanetary Society* 51 (August 1998): 301–308; Jeffrey T. Richelson, "Undercover in Outer Space: The Creation and Evolution of the NRO, 1960–1963," *International Journal of Intelligence and Counterintelligence* 13:3 (2000): 301–344.

5. Tom D. Crouch, *Lighter Than Air: An Illustrated History of Balloons and Airships* (Baltimore, MD: Johns Hopkins University Press, 2009); F. Stansbury Haydon and Tom D. Crouch, *Military Ballooning during the Early Civil War* (Baltimore, MD: Johns Hopkins University Press, 2000).

6. Lt. F. E. Humphreys, "The Wright Flyer and its Possible Uses in War," *Journal of the United States Artillery* 33:2 (1910): 99–107.

7. Lee Kennett, *The First Air War: 1914–1918* (New York: Free Press, 1991); John H. Morrow Jr., *The Great War in the Air: Military Aviation from 1909–1921* (Washington, DC: Smithsonian University Press, 1993); Richard P. Hallion, *Rise of the Fighter Aircraft, 1914–1918* (Baltimore, MD: The Nautical and Aviation Press, 1984).

8. Quote from "Aerial Reconnaissance in World War Two," Euro Group Gazette, November 19, 2003, available on-line at http://www.euro-downloads.com /gazette/G221.htm, accessed May 2, 2012 4:17 P.M. See also William E. Burrows, *By Any Means Necessary: America's Secret Air War in the Cold War* (New York: Farrar, Straus and Giroux, 2001).

9. Dino A. Brugioni, *Eyes in the Sky: Eisenhower, the CIA and Cold War Aerial Espionage* (Annapolis, MD: Naval Institute Press, 2010).

10. Glenn Hastedt, "Reconnaissance Satellites, Intelligence, and National Security," in Steven J. Dick and Roger D. Launius, eds., *Societal Impact of Spaceflight* (Washington, DC: NASA SP-2007, 4811), 370.

11. Roger D. Launius et al., "Spaceflight: The Development of Science, Surveillance, and Commerce in Space," *Proceedings of the IEEE* 100 (May 13, 2012): 1785–1818, on 1810.

12. On the many US difficulties in obtaining accurate strategic intelligence information on the Soviet Union see Lawrence Freedman, *US Intelligence and the Soviet Strategic Threat* (London: Macmillan Press, 1977); John Prados, *The Soviet Estimate: U.S. Intelligence Analysis & Russian Military Strength* (New York: Dial Press, 1982).

13. Frank Madden, "The Corona Camera System: Itek's Contribution to World Security," *Journal of the British Interplanetary Society* 52 (1999): 375–380.R. Cargill Hall, "Postwar Strategic Reconnaissance and the Genesis of CORONA," in Day et al., eds., *Eye in the Sky*, 86–118; R. Cargill Hall, "Origins of U.S. Space Policy: Eisenhower, Open Skies, and Freedom of Space," in John M. Logsdon, gen. ed., *Exploring the Unknown: Selected Documents in the History of the U.S. Civil Space Program, Volume 1*, 213–229; Tom Crouch, *The Eagle Aloft: Two Centuries of the Baloon in America*. Washington, DC: Smithsonian Institution Press, 1983, 644–649.

14. Col. Andrew J. Goodpaster, "Memorandum of Conference with the President, November 15, 1956, 2:30 PM," with Secretary Hoover, Admiral Radford, Mr. Allen Dulles, Mr. Richard Bissell, and Col. Goodpaster, DDE Library, Abilene, KS; Curtis Peebles, "The United States Air Force and the Military Space Program,"3 [unpublished paper in possession of author].

15. Coy F. Cross, "The U-2: From Francis Gary Powers to Kosovo," 3, 2002 [unpublished in possession of author].

16. Gregory W. Pedlow and Donald E. Welzenbach, *The CIA and the U-2 Program* (Langley, VA: CIA Center for the Study of Intelligence, 1998), 2–4.

17. "Soviet Note of December 15," *Department of State Bulletin,* document 91–1 in Appendix II of R. Cargill Hall and Clayton Laurie, eds., *Early Cold War Overflights: Symposium Proceedings* (Washington, DC: Office of the Historian, National Reconnaissance Office, 2003), Vol. 2, 573.

18. David F. Winkler, *Cold War at Sea: High-seas Confrontation Between the United States and the Soviet Union* (Annapolis, MD: Naval Institute Press, 2000); Don Oberdorfer, *From the Cold War to a New Era: The United States and the Soviet Union, 1983–1991* (Baltimore, MD: Johns Hopkins University Press, 1998); Seymour M. Hersh, *The Target is Destroyed* (New York: Random House, 1986); Richard William Johnson, *Shootdown: Flight 007 and the American Connection* (New York: Viking, 1986);Alexander Dallin, *Black Box: KAL 007 and the Superpowers* (Berkeley: University of California Press, 1985).

19. Pedlow and Welzenbach, *The CIA and the U-2 Program,* 10; Chris Pocock, *The U-2 Spyplane* (Atglen PA: Schiffer Publishing Ltd., 2000), 5–9. See also Burrows, *By Any Means Necessary.*

20. The U-2 was augmented in 1962 by the SR-71 Blackbird, which proved too technologically difficult to maintain and was retired in 1990. Meantime, the SR-71 set a world speed record at 2,193.167 miles per hour in 1976 for a coast to coast flight, but its true top speed remains classified information.

21. Roger D. Launius, Robert W. Smith, and John M. Logsdon, eds., *Reconsidering Sputnik: Forty Years Since the Soviet Satellite* (Amsterdam: Harwood, 2000), 172–173.

22. Philip Taubman, *Secret Empire: Eisenhower, the CIA, and the Hidden Story of America's Space Espionage,* (New York: Alfred A. Knopf, 2003); Howard E. McCurdy, *Space and the American Imagination* (Washington, DC: Smithsonian Institution Press, 1997), 58–9; Robert A. Divine, *The Sputnik Challenge* (New York: Oxford University Press, 1993), 11–12.

23. Jay Miller, *Lockheed's Skunk Works* (Arlington TX: Aerofax, Inc., 1993), 84; Dennis R. Jenkins, *Lockheed U-2 Dragon Lady* (North Branch MN: Specialty Press Publishers and Wholesalers, 1998), 14–18.

24. See Michael R. Beschloss, *Mayday: Eisenhower, Khrushchev and the U-2 Affair* (New York: Harper & Row, 1986); Gerald K. Haines and Robert E. Leggett, eds., *Watching the Bear: Essays on CIA's Analysis of the Soviet Union* (Langley, VA: Center for the Study of Intelligence, 2003).

25. R. Cargill Hall and Robert Butterworth, *Military Space and National Policy: Record and Interpretation* (Washington, DC: George C. Marshall Institute, 2006), 31.

26. The Report to the President by the Technological Capabilities Panel of the Science Advisory Committee, Vol. II, *Meeting the Threat of Surprise Attack* (Washington, DC, February 14, 1955), 151. This declassified Top Secret report can be found in the records of the Dwight D. Eisenhower administration for 1952–61 in the Office of the Special Assistant for National Security Affairs, NSC Policy Papers, Box 16, Folder NSC 5522, Technological Capabilities Panel, Eisenhower Presidential Library.

27. Quoted in Howard Jones, *Crucible of Power: A History of American Foreign Relations from 1945* (Lanham, MD: Rowman & Littlefield, 2009), 80.

28. Quoted in Walter A. McDougall, *The Heavens and the Earth: A Political History of the Space Age* (New York: Basic Books, 1985), 134; an abridged version, less the reference to military satellites, appears in "Memorandum of a Conference, President's Office, White House, Washington, October 8, 1957, 8:30 a.m.," *Volume XI* [347], 755–56.

29. These arguments have been effectively made in Rip Bulkeley, *Sputniks Crisis and Early United States Space Policy* (Bloomington: Indiana University Press, 1991); Divine, *Sputnik Challenge*; Dwayne A. Day, "New Revelations About the American

Satellite Programme Before Sputnik," *Spaceflight* 36 (1994): 372–373; R. Cargill Hall, "Origins of U.S. Space Policy," 213–229; Roger D. Launius, "Eisenhower, Sputnik, and the Creation of NASA: Technological Elites and the Public Policy Agenda," *Prologue: Quarterly of the National Archives and Records Administration* 28 (1996): 127001E43; R. Cargill Hall, "Earth Satellites: A Few Look by the United States Navy," in R. Cargill Hall, ed., *Essays on the History of Rocketry and Astronautics: Proceedings of the Third through the Sixth History Symposia of the International Academy of Astronautics* (San Diego, CA: Univelt, Inc., 1986), 253–78; R. Cargill Hall, "Eisenhower, Open Skies, and Freedom of Space," IAA-92–0184, paper delivered on September 2, 1992, to the International Astronautical Federation, Washington, DC; Derek W. Elliott, "Finding an Appropriate Commitment: Space Policy Under Eisenhower and Kennedy," Ph.D. Diss., George Washington University, 1992; R. Cargill Hall, "The Eisenhower Administration and the Cold War: Framing American Aeronautics to Serve National Security," *Prologue: The Journal of the National Archives* 27 (Spring 1995): 61–70; and several essays in Roger D. Launius, ed., *Organizing for the Use of Space: Historical Perspectives on a Persistent Issue* (San Diego, CA: Univelt, Inc., AAS History Series, Vol. 18, 1995).

30. For a lengthy early analysis of Eisenhower's Open Skies proposal see Walt W. Rostow, *Open Skies: Eisenhower's Proposal of July 21, 1955* (Austin: University of Texas Press, 1982).

31. Launius, et al., "Spaceflight," 1810.

32. John Lewis Gaddis, *The Long Peace: Inquiries Into the History of the Cold War* (New York: Oxford University Press, 1987), 203–205.

33. Lt. Col. Larry K. Grundhauser, "Sentinels Rising: Commercial High-Resolution Satellite Imagery and Its Implications for US National Security," *Aerospace Power Journal* 12 (Winter 1998): 61–80, on 76. See also, Peter D. Zimmerman, "Remote-Sensing Satellites, Superpower Relations, and Public Diplomacy," in Michael Krepon et al., eds., *Commercial Observation Satellites and International Security* (London: Macmillan, 1990), 34.

34. United Nations General Assembly Resolution 1721 (XVI), adopted on December 20, 1961.

35. "Treaty on Principles Governing the Activities of States in the Exploration and Use of Outer Space, Including the Moon and Other Celestial Bodies," Signed at Washington, London, Moscow, January 27, 1967, Article I, available on-line at U.S. State Department, http://www.state.gov/t/ac/trt/5181.htm, accessed October 10, 2006.

36. Dwayne A. Day, "Corona: America's First Spy Satellite Program," *Quest* 4 (1995): 4–21; Kevin C. Ruffner, ed., *Corona: America's First Satellite Program* (Washington, DC: CIA Center for the Study of Intelligence, 1995).

37. Hall, "Origins of U.S. Space Policy," 222–224.

38. R. Cargill Hall, *Samos to the Moon: The Clandestine Transfer of Reconnaissance Technology between Federal Agencies* (Washington, DC: National Reconnaissance Office, 2001), 2; R. Cargill Hall, "Postwar Strategic Reconnaissance and the Genesis of Corona," chap. 4 in Day, et. al., eds. *Eye in the Sky*, 110.

39. Robert L. Perry, "A History of Satellite Reconnaissance, Volume I—CORONA," October 1973, National Reconnaissance Office, pp. 219–20, copy in possession of author.

40. Ibid., 214.

41. William B. Scott, "Two-Stage-to-Orbit 'Blackstar' System Shelved at Groom Lake?" *Aviation Week & Space Technology*, March 5, 2006, available at: http://www.aviationweek.com/aw/generic/story_generic.jsp?channel=awst&id=news/030606p1.xml, accessed February 19, 2010.

42. Jeffrey T. Richelson, *America's Space Sentinels: DSP Satellites and National Security* (Lawrence: University Press of Kansas, 1999).

43. Paul B. Stares, *The Militarization of Space: U.S. Policy, 1945–1984* (Ithaca, NY: Cornell University Press, 1985), 81; Curtis Peebles, *Battle for Space* (New York: Beaufort Books, 1983), 83–92; Lt. Col. Bruce M. Deblois, "Space Sanctuary: A Viable National Strategy," *Aerospace Power Journal* 12 (1998): 41–57.

44. Lt. Col. Eric Nedergaard, "The F-15 ASAT—The Invitation to Struggle Accepted," National War College Study, December 10, 1990, 3–8 and 10–12; Marcia S. Smith, "ASATs: Antisatellite Weapons Systems," Congressional Research Service Issue Brief, December 7, 1989; Department of Defense, *Soviet Military Power 1990* (Washington, DC: Government Printing Office, September 1990), 59–60; Curtis Peebles, *High Frontier: The U.S. Air Force and the Military Space Program* (Washington, DC: Air Force History and Museums Program, 1997), 67.

45. Dwayne A. Day, "Blunt Arrows: The Limited Utility of ASATs," Space Review, available on-line at http://www.thespacereview.com/article/388/1, accessed October 11, 2006.

46. See Donald R. Baucom, "The Rise and Fall of Brilliant Pebbles," *Journal of Social, Political and Economic Studies* 29 (2004): 145–190; Donald R. Baucom, *The Origins of SDI: 1944–1983* (Lawrence, KA: University Press of Kansas, 1992); Frances Fitzgerald, *Way Out There in the Blue: Reagan, Star Wars, and the End of the Cold War* (New York: Simon and Schuster, 2000).

47. Peter Anson and Dennis Cummings, "The First Space War: The Contribution of Satellites to the Gulf War," *Royal United Services Institute Journal* 136 (1991): 45–53; Steven Lambakis, "The World's First Space War," *Orbis* 39 (1995): 417–413; Scott A. Weston, "Examining Space Warfare: Scenarios, Risks, and US Policy Implications," *Air & Space Power* 23:1 (2009): 73–82; Darren Huskisson, "Protecting the Space Network and the Future of Self-Defense," *Astropolitics: The International Journal of Space Politics & Policy* 5, No. 2 (May 2007): 123–43; Howard Kleinberg, "On War in Space," *Astropolitics: The International Journal of Space Politics & Policy* 5, Issue 1 (January 2007): 1–27.

48. John Poindexter, "Overview of the Information Awareness Office," remarks as prepared for delivery by Dr. John Poindexter, Director, Information Awareness Office, at the DARPATech 2002 Conference, August 2, 2002, available on-line at http://www.fas.org/irp/agency/dod/poindexter.html, accessed June 15, 2012; Gina Marie Stevens, "Privacy: Total Information Awareness Programs and Related Information Access, Collection, and Protection Laws," Congressional Research Service, RL31730, March 21, 2003.

49. Shane Harris, *The Watchers: The Rise of the Surveillance State* (New York: Penguin Books, 2001), 357.

50. Elizabeth B. Bazan, "The Foreign Intelligence Surveillance Act: An Overview of the Statutory Framework and Recent Judicial Decisions," Congressional Research Service, RL30465, April 2005; Peter P. Swire, "The System of Foreign Intelligence Surveillance Law," *George Washington Law Review* 72 (2004), available at: http://papers.ssrn.com/sol3/papers.cfm?abstract_id=586616, accessed June 15, 2012; David Alan Jordan, "Decrypting the Fourth Amendment: Warrantless NSA Surveillance and the Enhanced Expectation of Privacy Provided by Encrypted Voice over Internet Protocol," *Boston College Law Review* 47:1 (2006): 1–42; U.S. Department of Justice White Paper on NSA Legal Authorities, "Legal Authorities Supporting the Activities of the National Security Agency Described by the President," January 19, 2006, available at: http://fll.findlaw.com/news.findlaw.com/hdocs/docs/nsa/dojnsa11906wp.pdf, accessed May 15, 2012.

51. Harris, *The Watchers*, 357.

52. Nicholas Johnson, "Space Traffic Management: Concepts and Practices," *Space Policy* 20 (2004): 79–85.
53. Phillip J. Baines, "Prospects for "Non-Offensive" Defenses in Space," *Center for Nonproliferation Studies Occasional Paper No. 12*, 2004, 31.
54. Malcolm Mowthorp, "US Military Space Policy, 1945–92," *Space Policy* 18 (2002): 25–36, on 25.
55. Lt. Col. Bruce M. DeBlois, "Space Sanctuary: A Viable National Strategy," *Airpower Journal* 12 (Winter 1998): 41–57.
56. Donald H. Rumsfeld, et al., *Report of the Commission to Assess United States National Security Space Management and Organization* (Washington, DC: Government Printing Office, 2001), x.
57. Everett C. Dolman, "U.S. Military Transformation and Weapons in Space," *SAIS Review* 26 (Winter-Spring 2006): 163–74, quote on 163. See also Everett C. Dolman, *Astropolitik: Classical Geopolitics in the Space Age* (London, UK: Frank Cass, 2002).
58. Gen. Lance W. Lord, "Space Superiority," *High Frontier* 1 (Winter 2005): 4–5, on 4.
59. Michael Krepon and Michael Katz-Hyman, "The Responsibilities of Space Faring Nations," *Defense News*, October 16, 2006.
60. Charles V. Peña and Edward L. Hudgins, "Should the United States 'Weaponize' Space? Military and Commercial Implications," *Policy Analysis* 427: 1–24, on 16–17.
61. United Nations Institute For Disarmament Research, "Outer Space and Global Security, Geneva, November 26–27, 2002, Conference Report," 2–3.
62. Bronwen Maddox, "America Wants it All—Life, the Universe and Everything," *London Times*, October 19, 2006.
63. "Gore Condemns Bush Space Policy," *Popular Science*, October 20, 2006, available at: http://popsci.typepad.com/popsci/2006/10/gore_space_poli.html, accessed October 28, 2006.
64. Krepon and Katz-Hyman, "The Responsibilities of Space Faring Nations."
65. Jim Burke, e-mail to beethakore@yahoo.co.uk, et al., "US Space Policy Debate," October 19, 2006.
66. Robert L. Park, "Empire: President Bush Approves a New National Space Policy," *What's New*, October 20, 2006.
67. "Farewell Radio and Television Address to the American People," January 17, 1961, *Papers of the President, Dwight D. Eisenhower 1960–61* (Washington, DC: Government Printing Office, 1961), 1035–1040.

Chapter 8

Serendipitous Outcomes in Space History: From Space Photography to Environmental Surveillance

Sebastian Vincent Grevsmühl

On February 8, 1962, the US Navy, in collaboration with the US Weather Bureau and the Canadian government, launched a major observation effort "to correlate observations of the ice conditions in the Gulf of St. Lawrence made from surface ships and aircraft with those made from the TIROS [Television Infrared Observation] satellite."[1] Observation correlation in the context of satellite remote sensing meant two things. First of all, it implied learning how to look at the images provided by the first meteorological satellite program in order to use them in scientific studies. In order to make sense of the pictorial evidence, these images had to be correlated with other, better know "topographies of knowledge,"[2] such as aerial photography, which had already become fully operational during World War I. Secondly, observation correlation required cooperation between major Cold War military and civilian organizations, such as the US Navy and the US Weather Bureau. Their participation thus reveals that these correlation studies had hidden surveillance ambitions and were sponsored not just in light of benefits to scientific knowledge but also because of a national security imperative.

Historians of twentieth-century science and technology have yet to fully explore the history of National Aeronautical and Space Administration's (NASA) satellite programs for environmental surveillance. Pamela Mack's pioneering study of the Landsat satellite project of the late 1960s and 1970s has shown that during the Cold War satellites enabled to gain an increasingly global picture of environmental conditions. Unfortunately, her research was not followed by many comparable in-depth studies. Probably one of the most notable exceptions is Erik Conway's *Atmospheric Science at NASA* and, to a lesser degree, Henry Lambright's *NASA and the Environment: The Case of Ozone Depletion*.[3] By contrast, following Ron Doel's pioneering work, historians have covered the subject of military patronage of Cold War environmental sciences quite extensively.[4] Some scholars have examined the surveillance implications of early space photography and satellite imagery.[5] National Reconnaissance Office (NRO) historian Cargill Hall contributed particularly strongly to our current understanding of the close cooperation, yet also the severe tensions that existed between the NRO and the American space agency all along the Cold War period.[6] Clandestine technology transfer from military to nominally civilian institutions was common practice in the United States during the Cold War, as John Cloud and Dwayne Day have shown.[7]

In this chapter I pay closer attention to the actual mediation processes involved in producing novel environmental knowledge. I examine how information was gathered and interpreted and how the conclusions were drawn. One lesson to be taken from the following reflections on early missile and satellite technologies is that from the very beginning NASA managed to attract the attention of very different communities to the usefulness of remote environmental observations. We also learn that remote sensing imaging proved to be a technology with a far wider range of applications than those their inventors had in mind. For instance, V-2 photographs were conceived to understand the motion of rockets along their trajectory, but they also quickly attracted the interest of geographers and especially meteorologists who began to use the images to gain a better understanding of environmental phenomena. In a quite similar way, images obtained with the first US meteorological satellite TIROS served civilian as well as military goals. They helped convey new information on cloud coverage and atmospheric systems that was of interest not only to meteorologists but also to military planners, allowing for better scheduling of photographic reconnaissance sorties. Moreover, in the absence of clouds, the satellite images themselves could potentially reveal sensitive terrain information such as snow and ice cover.

This chapter argues that the discovery that these images could be used differently from what they had been originally designed for was often "serendipitous."[8] I also show that these novel applications of remote sensing imagery marked an important transition from their use in military research programs, devoted to improving weapons and surveillance of enemy forces, to their utility in "environmental surveillance"[9] studies.

Early Space Photography and Serendipity

One of the most striking elements in early history of space photography is that at the outset its utility was somewhat narrowly defined and, eventually, its users gained new insights on its potential for the advancement of environmental analysis. At the time of the early postwar rocket flights, leading scientists from both sides of the Iron Curtain claimed that observation technologies were narrowly conceived for operational use and "photography was rarely the main purpose of a flight."[10] For example, early photographs obtained on V-2 flights during the late 1940s and early 1950s at the US Army Ordnance's White Sands Proving Ground in New Mexico were taken in order "to acquire a better knowledge of various motions executed by the missile in going through the upper atmosphere."[11] The realization was in other words the result of serendipity, as the rocket camera was not deliberately directed toward the Earth or its features so as to image them, but rather toward the rocket's trajectory in order to reveal its path during the flight.

Various other types of instruments and detectors were also flown to analyze other characteristics of the missile system and its interaction with the medium through which it travelled. These included devices to monitor the influence of cosmic rays, disruption to telecommunications due to rocket exhaust, and other effects due to the missile's discharge.[12] As veritable flying laboratories, V-2 rockets and especially the so-called *Aerobees* (the first large vector for atmospheric research in the United States) were also fitted with detection devices and automatic recorders. Photographic equipment was in other words only one of many onboard sensors used to gather valuable data on the flight. However, when the routine rocket recording operations returned a wealth of numeric data and other material artifacts (including photographs), those scientists who had access to them realized that these could find application in other research fields. Rocket photographs showed important features of the Earth (including its curvature and jet streams) and could thus be reutilized in the context of meteorological research or in studies focusing on the whole Earth.

The success of V-2, Aerobee, and Viking photography is even more remarkable considering that most of those spectacular images were not only unexpected, but that missile experts attempted to fire the rockets during clear weather conditions, in order to prevent problems during the launch and ensure the safe reception of valuable information on the missiles' trajectory upon the vehicle's reentry.[13] What the specialists working at White Sands initially considered a nuisance, that is, the presence of meteorological systems in proximity of the launching site, eventually enabled "the most spectacular use of photography in connection with rocket research."[14] Analysts of the RAND Corporation (the think tank with close ties to the US Air Force) were also well aware of the accidental nature of rocket photographs, and RAND's first feasibility study on meteorological satellites openly acknowledged that the new information on cloud formations were gained "from data which were not originally gathered for this purpose."[15]

Photographs were not the only outcome of an operational use that eventually found different applications. Spectrographs flown aboard V-2s in the context of Army Ordnance's 1946 rocket program are another good example. They excited not only astronomers interested in solar spectra but also meteorologists and geophysicists studying absorption processes in the upper atmosphere. Moreover, the first solar spectrogram captured above the ozone layer provoked even broad acclaim not only among specialists but also in the national and international press.[16] Yet, as David DeVorkin has shown, it is important to mention that this considerable success did not rely at the time on the expertise of solar physicists or any of the leading researchers in atmospheric physics. It was Richard Tousey, a National Research Laboratory (NRL) specialist in laboratory optical techniques and the limits of vision, who designed the V-2 spectrograph. In fact, the vast majority of US researchers engaged in upper atmosphere research were not part of "traditional" academic communities. Trained in military laboratories during World War II, these radio and radar engineers, technicians, and optics specialists aligned scientific research to a national security agenda by promoting novel research with the aim of improving defense systems.[17]

Tousey's instrument was in other words a component of science experiments that, while nominally civilian, were tightly aligned to the interests of the US armed forces in preparing for a nuclear conflict. Improving ballistic missile trajectories was an imperative in building up reliable missile systems, which was the ultimate guiding objective of all the V-2 science experiments of the early

Cold War.[18] One major "side effect" was the considerable gain in environmental knowledge. Indeed, as Ron Doel has argued, it is certainly not an exaggeration to state that the guided missile helped tremendously to "constitute the physical environmental sciences in the US after 1945."[19]

The Mobilization of the First Space Photographs in the Context of Meteorological Research

Rocket photography had a significant impact in the advancement of meteorology in the United States and helped its most prominent experts to experiment with photographic materials blending together a variety of different techniques. It also made them eager to promote new satellite programs for weather reconnaissance. Already by the end of the 1940s, the White Sands rocket program had produced spectacular photographs, including some showing the Earth's curvature and a variety of meteorological systems above our planet. Those pictures (disseminated also via the popular press, such as in National Geographic[20]) helped to promote rocket imagery well beyond military departments and to ensure space photography became adopted in the running of other scientific endeavors.[21]

The photographs seemed to have promise especially for meteorological research and stimulated plans to launch a satellite mainly devoted to taking images from space to know more about jet streams and other large-scale atmospheric phenomena and processes.[22] Some NASA experts were wary of this solution due to the enormous complexity of the task ahead.[23] But others, such as the meteorologist Harry Wexler (the US Weather Bureau's observer on the panel overseeing the V-2 program at White Sands Proving Ground), were determined to push the scientific exploits to be derived from rocket experiments. Wexler was one of several US science administrators (also including Joseph Kaplan, Lloyd Berkner, James Van Allen, Homer Newell, and Fred Singer) who saw the development of a satellite as critical to the advancement of meteorology in the United States.[24] Yet they succeeded only when US military agencies offered funding for the endeavor. This was mainly because of their interest in technologies that drastically improved surveillance and the planning of military operations by providing up-to-date meteorological information. Already a 1946 report by the RAND Corporation argued for the feasibility of satellites and noted unambiguously that "perhaps the two most important classes of observation which can be made from such a satellite are the spotting of the points of impacts of bombs launched by us, and the observation of weather conditions over enemy territory."[25]

Aware that support existed for his plans, Wexler vigorously campaigned for a US weather satellite. In May 1954, he presented his views to a large audience gathered on occasion of the Third Symposium on Space Travel held at Hayden Planetarium (New York). For the first time Wexler's vision of a satellite Earth observation program emerged and he used numerous slides of photographs "accidentally" showing meteorological systems obtained with the cameras installed on the V-2. Two years later, he presented his ideas again at the Third Annual Conference of the American Astronautical Society. Besides a rather rudimentary photographic mosaic already used during his previous speech, Wexler now showed a remarkably sophisticated composite image, which Otto E. Berg of the Naval Research Laboratory (NRL) had pieced together with the help of 16-mm photographs obtained in October 1954 on one of the Aerobee-Hi rocket flights (Figure 8.1). Made out of more than 100 color photographs, enlarged and mounted on a sphere, Berg's composite picture was indeed an impressive visual argument for the need of satellite-based synoptic weather observations.[26]

Yet the mosaic was a serendipitous outcome of NRL's missile program and nobody could anticipate that rocket photographs would be used for this purpose. Few also believed it possible to visualize for instance a tropical storm that, in fact, was now visible near the Gulf of Mexico in the upper left of the mosaic. Berg's photomontage clearly exemplified the enormous potential of satellite photography for meteorological analysis. This is also the reason why the article that Berg published together with Lester Hubert (US Weather Bureau) on the mosaic concluded that "[t]he possibilities suggested by this accidental rocket reconnaissance of a tropical storm are tremendous."[27] In September 1955 the popular press celebrated Berg's "portrait of the Earth" and a double color page was published in the magazine *Life*.[28] The mosaic thus generated interests in satellite photography well beyond military circles. The serendipitous finding that clouds could be continuously photographed from space made Wexler even more enthusiastic. Presumably drawing on the parlance of military personnel reconnoitering Soviet forces, he now envisaged that a weather satellite could function as a "storm patrol."[29]

The photographic mosaic, however, also demonstrated the advantages of combining different techniques of analysis of pictorial evidence consisting of composing images and then treating them with a variety of meteorological methods.

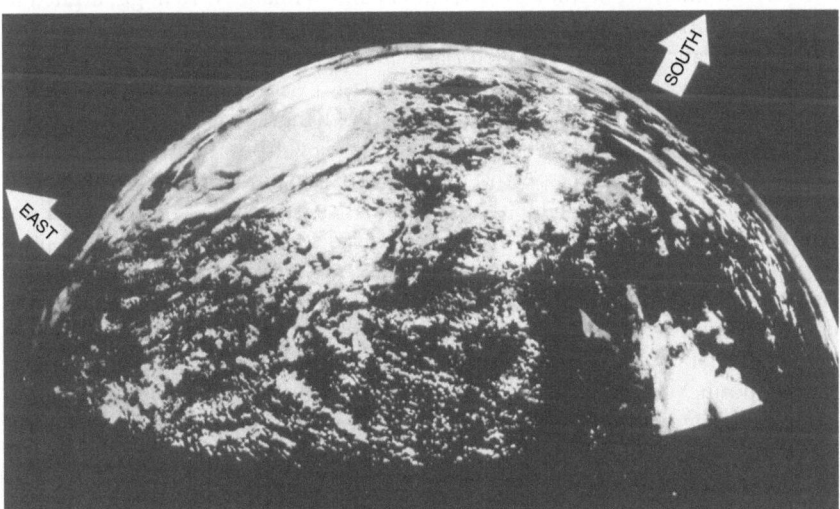

Figure 8.1 Black-and-white photomontage of more than 100 individual photographs obtained during a US Navy *Aerobee* flight in 1954 and mounted on a sphere by NRL scientist Otto Berg. A large whirlpool of cloud left over from a tropical cyclone above the Gulf of Mexico, discovered by chance, is visible in the upper left of the image

Source: NOAA image library.

The whole complexity of the synoptic weather situation, especially, as in the case of Berg's mosaic, due to the presence of several small vortices, proved to be undetectable when relying solely on the standard contour analysis traditionally employed in meteorology. Indeed, as Berg and Hubert point out in their article, three vortices were "sufficiently small to be overlooked in the routine contour analysis" as they were below "the detection threshold of [the] data network."[30] The data points that the meteorologists had at their disposal were in other words still too sparse in order to be able to draw contour lines allowing to reveal the three circular structures that were however clearly visible in Berg's mosaic.[31]

Rocket photographs revealed, in other words, a whole new complexity in the state of the atmosphere. They also helped visualize new weather systems and patterns such as for example cyclone structures that explained in Berg's case the actual cause of a local high-precipitation phenomenon, as normally known from the tropical regions, by simply visualizing their presence.[32] Yet the photographs could not represent a goal in their own right. Many elements of the photographic mosaic could only be of utility if analyzed with traditional meteorological methods. Therefore, Berg's recent mosaic represented a good opportunity to correlate novel pictorial data with more traditional methods of meteorological analysis. The example clearly shows that each visualization medium possesses its own "epistemic limits" according to what I have called elsewhere an "epistemic topography."[33] Rocket and satellite photographs were breaking new ground in meteorological work mainly because they enabled the observation of large-scale meteorological systems from *above* rather than below, allowing therefore a complete rethinking of the global atmospheric system.

The new epistemic topography called for a new visual language.[34] In 1951, two RAND analysts, Stanley M. Greenfield and William W. Kellogg, argued in a report for introducing a whole new method of cloud classification based on the visual appearance of clouds in pictorial evidence from rockets. They also stated that the new visual language ought to replace existing classification criteria, since observing clouds from *above* dramatically changes shapes and patterns and they often do not coincide with the traditional view from the ground. In other words, the inversion of perspective and the broader, synoptic viewpoint created fundamentally new knowledge that traditional classification schemes could not produce.[35] Unsurprisingly, the most experienced scientists involved in V-2 photographic analysis concluded at the time that "[t]he manner in which clouds or a cloud formation seen from above and from below coincide is still far from being definitely known."[36]

However, by combining different types of knowledge, by (often literally) superimposing traditional meteorological analysis and visualization of cloud cover, it slowly became possible to identify new large-scale phenomena and to correlate them with better-established local knowledge.[37] Indeed, the example of cloud photography shows very well the necessity of integrating different kinds of epistemic topographies, which assisted in the development of meteorology all through the Cold War.[38]

By experimenting with rocket photographs and correlating visual data with other techniques Wexler, Berg, and other prominent US meteorologists could set a new agenda in meteorological studies in the United States, a rationale based on environmental surveillance. This was an agenda that, quite evidently, drew on existing Cold War challenges in terms of both method and instruments. Methodologically, it promoted reconnaissance as a useful way to attack weather analysis and in terms of instrumentation it advocated reusing equipment originally developed for military research purposes. Yet Wexler even envisioned the advantages to be derived from a satellite devoted to meteorology thanks to the assistance of a professional painter, as we shall now see.

Harry Wexler and the Invention of a First Satellite Icon

The integration of different topographies of knowledge in the field of meteorology was one of Harry Wexler's main objectives. Although enthusiastic about Berg's photographic mosaic, he was aware that rocket photography was far less effective than satellite photography. In the long run, only satellite observations of the atmosphere would allow a synoptic and continuous production of weather data. Wexler was also aware of the great power pictures could bear on funding institutions and decision makers. Indeed, as David DeVorkin noted, the offices of Wexler superior's were always "well adorned with photographs of storms and cloud systems taken from space."[39] In 1954 he thus sought to commission a painting from an unknown artist who was "stimulated by such chance photos from research rockets."[40] Also, Wexler was well placed to judge how impressive the natural power of meteorological phenomena could be. As a former member of the Meteorology Division in the Army Air Forces, in 1944 he actually flew across a hurricane that would later be famously called for its extreme violence the "Great Atlantic Hurricane."[41]

The painting ordered by Wexler shows a hypothetical view of parts of the Earth and its atmosphere at an altitude of about 6,400 kilometers above Amarillo, Texas. Due to its circular framing, some cultural historians have referred to the painting as a precursor of the famous blue marble.[42] Yet an observer would have to travel at least double the distance Wexler had chosen for this hypothetical view of the Earth's atmosphere in order to see the curvature of planet Earth appear in the way that it was portrayed by the *Apollo* astronauts.[43]

The painting was realized with the support of numerous new scientific insights, as Wexler explained during his talk at the Third Symposium on Space Travel. On that occasion the painting acquired a new status as a "truly scientific image," even if its evident "constructed" character strikes the contemporary observer.[44] Most of the depicted weather systems were quite "realistically" anticipated and showed a truly remarkable correspondence to high-altitude cloud photographs as data from TIROS and other programs would reveal a few years later.[45]

All the elements depicted were carefully elaborated on the canvas. For example, the continental parts of the painting are represented taking into account "reflectivity of sunlight" and the "scattering and depleting effects on the passage of light through the Earth's atmosphere."[46] Furthermore, the image is dominated by very different kinds of clouds, ranging from "a cyclone family of three storms" to "a hurricane" and cloud streets, to which "albedo values were assigned [...] and their brightnesses [sic] computed."[47] Yet the probability for each of these different cloud types to appear simultaneously was low: "the trade cumuli could undoubtedly be observed on almost any day and others, such as the hurricane, seen only rarely."[48]

The unknown artist, thanks to Wexler's guidance, was thus able to unite in one single picture the greatest possible number of different cloud types and produce a sort of condensed cloud "atlas." This explains Wexler's interest in this painting as it enabled to establish a veritable *visual typology* of Earth's "atmospheric systems" seen from space. Photographic naturalism, even if it had been at Wexler's disposition, would never have sufficed to reveal what Wexler wanted to show to his public.[49] Moreover, the choice of a *perpendicular* perspective— introducing the theme of the planisphere[50]—seems indispensable in underlining Wexler's intention to transform the sky illustrations into quantifiable and clearly identifiable scientific objects, reinforcing his message that these instruments ought soon to be built.

Throughout the 1950s and 1960s, scientists in the United States, but also the broad public, hoped that these new technologies would one day even allow efficient "weather control."[51] It comes therefore as no surprise that Wexler's innovative understanding of weather analysis and observation also relied on the support of military organizations to become reality. If NASA's TIROS program was conceived as a "storm patrol,"[52] then it was also a means to a Cold War surveillance end, as we shall now see.

Towards an Integration of Data: TIROS, the First Meteorological Satellite Program

The earliest RAND report on meteorological satellite surveillance unambiguously stated that "in the event of armed conflict there will be large regions of the world from which it will be impossible to obtain weather information by normal means. Owing to the fact that the success of any aerial reconnaissance depends, to a large extent, on [...] knowledge of the weather conditions over the target, the lack of this information will be felt more and more as any planned air offensive progresses. Systematic weather reconnaissance by some unconventional means must therefore be undertaken."[53] When Wexler's plans were examined by prominent military agencies in the United States, what caught their attention was not the potential of space photography for meteorological research alone, but also for surveillance operations.

The integration of surveillance technologies, such as photographic and television cameras, into the satellite payload seemed to have promise in terms of boosting the capacity to reconnoiter enemy territories, and was therefore worth significant investments. As a consequence, the US Air Force and the US Army financed research on different satellite observation technologies.[54] If military objectives, such as improving the performance of rockets, gave leeway to developing important areas of civilian research, including the application of television and photography to meteorological programs, then these programs offered cover for furthering other military projects, including surveillance satellites. Even knowledge of cloud coverage in itself provided highly relevant intelligence information. It allowed for better coordination of traditional photographic reconnaissance flights, especially since cloud coverage posed a major threat to their successful accomplishment.

It therefore made sense, at least from the US perspective, to vigorously promote free access to space to improve the ability to reconnoiter enemy territories.

The International Geophysical Year scientific satellite program provided therefore the historical and legal basis for what Cargill Hall rightly identified as "a stalking horse to establish the precedent of overflight in space for the eventual operation of military reconnaissance satellites."[55] Indeed, as Walter McDougall has shown and as Roger Launius argues in his contribution to this book, US presidents Dwight D. Eisenhower and John F. Kennedy clearly understood the multiple benefits to be derived from granting free access to outer space. Eisenhower's controversial "Open Skies" proposal and Kennedy's plea for "peaceful uses of outer space" both aimed to defuse any possible objections against satellite reconnaissance.[56]

In the field of meteorology, the official and well-known outcome of this strategy was the initiation under the Kennedy administration of discussions with the Soviet Union on a joint meteorological satellite program, united under the aegis of the World Meteorological Organization. Approved in 1963 as the World Weather Watch and still in operation today, the program, with its subsystems, coordinates meteorological observation efforts to provide weather services in all countries.[57]

However, the free access to space policy also served military reconnaissance and surveillance interests. Many scientific satellite programs, some similar to TIROS, offered cover for spy missions. For example, NASA's Discoverer program—officially announced to the public as a research program dedicated to examining and reporting on the space environment, including most notably biomedical experiments with mice and monkeys—carried as its main payload the first US photoreconnaissance satellite camera system used as part of the highly classified CORONA program.[58] Also, in the 1960s meteorological research satellites were routinely used as cover for military surveillance. For instance, between 1962 and 1994 the US Air Force Defense Meteorological Satellite Program (DMSP) deployed a whole set of meteorological satellites, largely identical to their civil counterparts of the TIROS program.[59]

The TIROS meteorological satellites were themselves born out of a surveillance satellite program. TIROS had an immediate military precursor known as US Army project Janus that was initially conceived as a reconnaissance project developed for spying on Soviet territory.[60] The imaging technology was based on a study from the Radio Corporation of America (RCA), a proposal initially presented to the US Air Force for their secret reconnaissance program known as WS-117L. After the bid was lost to Lockheed, RCA sold the idea to the Army where the work was developed further. Shortly after the NASA was created, the program (then known as Janus 2) was transferred in April 1959 to the Space Agency and renamed TIROS. Within the TIROS program, NASA obtained the overall responsibility for engineering and launch and the US Weather Bureau had to oversee operation and data interpretation.[61]

This transfer to nominally civilian institutions is generally considered the beginning of a clear separation of military and civilian programs. Yet the TIROS program shows that throughout the Cold War, this distinction is highly ambiguous, if not to say artificial. Indeed, a closer look at Cold War meteorology efforts erodes our confidence in the distinction between nominally military and nominally civilian uses. Despite important achievements of the DMSP, for example during the Cuban missile crisis in 1962, the Department of Defense also continued to make use of TIROS imagery. In fact, in order to fully assure DoD's need for an *operational* weather satellite, TIROS's two read-out sites at Fort Monmouth and on Hawaii were permanently staffed with "teams composed of meteorologists from the Weather Bureau and Department of Defense agencies" and a "Navy meteorologist was stationed at Ft. Monmouth during the entire operational period of TIROS I."[62] For the same purpose, as documented by the agreement that officially transferred the program to NASA, a joint DoD-NASA advisory group was put in place and substantial DoD funding allocated to the program in order to assure full cooperation.[63]

In terms of technology, it is true that the TIROS meteorological satellites had relatively low-resolution central reconnaissance components because of the civilian use NASA was supposed to make of them.[64] But TIROS's reduced resolution enabled more information to be gained than expected. Both satellite cameras produced black-and-white images composed of distinct lines, making the *contrast* of the vidicon images one of their most important criteria.[65] The imaging technology was designed to better identify and visualize meteorological systems, but once the first satellite was operational other Earth features also became discernable.

This is a recurrent theme in the history of remote sensing: observation technologies produce a "surplus" of information, something not anticipated when they were designed. Following the reception of the first TIROS images, scientists started to become interested in phenomena that had little to do with meteorology. Some photographic mosaics portraying regions surrounding the Gulf of Saint Lawrence (Canada) showed numerous white spots on black background (Figure 8.2). More serendipitous discoveries were about to happen and

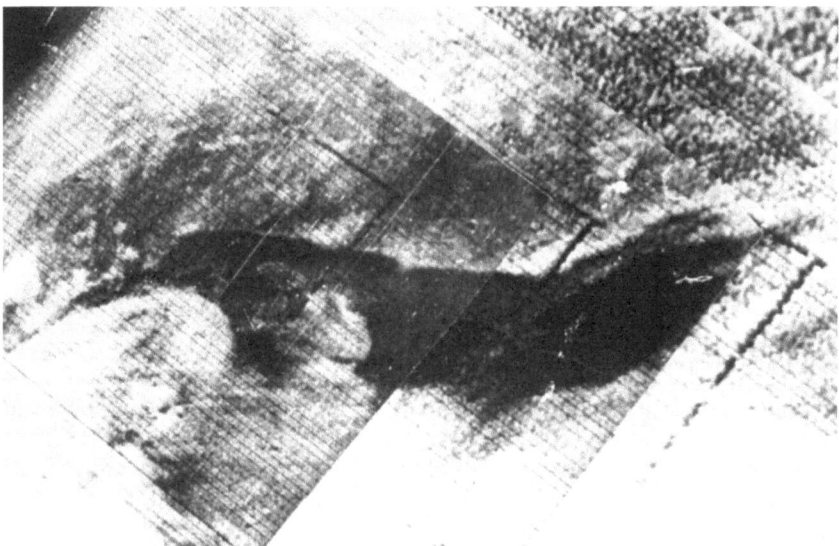

Figure 8.2 Composite image of oblique views taken from TIROS-I above the Gulf of Saint Lawrence on April 1, 1960. The image has enough contrast to allow distinguishing water, clouds and sea ice

Source: NOAA image library.

Harry Wexler eventually concluded that those white spots could be sea ice.[66] So in the 1960s, Wexler supervised a group of scientists at the US Weather Bureau that investigated in greater depth ice surveillance techniques—an application that touched upon, as I will show now, military, scientific, political, as well as economic questions.[67]

TIREC: Environmental Satellite Surveillance during the Cold War

The accidental reconnaissance of sea ice through a satellite designed to produce pictures of clouds eventually led to project TIREC. Launched in 1962 project TIREC aimed "to correlate observations of the ice conditions in the Gulf of Saint Lawrence made from surface ships and aircraft with those made from the TIROS satellite."[68] TIREC, an acronym for TIROS Ice Reconnaissance, had as main objective the development of "procedures and techniques for interpreting satellite readouts of ice formation."[69] It was in many ways an unforeseen by-product of the first American weather satellite program, a serendipitous application of remote cloud observation, initially not included in the list of possible weather satellite uses.

At the end of the first two experimental phases, TIREC confirmed, even without a refined resolution, what Wexler had foreseen. It proved indeed possible to distinguish clouds from ice and to produce maps of sea ice. The comparison of images taken at different times as well as their geometric correction (for example, image distortion resulting from the lens of the camera had to be corrected) helped in achieving this task.

"Measurement of ice and snow cover," as a report to the US Congress in 1962 deliberately vaguely put it, had indeed "also [...] a military application."[70] Lavishly endowed with military funding, TIREC eventually helped to more firmly establish environmental surveillance as a way to establish control over polar regions. Sea ice maps obtained through correlation of satellite data allowed for improved coordination of military and scientific logistics in the cold regions. They also granted otherwise unavailable information on access to strategic sites via polar transit areas. As one representative of the Canadian Defense Board noted, sea ice observations were especially important to identify places where submarines may emerge.[71]

Ice observations were also of "direct economic interest," a major motivation for putting TIREC in place.[72] In a study on the nonmeteorological uses of TIROS and the Nimbus satellite programs, Fred Singer (better known today for his global warming skeptic declarations[73]) argued that meteorological satellites could have saved Canadian and US administrations, for the year 1961 alone, no less than $1.7 million traditionally put aside for logistical operations and the planning of navigable waterways, a figure geologist and geophysicist Paul Lowman of Goddard Space Flight Center rightly considered "misleadingly conservative" as further savings could be derived from using the same system in Western Europe, Russia, and especially Antarctica.[74] TIROS data also offered invaluable support to the US Navy in Antarctic exploration and helped to strengthen its role as principal logistics operator at the South Pole.[75]

Climate scientists benefitted from satellite-based ice observations and correlation studies too. These proved decisive in establishing local temperature trends (most notably warming trends), as well as water circulation estimates, all factors that became important elements of modern climate change interpretations and predictions.[76] For example, synoptic ice observations allowed revealing that the Antarctic continent doubles its size each winter, growing well beyond Europe's surface area by attaining about 34 million square kilometers thanks to the accumulation of sea ice. Indeed, from the 1960s onward, sea ice accumulation as inferred from satellite images has been routinely used to assess global climatic changes.[77]

Thus Project TIREC pioneered Cold War environmental surveillance. The reasons behind its development were not just scientific. Economic, logistic, strategic, and geopolitical ambitions were equally relevant.[78] Soon after Project TIREC was established, a conference on satellite ice studies was held in Washington, DC. By looking at the agencies and institutions represented, one understands how important these studies had now become to civilian and military organizations alike. Representatives of Canada's Defence Research Board and its Joint Photographic Intelligence Center, together with the Royal Canadian Air Force were present. Among participants from the United States, the Weather Bureau, NASA, the Department of Transport, and the National Science Foundation (US Antarctic Program) had all sent delegates. Among the 51 participants, only four can be identified as coming from universities and not holding any official military accreditation.[79] These numbers show very well that at the very beginning of satellite ice observations—they coincide practically with the very birth of the first meteorological satellite program—military concerns clearly went hand in hand with scientific and economic interests.

Conclusion

We should think of the example of space photography as shedding new light on how military patronage made it possible to realize the inner surveillance potential of a range of imaging technologies—a potential that was not foreseen when photographic cameras were first installed on rockets.

In order to understand the history of satellite imagery we need to reconsider the importance that accidental discoveries played in key moments of its unfolding. The use of satellite photography for meteorological purposes was the unintended outcome of attempts to monitor the path of ballistic missiles during their flight. The application of satellite cameras to monitor sea ice was also unforeseen, as photographs produced to "patrol" storms eventually captured more details than expected; including sea ice formations. Thus serendipity proved to be a crucial heuristic element in the contemporary analysis of earth features such as clouds and polar ice caps. Numerous case studies in history of science have shown the diverse ways in which instruments and technologies may find new uses, most notably through the exploitation, by chance, of some of their "hidden" characteristics and properties. Turchetti et al. went a step further and showed in their case study on radio echo-sounding in Antarctica that "when accidents and errors become the subject of scientific enquiry, they can instigate a broader analysis of the range of applications associated with the experimental apparatus in use and in turn favor its adoption more remotely from its current domain."[80] Project TIREC similarly shows that beyond simple exploitation of hidden properties, systematic analysis of environmental surveillance imagery favored its adoption in other scientific domains.

Closely linked to this accidental discovery is the importance of material culture linked to the first meteorological satellite program at the beginning of the 1960s. The particularities of the visualization technology, most notably the poor resolution depending directly on the visual contrast, largely guided the use of satellite images. Within the context of the Cold War, this meant that a "civilian" observation technology, as soon as it aroused interest from the military, could more easily attract important funding, favoring the production of environmental knowledge with immediate utility for national security.

However, the adoption of the new remote sensing techniques that accidental findings enabled required much more than just serendipity. The military sponsorship of new projects such as TIROS and TIREC made it possible to realize what accidents had proven just as a possibility but needed substantial funding to show its real potential. This chapter has shown that support was given in light of the benefits to surveillance and military operations to be derived from satellite imagery. The synoptic view of the satellite could easily help decide whether aerial reconnaissance sorties over enemy territory would be efficient or not. Yet there was also a more generally added value of satellite photography in allowing assessments to be made about accessibility to sea ice covered areas, controlling navigable waterways and identifying regions which could be potentially used as hideout for enemy submersibles. So while the opportunity existed to explore what was serendipitously found through space photography, it was the Cold War urgency of improved surveillance that enabled to further explore the potential of these accidental discoveries. New knowledge of key earth features thus emerged in the search of more sophisticated methods to know about enemy forces.

Moreover the TIREC project shows that NASA was also actively engaged in environmental surveillance and continued to pursue it for several years after TIREC reached completion. For instance, photographs from the Mercury and Gemini space missions attracted great interest from a number of experts working on a number of different scientific disciplines. In particular, during the 1960s, geographers, geologists, and hydrologists took advantage of the new possibilities that remote sensing offered them and used space photography in a variety of projects favoring, for instance, a more systematic monitoring of natural resources. Oceanographers also profited from imagery produced during space programs, as in the case of TIROS, profiting from the growing spectrum of new technologies for environmental exploration and surveillance.[81] Indeed, it is not an exaggeration to suggest that the interest that space photography generated was a decisive factor in the adoption of satellite-based environmental *remote sensing* techniques in the 1970s, including those aimed to more accurately ascertain the availability of natural resources and vital aspects of global environmental change.[82]

Notes

1. Abraham Schnapf, "The TIROS Global System," *Annals of the New York Academy of Sciences* 134 (1965): 149–166, citation 156.
2. Sebastian Vincent Grevsmühl, "Epistemische Topografien. Fotografische und radartechnische Wahrnehmungsräume," in Ingeborg Reichle, Steffen Siegel,

Achim Spelten (eds.), *Verwandte Bilder. Die Fragen der Bildwissenschaft* (Berlin: Kadmos, 2007), 263–279.

3. Pamela E. Mack, *Viewing the Earth: The Social Construction of the Landsat Satellite System* (Cambridge, MA: MIT Press, 1990); Erik M. Conway, *Atmospheric Science at NASA: A History* (Baltimore: Johns Hopkins University Press, 2008); W. Henry Lambright, *NASA and the Environment: The Case of Ozone Depletion* (Washington, DC: NASA, 2005). See also: Pamela E. Mack and Ray A. Williamson, "Observing the Earth from Space," in John M. Logsdon (ed.), *Exploring the Unknown: Selected Documents in the History of the U.S. Civil Space Program, Vol.III: Using Space* (Washington, DC: NASA, 1998), 155–177; John H. McElroy and Ray A. Williamson, "The Evolution of Earth Science Research from Space: NASA's Earth Observing System," in John M. Logsdon, Stephen J. Garber, Roger D. Launius, and Ray A. Williamson (eds.), *Exploring the Unknown: Selected Documents in the History of the U.S. Civil Space Program, Vol.VI: Space and Earth Science* (Washington, DC: NASA, 2004), 441–473, and finally Erik M. Conway, "Drowning in Data: Satellite Oceanography and Information Overload in the Earth Sciences," *Historical Studies in the Physical and Biological Sciences* 37 (2006): 127–151.

4. See among others: Ronald E. Doel, "Constituting the Postwar Earth Sciences: The Military's Influence on the Environmental Sciences in the USA After 1945," *Social Studies of Science* 33 (2003): 635–666. See in the same issue: Kai-Henrik Barth, "The Politics of Seismology: Nuclear Testing, Arms Control and the Transformation of a Discipline," *Social Studies of Science* 33 (2003): 743–781. Jacob D. Hamblin, *Oceanographers and the Cold War: The Disciples of Marine Science* (Seattle: University of Washington Press, 2005); Jacob D. Hamblin, *Arming Mother Nature: The Birth of Catastrophic Environmentalism* (Oxford: Oxford University Press, 2013); John Cloud and Keith C. Clarke, "Through a Shutter Darkly: The Tangled Relationships Between Civilian, Military, and Intelligence Remote Sensing in the Early U.S. Space Program," in Judith Repps (ed.), *Secrecy and Knowledge Production,* Peace Studies Program Occasional Paper no. 23 (Ithaca: Cornell University, 1999), 36–56; John G. Cloud, "American Cartographic Transformations During the Cold War," *Cartography and Geographic Information Science* 29 (2002): 261–282; Simone Turchetti, Katrina Dean, Simon Naylor and Martin Siegert, "Accidents and Opportunities: A History of the Radio Echo-Sounding of Antarctica, 1958–79," *British Journal of the History of Science* 41 (2008): 417–444.

5. I adopt here a common, although arbitrary definition of "space imagery," which postulates that any activity above 100 kilometers altitude may be considered an (outer) *space* activity. Early V-2 flights (launched around 1947) went beyond 100 kilometers altitude and may be counted therefore amongst the first space activity. For this common definition, see for example: Alain Dupas, *La nouvelle conquête spatiale* (Paris: Odile Jacob, 2010), 95.

6. See for example: R. Cargill Hall, *A History of the Military Polar Orbiting Meteorological Satellite Program* (Chantilly, VA: NRO Office of the Historian, 2001); R. Cargill Hall, "Origins of U.S. Space Policy: Eisenhower, Open Skies, and Freedom of Space," in John Logsdon et al. (eds.), *Exploring the Unknown: Selected Documents in the History of the U.S. Civil Space Program, Vol.I: Organizing for Exploration* (Washington, DC: NASA History Division, 1995), 213–229.

7. For example, NRO reconnaissance technology was secretly transferred to NASA's Lunar Orbiter program and the US Geological Survey acquired both CORONA photography and technologies to best use the photography for mapping purposes. See especially: John Cloud, *Hidden in Plain Sight: CORONA and the Clandestine*

Geography of the Cold War, PhD diss. (Santa Barbara: University of California, 1999), 256–264; R. Cargill Hall, *SAMOS to the Moon: The Clandestine Transfer of Reconnaissance Technology Between Government* Agencies (Chantilly, VA: NRO Office of the Historian, 2001); Dwayne A. Day, "Mapping the Dark Side of the World, Part 1: The KH-5 ARGON Geodetic Satellite," *Spaceflight* 40 (1998): 264–269; Dwayne A. Day, "Mapping the Dark Side of the World, Part 2: Secret Geodetic Programmes After ARGON," *Spaceflight* 40 (1998): 303–310.

8. "Serendipity" refers to the process of accidental discovery in science. In many cases, research questions are oriented toward other goals and uses than those that the actual discovery reveals. See for example: Robert K. Merton and Elinor Barber, *The Travels and Adventures of Serendipity: A Study in Sociological Semantics and the Sociology of Science* (Princeton: Princeton University Press, 2004); Royston Roberts, *Serendipity: Accidental Discoveries in Science* (New York: Wiley, 1989).

9. The notion "environmental surveillance" refers to all observation, measurement, and detection technologies, which allow to describe on a more or less global-scale environmental phenomena.

10. See on this point: B.V. Vinogradov, "Kosmicheskaya fotografiya dlya geografiicheskogo izucheniya Zemli," *Izvestiya Vsesoyuznogo Geografiicheskogo Obshchestva* 98 (1966): 101–111; translation (by NASA): B.V. Vinogradov, *Space Photography for the Geographical Study of the Earth*, NASA TT-F-10246 (Washington, DC: NASA, 1966). The citation is from the Anglo-Canadian geographer Brian Bird: J. Brian Bird and A. Morrison, "Space Photography and Its Geographical Applications," *Geographical Review* 54 (1964): 463–486, citation 468.

11. Thor Bergstralh, "Photography from the V-2 at Altitudes Ranging up to 160 Kilometers," in Homer E. Newell and Joseph W. Siry (eds.), *Upper Atmosphere Research Report no. IV*, NRL report R-3171 (Washington, DC: Naval Research Laboratory, 1947), 119–130, citation 119; Bergstralh's chapter is a reprint of a report that was published beforehand individually as NRL report R-3083, 1947.

12. David DeVorkin cites a large range of applications in his excellent book: David DeVorkin, *Science with a Vengeance: How the Military Created the US Space Sciences After World War II* (New York: Springer, 1992). A few NRL (Naval Research Laboratory) reports document also very well those efforts; one in particular will be discussed later.

13. A good overview is given by Homer Newell in his chapter on "High-altitude Photography," in Homer E. Newell, *High Altitude Rocket Research* (New York: Academy Press, 1953), 283–288, especially 283–284. See also: Robert Poole, *Earthrise: How Man First Saw the Earth* (New Haven, CT: Yale University Press, 2008), 64.

14. Newell, *High Altitude Rocket Research*, 283.

15. Stanley M. Greenfield and William W. Kellogg, *Inquiry Into the Feasibility of Weather Reconnaissance From a Satellite Vehicle*, report no. R-365 (Santa Monica: RAND Corporation, 1951), 14.

16. *The Washington Post* reproduced two of the spectrograms on its front page of its October 30, 1946, issue. Other newspapers adopted enthusiastically the story, see: DeVorkin, *Science with a Vengeance*, 143–144.

17. DeVorkin, *Science with a Vengeance*, 1–6.

18. See: ibid., especially Chapter 9. See also: Ronald E. Doel, "Quelle place pour les sciences de l'environnement physique dans l'histoire environnementale," *Revue d'histoire moderne et contemporaine* 56 (2009): 137–164, here 149.

19. Ronald E. Doel, "Constituting the Postwar Earth Sciences: The Military's Influence on the Environmental Sciences in the USA After 1945," *Social Studies of Science* 33 (2003): 635–666, here 638.

20. See: Ryan Edgington, "An 'All-seeing Flying Eye': V-2 Rockets and the Promises of Earth Photography," *History and Technology* 28 (2012): 363–371.

21. Meteorological analysis derived from rocket imagery, illustrated by V-2 photographs, was first suggested by USAF meteorologist D. L. Crowson, "Cloud Observations From Rockets," *Bulletin of the American Meteorological Society* (1949): 17–22.

22. Jet stream analysis became next to cyclone tracking one of the major application fields. For an example of visual jet stream analysis, see: US Navy, *Weather Analysis from Satellite Observations* (Norfolk: Navy Weather Research Facility, 1960), 30–31.

23. See: Pamela E. Mack and Ray A. Williamson, "Observing the Earth from Space," in Logsdon, *Exploring the Unknown Vol.III*, 155–177, here 156.

24. See on this point in more detail especially: Roger D. Launius, "What Are Turning Points in History, and What Were They for the Space Age?" in Steven J. Dick and Roger D. Launius (eds.), *Societal Impact of Spaceflight* (Washington, DC: NASA History Division, 2007), 19–39, here 31.

25. Douglas Aircraft Company, *Preliminary Design of an Experimental World-Circling Spaceship*, report no. SM-11827 (Santa Monica: Douglas Aircraft Company, 1946), 11.

26. The original color photomontage is today part of the National Air and Space Museum collection, NASM object A19620042000. For a short description and black-and-white reproduction of the mosaic, see also: Edgar M. Cortright, *Exploring Space with a Camera* (Washington, DC: NASA, 1968), 4–5.

27. For a detailed description, see: Lester F. Hubert and Otto Berg, "A Rocket Portrait of a Tropical Storm," *Monthly Weather Review* 83 (1955): 119–124.

28. See: Anon., "A 100 Mile High Portrait of Earth," *Life Magazine* 39 (5 September 1955), 10–11.

29. See: Harry Wexler, "Observing the Weather from a Satellite Vehicle," *Journal of the British Interplanetary Society* 7 (1954): 269–276, citation 269. The photographic mosaic also appeared in publications of the US Weather Bureau, promoting the use of satellite vehicles in meteorology. See for example: US Department of Commerce Weather Bureau, *Meteorological Satellites — Global Weather Observers* (Washington, DC: Department of Commerce, 1959).

30. Hubert and Berg, "A Rocket Portrait of a Tropical Storm," 122.

31. Contouring, in the atmospheric sciences jargon, means connecting data points of equal value. In general, contour analysis allows to understand how different geophysical variables change in time and space.

32. See: Hubert and Berg, "A Rocket Portrait of a Tropical Storm," 122.

33. The notion refers to the nature of instrumental logic which is proper to each visualization technology. Each visualization technology allows to see the world in a different way and it produces therefore knowledge that is specific to the instrumentation mobilized. See on this point in more detail: Grevsmühl, "Epistemische Topografien," 263–279.

34. Martin Rudwick was among the first historians who insisted on the importance of a common visual language in the making of a scientfic discipline, which was in his case geology around 1830, see: Martin Rudwick, "The Emergence of a Visual Language for Geological Science 1760–1840," in Martin Rudwick (ed.), *The New Science of Geology: Studies in the Earth Sciences in the Age of Revolution* (Aldershot, Burlington: Ashgate, 2004 [1976]), 149–195.

35. Greenfield and Kellogg, *Inquiry Into the Feasibility of Weather Reconnaissance*, 22. Richard Hamblyn has shown how Luke Howard introduce in England, in 1802, the vocabulary (which became general standard) to describe clouds mostly in function

of their appearance—as seen from the ground—and altitude: Richard Hamblyn, *The Invention of Clouds: How An Amateur Meteorologist Forged the Language of the Sky* (London: Picador, 2001). However, a new descriptive section concerning specifically cloud classification as seen from *above* was included only in 1975 in the standard atlas meteorologist use. The photographic volume accompanying the first, descriptive volume took even longer to prepare—it appeared only in 1987: World Meteorological Organization, *International Cloud Atlas: Manual on the Observation of Clouds and Other Meteors*, WMO-no. 407, vol. 1 (text), vol. 2 (photographs) (Geneva: WMO, 1987 [1975]). Howard's classification system remains the main standard for small-scale phenomena. Leopold Kletter mentions some of the new cloud types which may only be observed from a satellite: Leopold Kletter, "Die praktische Auswertung der Bildsendungen der Wettersatelliten," *Schriften des Vereins zur Verbreitung naturwissenschaftlicher Kenntnisse* 110 (1970): 23–35.

36. This point is discussed in more detail, well before the launch of TIROS, in: William K. Widger and Chan N. Touart, "Utilization of Satellite Observations in Weather Analysis and Forecasting," *Bulletin of the American Meteorological Society* 38 (1957): 521–533.

37. For a typical example of a mosaic of TIROS photographs superimposed upon surface analysis and compared with conventional analysis, see: US Navy, *Weather Analysis from Satellite Observations*, 44.

38. The question of integration is for example discussed in more detail in a historical perspective in: Amy Dahan, "Putting the Earth System in a Numerical Box? The Evolution from Climate Modeling toward Global Change," *Studies in History and Philosophy of Modern Physics* 41 (2010): 282–292. Today, the high level of integration, which was put into place in climatology, meteorology's complementary discipline, is seen with quite some reticence.

39. DeVorkin, *Science with a Vengeance*, 145.

40. Harry Wexler, "The Satellite and Meteorology," *Technical Session Preprints of the American Astronautical Society*, Preprint no. 104254 (1956): 1–15, citation 8. On the painting, see especially the important work of Jim Fleming: James R. Fleming, "A 1954 Color Painting of Weather Systems as Viewed From a Future Satellite," *Bulleting of the American Meteorological Society* 88 (2007): 1525–1527; James R. Fleming, "Earth Observations from Space: Achievements, Challenges, and Realities," in Steven J. Dick (ed.), *NASA's First Fifty Years: Historical Perspectives* (Washington, DC: NASA History Division, 2010), 543–562, especially 548–549; James R. Fleming, "Polar and Global Meteorology in the Career of Harry Wexler," in James R. Fleming, Roger D. Launius and David H. DeVorkin (eds.), *Globalizing Polar Science: Reconsidering the International Polar and Geophysical Years* (New York: Palgrave, 2010), 225–241, here 232. Wexler's color painting is also discussed in: P. Krishna Rao, *Evolution of the Weather Satellite Program in the U.S. Department of Commerce—A Brief Outline*, NOAA technical report NESDIS 101 (Washington, DC: Department of Commerce, NOAA, 2001), 8.

41. For biographical information on Wexler, I refer to Wexler's file in the archives of the International Geophysical Year: National Academy of Sciences, Archives of the IGY, "Series 13: IGY Personnel—Antarctic: Wexler, Harry, 1955–1957."

42. Alexander Linke and Dominique Rudin, "The Earth as Seen from Apollo 8 in Space," *Rheinsprung 11: Zeitschrift für Bildkritik* 1 (2011): 147–156.

43. Robert Poole traces this history in great detail in: Robert Poole, *Earthrise: How Man First Saw the Earth* (New Haven: Yale University Press, 2008).

44. See: Wexler, "Observing the Weather from a Satellite Vehicle," 269–276. The painting is reproduced as supporting document at the end of chapter 2 in: Logsdon, *Exploring the Unknown Vol.III*, 177–183.

45. William K. Widger, "Satellite Meteorology—Fancy and Fact," *Weather* 16 (1961): 47–55.

46. Wexler, "The Satellite and Meteorology," 10.

47. Ibid., 10.

48. Ibid., 12.

49. The scientific painting is therefore part of a larger historical development that may be observed all along the age of mechanical objectivity in which the principal function of the painting could only survive in scientific niches where classification and typology were indispensable; see: Peter Galison, "Judgment against Objectivity," in Caroline Jones and Peter Galison (eds.), *Picturing Science, Producing Art* (New York, London: Routledge, 1998), 327–359.

50. The term planisphere refers to the map-like view of the Earth as a result of the convergence of photography and flight during the nineteenth century, as put into pratice by early balloonists-photographers such as Nadar. The initial objective of Nadar's perpendicular photography was to elaborate *maps*, one of two common meanings of the term planisphere. See on this point in more detail: Sebastian Grevsmühl, La Terre vue d'en haut: l'invention de l'environnement global (Paris: Seuil, 2014); Sebastian Grevsmühl, *A la recherche de l'environnement global: De l'Antarctique à l'Espace et retour*, PhD thesis in history of science (Paris: Ecole des Hautes Etudes en Sciences Sociales, 2012), 240–243.

51. Weather control was until the early 1970s a central theme in meteorology, see for example: Jacob Hamblin, *Arming Mother Nature: The Birth of Catastrophic Environmentalism* (New York: Oxford University Press, 2013); James R. Fleming, *Fixing the Sky. The Checkered History of Weather and Climate Control* (New York: Columbia University Press, 2010); Chunglin Kwa, "The Rise and Fall of Weather Modification: Changes in American Attitudes toward Technology, Nature, and Society," in Clark A. Miller and Paul N. Edwards (eds.), *Changing the Atmosphere: Expert Knowledge and Environmental Governance* (Cambridge, MA: MIT Press, 2001), 135–165.

52. Wexler, "Observing the Weather from a Satellite Vehicle," 269. Especially TIROS-III, launched during the US hurricane season, would acquire that reputation: Leopold Kletter, "Meteorologische Satelliten erforschen das Weltwetter," *Schriften des Vereins zur Verbreitung naturwissenschaftlicher Kenntnisse* 103 (1963): 1–17, here 7.

53. Greenfield and Kellogg, *Inquiry Into the Feasibility of Weather Reconnaissance*, 1. See also: Hall, *A History of the Military Polar Orbiting Meteorological Satellite Program* as well as: Stephen B. Johnson, "The Political Economy of Spaceflight," in Steven J. Dick and Roger D. Launius (eds.), *Societal Impact of Spaceflight* (Washington, DC: NASA History Division, 2007), 141–191, especially 171.

54. Merton E. Davies and William R. Harris, *RAND's Role in the Evolution of Balloon and Satellite Observation Systems and Related U.S. Space Technology*, RAND report R-3692-RC (Santa Monica: RAND Corporation, 1988), 23. A detailed description of the historical dynamics between government bodies and the different military members can be found in: Hall, "Origins of U.S. Space Policy," 213–229. See also: Dwayne A. Day, "Invitation to Struggle: The History of Civilian-Military Relations in Space," in John M. Logsdon (ed.), *Exploring the Unknown: Selected Documents in the History of the U.S. Civilian Space Program. Vol.II: External Relationships*, NASA SP-4407 (Washington, DC: NASA, 1996), 233–270.

55. Hall, "Origins of U.S. Space Policy," 222.

56. Walter A. McDougall, *The Heavens and the Earth: A Political History of the Space Age* (New York: Basic Books, 1985), 117–118, 349–352. See also: Paul N.

Edwards, "Meteorology as Infrastructural Globalism," *Osiris* 21 (2006), 229–250, especially 245. See also Launius in this collection.

57. For a brief historical overview on World Weather Watch, see: James R. Rasmussen, "Historical Development of the World Weather Watch," *WMO Bulletin* 52 (2003): 16–25.

58. CORONA was the codename of the first US photoreconnaissance satellite system that successfully returned during a period of twelve years (1960–1972) a large number of film capsules that were collected midair by specially equipped cargo planes and subsequently developed and interpreted by a USAF and CIA personnel. For a concise overview, see: David C. Arnold, "Corona," in Stephen B. Johnson (ed.), *Space Exploration and Humanity: A Historical Encyclopedia*, vol. 2 (Santa Barbara: ABC-CLIO, 2010), 873–875. See also: Dwayne A. Day, "The Development and Improvement of the CORONA Satellite," in Dwayne A. Day, John M. Logsdon and Brian Latell (eds.), *Eye in the Sky: The Story of the Corona Spy Satellites*, (Washington, DC: Smithsonian Institution Press, 1998), 48–85, here 53. Many now declassified records relating to the CORONA program and other formerly secret US photoreconnaissance programs may be accessed on the NRO website: http://www.nro.gov/FOIA/declass/index.html.

59. For a concise overview, see: James C. Mesco, "Defense Meteorological Satellite Program," in Johnson, *Space Exploration and Humanity*, vol. 2, 873–875.

60. See: Davies and Harris, *RAND's Role in the Evolution of Balloon and Satellite Observation Systems*, 25; Day, "Invitation to Struggle," 243 and 246.

61. See: NASA, *TIROS: The System and its Evolution*, NASA technical report NASA-TM-X-56696 (Washington, DC: NASA, 1965), 4. For an overview, see: David Leverington et al., "Tiros," in Johnson, *Space Exploration and Humanity*, vol. 1, 360–363.

62. US Navy, *Weather Analysis from Satellite Observations*, 32.

63. Report to Congress from the President of the United States January 1, to December 31, 1959, p. 26. See the DoD-NASA agreement in: John M. Logsdon (ed.), *Exploring the Unknown, Vol. III: Using Space*, 203–204.

64. RCA deliberately downgraded the TIROS resolution for the simplified civilian version, see: Richard Leroy Chapman, *A Case Study of the United States Weather Satellite Program: The Interaction of Science and Politics*, PhD thesis in political sciences (Syracuse: Syracuse University, 1967); cited in Conway, *Atmospheric Science at NASA*, 27. Information confirmed in: NASA, *TIROS: The System and its Evolution*, 7.

65. For a more detailed technical description, see the explanation of the TIROS I television system in the NSSDC Master Catalog: http://nssdc.gsfc.nasa.gov/nmc/experimentDisplay.do?id=1960–002B-01 (accessed October 24, 2013). In the case of TIROS the term satellite *photography* can be misleading. The TIROS imaging system was not based on photography but on vidicon tubes. The vidicon images were relayed to one of the two ground stations where they were visualized on a TV-screen that was eventually photographed by a 35mm camera.

66. See the contribution of Robert Popham of the US Weather Bureau to the first conference on sea ice: Maurice D. Baliles and Herbert Neiss (eds.), *Conference on Satellite Ice Studies*, Meteorological Satellite Laboratory report no. 20 (Washington, DC: US Weather Bureau, 1963), 23.

67. For the first sea ice report, see: David Q. Wark and Robert W. Popham, "Tiros I Observations of Ice in the Gulf of St. Lawrence," *Monthly Weather Review* 88 (1960): 182–186. For an analysis of the first scientific advances in this young domain, see: David Q. Wark and Robert W. Popham, "The Development of Satellite Ice Surveillance Techniques," in Harry Wexler and James E. Caskey

(eds.), *Proceedings of the First International Symposium on Rocket and Satellite Meteorology* (Amsterdam: North-Holland, 1963), 415–418.

68. Schnapf, "The TIROS Global System," 156.

69. NASA, *Astronautical and Aeronautical Events of 1962: Report of the National Aeronautics and Space Administration to the Committee on Science and Astronautics, U.S. House of Representatives, Eighty-eighth Congress, First Session* (Washington, DC: Government Printing Office, 1963), 14.

70. Library of Congress, *Meteorological Satellites, Staff Report Prepared for the Use of the Committee on Aeronautical and Space Sciences United States Senate, Eighty-seventh Congress, Second Session* (Washington, DC: Government Printing Office, 1962), 39.

71. As in many other cases, the production of reliable maps demanded nevertheless the mobilization of traditional aerial reconnaissance. See the report: Defense Research Board, *Project TIREC, February-April 1962, Preliminary report by the Canadian Participating Agencies, Canada*, Defense Research Board, Directorate of Physical Research (Geophysics), Report No. Misc. G-11, February 1963. A similar argument is made in the following study: Baliles and Neiss, *Conference on Satellite Ice Studies*, 90.

72. Baliles and Neiss, *Conference on Satellite Ice Studies*, 26.

73. Before becoming climate change skeptic, Fred Singer fought other highly ambiguous ideological battles, such as the acid rain debate, the ozone depletion controversy as well as health dangers associated to passive smoking, defending mostly a position contradicting general scientific consensus, see: Naomi Oreskes and Erik M. Conway, *Merchants of Doubt* (New York: Bloomsbury Press, 2010). See also: Naomi Oreskes and Erik M. Conway, "Challenging Knowledge: How Climate Science Became a Victim of the Cold War," in Robert N. Proctor and Londa Schiebinger (eds.), *Agnotology: The Making and Unmaking of Ignorance* (Stanford: Stanford University Press, 2008), 55–89.

74. S. Fred Singer and Robert W. Popham, "Non-meteorological Observations from Weather Satellites," *Aeronautics and Aerospace Engineering*, 1 (1963): 89–92; Paul D. Lowman, *A Review of Photography of the Earth from Sounding Rockets and Satellites*, NASA TN D-1868 (Washington, DC: NASA, 1964), 18.

75. From 1961 onward, satellite data played for the US Navy a major role in the planning of Antarctic logistics, see: NASA Space Applications Program Office, *A Survey of Space Applications*, NASA SP-142 (Washington, D.C.: NASA, 1967), 90; Alan T. Waterman, "New Horizons for the Atmospheric Sciences," *Annals of the New York Academy of Sciences*, 95 (1961): 688–696, especially 689. See also: Radio Corporation of America, *TIROS: A Story of Achievement*, AED P-5167A (Princeton: RCA, February 1964).

76. See for a telling example: David W. J. Thompson and Susan Solomon, "Interpretation of Recent Southern Hemisphere Climate Change," *Science* 269 (2002): 895–899.

77. One may note that law theorists also have analyzed this phenomenon since sea ice makes it extremely difficult to establish continental limits and territorial extension into the sea: Christopher C. Joyner, *Antarctica and the Law of the Sea* (Dordrecht: Martinus Nijhoff, 1992), 14–16. "Sea ice" is therefore, at least from a law viewpoint, an object incredibly difficult to define.

78. TIREC was therefore a key source of environmental information in relation to Cold War concerns in fields like geophysics, seismology, oceanography, cartography and glaciology.

79. See the list of participants in: Baliles and Neiss, *Conference on Satellite Ice Studies*, iv–vi.

80. Turchetti et al., "Accidents and Opportunities," 419.
81. For the large spectrum of applications developed in oceanography from the very beginning of the Space Age, see the conference report: Gifford C. Ewing (ed.), *Oceanography from Space, Proceedings of the Conference on the Feasibility of Conducting Oceanographic Explorations from Aircraft, Manned Orbital and Lunar Laboratories. Held August 24–28 at Woods Hole, Massachusetts, USA*, ref. no.65–10 (Woods Hole: Woods Hole Oceanographic Institution, 1965). For two examples of very specific uses outside of the initial application context, see: Alastair Morrison and M. Christine Chown, *Photography of the Western Sahara Desert from the Mercury MA-4 Spacecraft*, NASA Contractor report CR-126 (Montreal: Quebec, 1964); Paul M. Merifield and James Rammelkamp, *Terrain in Tiros Pictures*, contract no. NAS 5–3390, LR 17848 (Burbank: Lockheed-California Company, 1964). On scientific applications of photographs of the Gemini program, see for example: Paul D. Lowman, James A. McDivitt, and Edward H. White, *Terrain Photography on the Gemini IV Mission: Preliminary Report*, NASA Technical Note D-3982 (Washington, DC: NASA, 1967).
82. See the contribution of two representative of the US Geological Survey: William A. Fischer and Charles J. Robinove, "A Rationale for a General Purpose Earth Resources Observation Satellite," in *Proceedings of University of Washington Remote Sensing Symposium* (Washington, DC: Washington University, 1968), 36.

Section V

From Surveillance to Environmental Monitoring

Chapter 9

Observing the Environmental Turn through the Global Environment Monitoring System

Soraya Boudia

The Global Environment Monitoring System (GEMS), currently involving the collaboration of 140 countries across the world, was conceived during the late 1960s within the wider context of growing environmental consciousness—and the availability of instruments and programs that could provide a truly global data set. Under the aegis of the United Nations Environment Programme (UNEP), its mission is to collect and process data from transnational monitoring systems observing the effects of human activities on the environment and health. GEMS is based on the expertise and infrastructure of specialized United Nations (UN) agencies, and encompasses a number of scientific and professional communities.

By exploring the origins of GEMS, this chapter sheds light on three overlapping issues concerning the expansion of geoscientific research during the Cold War. The first is the establishment of what Geoffrey Bowker, Susan Leigh Star, and others have described as "global infrastructures,"[1] that is, transnational technical infrastructures such as telecommunication or transport networks[2] that have facilitated globalization.[3] GEMS is itself one such infrastructure, while also being a product of a set of existing structures for surveillance and monitoring that grew tremendously during and after World War II. New technologies for remote sensing (including for instance sonar and radar) extended the capabilities and the scope of networks monitoring the environment that dated back to the early twentieth century. The birth of new technologies, from computers to satellites, allowed the collection and processing of data on an unprecedented scale. The Cold War stimulated the development of novel technological systems, while also providing geoscientists with powerful new data sets and research possibilities.[4]

The global nature of these technologies and networks relates both to the scope of their implementation, on a scale previously unprecedented, and to the type of information and images they produced of the earth. This is what Paul Edwards has labeled "infrastructural globalism."[5] Studies of these global infrastructures have highlighted the role of scientific techniques in the definition of the earth as a single entity, a globe that can be investigated, measured and visually represented as a whole.[6] In these different technical and scientific operations, a global perspective became increasingly common, particularly when events such as the International Geophysical Year (IGY, 1957–58) showed the potential for worldwide data collection. The convergence of these operations and policies, as much as developments in any specific scientific discipline, contributed to forming the idea of the earth as a global environment consisting of interdependent ecosystems.

Secondly, GEMS both reflected and contributed to the "environmental turn" in politics and society from the late 1960s. The publication of Rachel Carson's book *Silent Spring* in 1962 is widely credited with sparking interest in the destructive effect of pesticides and other industrial toxins upon the natural environment. As the decade progressed, consciousness of the impact of humans upon the earth increased. Frequently this was allied to progressive political and cultural movements, and to an expression of the politically divided planet as a potentially unified entity. But the environmental turn also had a harder edge. The Committee on the Challenges of Modern Society (CCMS), established by the North Atlantic Treaty Organization (NATO) in 1969, was in large part an acknowledgment that the legitimacy of the Cold War West would be undermined if it ignored the increasingly important issue of environmental responsibility.[7] The Club of Rome's 1972 report *The Limits to Growth* further established the environment as politically relevant. GEMS and its predecessors connected existing Cold War surveillance networks to the new environmental challenges, within an organizational mandate that stressed global human development. As the environment became a matter of mainstream political concern and the effects of humans upon it a potential threat to social stability, monitoring threats became a politically important matter.

Finally, this chapter follows Edwards in paying particular attention to what Martin Hewson has described as the final stage of "informational globalism," the conceptualization of monitoring infrastructures to know the global environment.[8] The term "globalization" is today both ubiquitous and regarded as an economic and political project rather than a neutral statement of fact. Yet it was in the 1970s, in the wake of decolonization and during an economic crisis due largely to rising oil prices, that globalization was first used to describe a world characterized by strong interdependencies.[9] A global perspective on the environment did not dictate either ignorance or transcendence of political divisions, but particularly when expressed through the medium of a global organization—the United Nations—the synergy between global environmental monitoring and global social and economic development was clear. The concept of globalization, concerning both markets and the environment, contributed to spread new ways of interpreting the changing world and political action at a global scale. Globalization in general, just like GEMS in particular, was a construction rather than an inevitable and natural development, created by actors with specific objectives.[10]

The Genesis of Global Environmental Monitoring

After World War II, a range of international organizations emerged with mandates covering diverse aspects of political, economic, cultural, and scientific activity around the world. These organizations invariably reflected contemporary power relations in addition to ideals of global collaboration. The UN, founded in San Francisco in 1945, was open to all sovereign states, but its membership was at first hardly universal—and the privileged status of five victorious wartime powers was recognized through a permanent veto in the organization's Security Council.[11] The UN soon included a raft of agencies such as the United Nations Educational, Scientific and Cultural Organization (UNESCO), the Food and Agriculture Organization of the United Nations (FAO), and the former International Meteorological Organization, reestablished as the World Meteorological Organization (WMO) in 1951 under UN auspices. Each of these bodies could not fully transcend the ideological divide of the early Cold War, especially as science in the Soviet Union remained almost entirely isolated from 1945 until after the death of Stalin in 1953.[12] International organizations did not produce global actions.

One of the first examples of truly global scientific action came with the International Geophysical Year (IGY) in 1957–58. British geophysicist Sydney Chapman and his US colleague Lloyd Berkner famously conceived this event at a gathering at the home of the American physicist James van Allen in 1950.[13] During the following seven years, Berkner worked tirelessly to bring about a major scientific event that also possessed significant value for US foreign policy.[14] The IGY's development into a truly global program of observation, monitoring, and exploration required considerable political energy under the aegis of the only recognized nongovernmental organization uniting scientists across the globe, the International Council of Scientific Unions (ICSU), and the WMO.[15] In the years after the IGY other major international scientific events took place, including the International Indian Ocean Expeditions (1960–65), the International Year of the Quiet Sun (1964–1965), and the International Biological Program (IBP, 1964–74). The WMO and ICSU jointly sponsored the Global Atmospheric Research Program (GARP, 1967–1982), which coordinated a series of large-scale experiments around the world to provide a stronger understanding of global atmospheric processes. Both scientist- and government-led international initiatives produced infrastructures at both the technological level—instruments and techniques for measuring environmental phenomena—and at the institutional level, through new networks between researchers and their patrons. Often they relied heavily on military funding. As Ron Doel has famously argued in the case of the United States, the potential to know and control the geophysical environment made it an attractive field to military patrons.[16] Nowhere was this clearer than in the studies of the effects of nuclear weapons (geophysical as well as biological), in which vast destructive power was accompanied by a study of its effects across huge geographic areas.[17]

Projects to investigate and monitor the earth's environment became more politically relevant with the rise of a "green" consciousness. Fears about the effects of radiation emerged as a powerful force during the 1950s.[18] Jacob Hamblin has documented the lack of widespread concern over dumping radioactive waste in the oceans during the 1950s,[19] but already in 1960 public fears in France—informed by oceanographers and the famous explorer Jacques Cousteau—helped prevent such a discharge from taking place in the Mediterranean. The 1962 bestseller *Silent Spring* focused public attention on the potential for chemicals such as dichlorodiphenyltrichloroethane (DDT) to impact on the environment, particularly birdlife.[20] The World Wildlife Fund was established in 1961 not to identify environmental dangers—which its founders felt were already clear—but rather to obtain funding to tackle them.[21] By 1969 public awareness of environmental protection as a political issue had grown to the point where NATO founded the CCMS, in which environmental issues were directly linked to the stability and progress of Western societies.[22] The establishment of the Environmental Protection Agency (EPA) in the United States followed shortly, while the European Economic Community began to develop a framework for environmental regulation too.[23] These initiatives also compelled organizations that, from the IGY onward, had pioneered international scientific collaboration to redefine their programs. For instance, the ICSU now appointed a Scientific Committee on Problems in the Environment (SCOPE) to put one of the IBP's key objectives, namely the study of human influence on the environment, "on its own permanent footing."[24]

Growing appreciation of the global nature of earth systems led to a matching appreciation for pollution and environmental contamination as global issues.[25] Pollution was no longer local but could now be understood as affecting the entire planet. The consequences were not only immediate; they could be felt decades after exposure or contamination, and over several generations. Transnational approaches to environmental issues were increasingly discussed, including reflections on how the combined scientific, political, and administrative character of the challenges required regulatory and institutional reconfigurations.[26] Nor was awareness of environmental threats always matched by practical knowledge of how to address them. Famously in March 1967 the SS *Torrey Canyon* sank near the British coast, releasing more than 90,000 tons of crude oil into Atlantic waters, which spread across an area 35 miles long and 20 miles wide. Lack of knowledge about how to respond to such an environmental crisis led the Royal Air Force and the Royal Navy to resort to burning off the fuel by setting it ablaze with rockets.[27] And it was only in 1968 that the Japanese government officially recognized the deleterious health effects of consuming fish contaminated by mercury released from chemical plants, even though the first symptoms of mercury poisoning among the population in the harbor city of Minamata had been recorded 20 years earlier.[28]

Realizing that environmental threats were becoming an important social, economic, and political issue, on December 3, 1968, the UN General Assembly voted unanimously to convene a Conference on the Human Environment. This marked a formal recognition that the peak global organization ought to play a role in addressing what was by now perceived as a global matter. The formal resolution described the environmental threats of the present as risks to "dignity" and "basic human rights" and linking the health of the environment to "sound economic and social development."[29] The dedicated "Conference on the Human Environment" would be convened in 1972, with the goal—as its name suggested—of examining the impact of environmental issues upon social and economic development. The international scientific and technical organizations beneath the UN umbrella were recognized as sources of advice while making clear that a wide variety of other agencies could contribute. Once it had been approved, the conference quickly formed a frame for other UN activities with relevance to the environment, such as its ongoing attempts to review issues related to marine pollution.[30]

The Conference on the Human Environment, which also became known as the "First Earth Summit," took place in Stockholm from June 5–16, 1972. Its organization was led by the Canadian entrepreneur Maurice Strong, who had long been an advocate of the UN and its operations. Part of the preparation for the conference involved Strong commissioning a report by the French-American biologist Rene Dubos and the British economist Barbara Ward and published in 1971, titled *Only One Earth*.[31] Like *The Limits to Growth*, published the following year, *Only One Earth* popularized ideas of environmental limits with consequences for human civilization in a manner that addressed a public beyond august experts. It drew extensively on the image of "Spaceship Earth," which Ward had done much to popularize in the late 1960s, emphasizing the fragile and limited nature of humanity's habitat.[32]

If Ward and Dubos helped provide the Stockholm Conference with a guiding vision of the human environment, the conception of a mechanism to monitor and assess the environment globally was yet to develop. This arose from a coming together of "green" environmental sentiments and more traditional management procedures in the organization of international scientific projects, with links of varying strength to the array of existing international scientific organizations. Carroll Wilson, professor of management at the Sloan School within the Massachusetts Institute of Technology (MIT), was a key figure in this effort. Wilson had served as the first general manager of the United States Atomic Energy Commission (AEC) and president of the Climax Uranium Company but like Strong, he had a long-standing interest in broader political issues. This led him to the Club of Rome, along with around 70 scientists, educators, economists, and industrialists from different countries,[33] and then to assemble a group of experts at MIT with backgrounds from the natural and social sciences in order to tackle environmental challenges. These experts spent July 1970 discussing a variety of problems from pollution to the finite nature of natural resources, and how those problems might be addressed. Funding came from US government agencies that had been sponsoring science throughout the Cold War both nationally and internationally, including the Ford Foundation, the Rockefeller Foundation, and the Sloan Foundation. Wilson subsequently published the results in a landmark report, the *Study of Critical Environmental Problems* (SCEP).[34] This document highlighted that an effective response to environmental threats was premised upon the edification of a global infrastructure for collecting environmental data. The report also called for a harmonization and standardization of existing structures and methods. Perhaps most importantly, its contributors stressed that "the global" was not only the geophysical scale at which information ought to be gathered, but also the political scale at which solutions should be conceived and adopted.

Even so, the SCEP report effectively called for a mobilization of those scientists and organizations that were already active in the field of global data collection because they had pioneered it since the IGY. In December 1970, Strong sought to implement the SCEP recommendations by commissioning the ICSU-sponsored SCOPE to draw up a report and recommend the design, parameters, and technical organization needed for a coherent global environmental monitoring system.[35] This implementation made visible the important connection between the new environmental monitoring plans and the Cold War surveillance. So did the involvement of prominent scientists such as Roger Revelle and Thomas Malone, each of whom played key roles in both SCOPE and the wider framework of Cold War geoscience. Revelle had served as director of Scripps Institution of Oceanography in La Jolla, California, from 1950 to 1964, overseeing the Institution's continued rise to one of the most important centers for marine research in the world (thanks in large part to military funding).[36] Revelle was personally involved with ocean and atmospheric monitoring research, including tracking fallout from the Bikini Atoll atomic tests, and a fixture on international oceanographic organizations before founding the Harvard University Center for Population Studies. The meteorologist Thomas Malone was a former chairman of the American Meteorological Society (1960–1961) and of the American Geophysical Union (1962–1964) in addition to being the ICSU vice-chairman.

Revelle and Malone both wanted SCOPE to stimulate large international collaborative projects, in line with the IGY tradition and the raft of organizations, including the Scientific Committee on Antarctic and Ocean Research (SCAR and SCOR), that they had contributed to establish. They convinced the Ford Foundation to fund an assessment of international studies and monitoring of the environment by gathering and analyzing available data on global and regional effects. At its first meeting in Madrid, in September 1970, SCOPE set up a commission to investigate the methodology of monitoring, including the selection of suitable parameters, to ensure comparability of methods and coordination of systems. In particular, the commission aimed to design an integrated monitoring system for air, water, soils, and humans, taking into consideration already existing activities. Another goal was to design a network of background stations far from population centers, while also considering the development of other methods for monitoring areas of high human impact.[37]

The 1971 SCOPE report on the planned global monitoring system formed the basis for discussion at the Stockholm conference of the following year. The report outlined the state of the art on environmental monitoring and assessment along with 18 recommendations. These included how to organize the overall monitoring system, including an analysis of how many and what type of observation stations would be needed and where they ought to be placed, plus what type of data ought to be collected in the first instance—prioritizing the study of pollutants such as carbon dioxide, lead, mercury and cadmium.[38] Finally, the report stressed the importance of promoting research and pilot studies for the development of new and more efficient monitoring systems. The SCOPE experts called for the integration of these data into a global system and proposed the creation of a new infrastructure for global environmental data collection, under the aegis of the UN.[39]

The Stockholm conference ratified the SCOPE recommendations effectively setting the goals of what would become GEMS, while also laying the groundwork for what would become the UN Environment Programme (UNEP).[40] Charged specifically with considering global environmental issues—and initially led by Maurice Strong—this body was headquartered in Nairobi. The choice of location reflected a growing sense that the environmental issues were important for the developing as well as the developed world, a sentiment that pervaded the Stockholm Conference as well as appearing in official UN documentation.

At the same time, UNEP—like SCOPE, and GEMS—was a product of the Cold War surveillance imperative in terms of both material and human infrastructures. The career path of Robert A. Frosch, who became assistant director of UNEP's scientific and technical projects via a stint as assistant secretary of the Navy for Research and Development, is indicative of this legacy. The American physicist's previous scientific endeavors placed him firmly among the Cold Warriors. Privy to secret US Navy research projects, including the global undersea surveillance network ARTEMIS, he also served as director of Nuclear Test Detection in the US project Vela Uniform.

The overarching project within which GEMS was located—Earthwatch—was outlined in 1972 and established at the first extraordinary meeting of the UNEP held in Geneva on June 12–22, 1973.[41] Drawing upon the SCOPE study, Earthwatch aimed to provide an efficient environmental assessment mechanism through monitoring, research, evaluation, and information exchange.[42] Financial resources for the program came from UNEP's initial endowment of $100 million over five years. The specific idea for GEMS emerged during an intergovernmental meeting on monitoring that took in Nairobi in 1974 as the monitoring component in the Earthwatch program.[43] GEMS would this aim to make assessments on a variety of environmental issues including atmospheric pollution and its impact on climate, contaminants in the food chain and agriculture, ocean pollution, and forecasting disasters such as earthquakes. Consistent with the SCOPE's recommendations, GEMS was not intended to collect and store data from scratch, but rather to work with existing data collection platforms through the so-called Information Referral System, a UNEP device to "encourage worldwide interchange of environmental information through the design, coordination, and operation of a system of referral to sources of environmental information and data."[44] In particular, it was expected that during its development both Earthwatch and its monitoring mechanism, GEMS, would draw extensively on the environmental programs of UN organizations, partly to avoid duplicating research and also to coordinate the ongoing exercises more effectively.

As the SCOPE report had suggested, integration as much as initiation was the key to constructing an effective global monitoring system. Earthwatch thus brought together work hitherto performed by over 30 governmental and nongovernmental organizations, including UN and ICSU agencies and committees.[45] But Joshua Howe has argued that Earthwatch was originally intended to have two complementary objectives: to coordinate, expand, and reorganize existing research and data collection; and to identify problems and issues requiring action on a global scale, thus defining guidelines for the development of new collaborative research.[46] The US delegates, inspired by the model of scientific cooperation based on the IGY experience as well as the collaborative, intergovernmental model of governance imposed in Antarctica from 1961, proposed the creation of a comprehensive international framework for the evaluation of environmental problems on a global scale.[47] Howe notes that American scientists soon came to view Earthwatch as a "clearinghouse" for existing initiatives, however, rather than a vehicle for more imaginative research. The first of its objectives—coordination—clearly trumped the second.[48]

Surveillance of the Environment and
Environmental Diplomacy

The most visible actors in the genesis of the GEMS were scientists. Revelle and Malone played an active role in promoting a global monitoring system. At the same time, their contribution reflected the political importance that the earth and atmospheric sciences had already acquired in the United States,[49] in addition to the growth of the ecosystem concept in academic ecology (and appreciation for the role both earth and life sciences could play in evaluating human effects).[50] Most importantly for the purpose of this chapter, American involvement in GEMS reflected the intense mobilization of the US scientific community in environmental policymaking at an international level.[51] Their involvement was encouraged by the US State Department in line with the Nixon administration's new "environmental diplomacy."[52]

The Vietnam conflict catalyzed anger against the United States's foreign policies both at home and abroad, undermining American capacity to effectively project ideological leadership.[53] On the domestic front, the Nixon administration saw environmental concern as a pragmatic political strategy for appealing to conservative, middle-class voters rather than student protesters: being green was not the sole province of the antiwar left.[54] On the international front, such concerns were a means of softening the association between science, warfare, capitalism, and environmental damage—particularly when the use of defoliants in Vietnam had led Swedish Prime Minister Olof Palme to speak of "ecocide" in South-East Asia.[55] The decidedly nongreen philosophy of mutually assured destruction allied science and technology to the means of global destruction (although surveillance systems could also provide mutual surveillance that potentially prevented strategic miscalculations). On the other hand, like the IGY over a decade earlier, protection of the environment through global monitoring could be pitched as an example of cooperation that bucked rather than exemplified military tensions. Nixon doubtless also recognized that if such a proposal came from the Western bloc, it had the power to strengthen the alliance while weakening ties in the Soviet alliance.[56] The environment thus became a vehicle for diplomacy in which scientists were essential actors, guarantors that global environmental monitoring was a serious and legitimate project in a symbolic sense in addition to obtaining knowledge with practical value for policymaking.

The Nixon administration's environmental diplomacy took several forms.[57] In January 1970, Secretary of State William Rogers created a new Office of Environmental Affairs, and asked all US ambassadors to promote environmental policies in the countries in which they were posted, and to inform the Department of State of all the initiatives and policies developed by foreign governments.[58] The administration ensured that environmental issues became a concern for North Atlantic Treaty Organization (NATO), which had already sponsored scientific research (particularly in oceanography) for military-strategic purposes, through a new Committee on the Challenges of Modern Society (CCMS).[59] The US government also fostered bilateral collaboration agreements with several countries, including Canada and even the Soviet Union and played a leading role in the Stockholm conference.[60]

Nevertheless, the broader dynamics of Cold War conflict pervaded the Stockholm Conference, as they did practically every formal international forum. Since the German Democratic Republic did not have a seat at the UN, it was not invited, prompting the Soviet Union and most Eastern Bloc states to boycott the conference.[61] And although developing countries were invited to contribute, echoing the concern for global economic development that featured in the original UN resolution, the floor of the Stockholm Conference remained a forum under the control of the First World. An amendment denouncing environmental degradation associated with expansionism, apartheid, colonialism, and racism, tabled by a representative of the United Republic of Tanzania and supported by a further 12 African states, was declared an "extraneous matter."[62] When the request to condemn the use of "chemical and biological agents in wars of aggression which degrade man and his environment" was put forward, the Vietnam-sore US delegation proposed to put it "in a wider framework," thus leading to the bland statement that "states must intensify efforts to maintain international peace" in the final resolution.[63]

Unsurprisingly, when discussions began in 1972 about the establishment of an integrated monitoring work of Earthwatch—GEMS—the negotiations were often fraught. Criticism soon surfaced within and from outside since the surveillance of environment, like environmental diplomacy, appeared more like a device in the hands of a few US and UN actors rather than a truly global mechanisms of assessment, management, and decision making. Representatives of the specialized UN organizations such as the WMO expressed reluctance to cede authority to an embryonic agency, especially when existing monitoring projects (such as the WMO's GARP program and the WMO Integrated Global Ocean Station System [IGOSS]) had been running satisfactorily for several years and would continue to do so irrespective of GEMS.[64] Ambition-driven rivalries are not uncommon in the world of international agencies, but in this case conflicts were exacerbated by the fact that despite its headquarters being in Nairobi and its mission embracing the developing world, sponsorship of the UNEP program came almost entirely from the US administration and the governments of another few developed countries.[65] It was thus difficult to envision GEMS as a sufficiently autonomous institution to warrant an overarching role within the existing institutional landscape.

GEMS was approved in 1972 and monitoring operations begun in 1976, after substantial planning discussions. Drawing on the SCOPE report, UNEP established five monitoring programs: on climate, long-range transport of pollutants, health-related, ocean-related, and on terrestrial resources.[66] The UNEP Intergovernmental Meeting on Monitoring that took place in Nairobi in 1974 further refined these goals, also listing the pollutants to be monitored in a specific "order of priority."[67] But this was an order of priority that did not align with the SCOPE study in prioritizing, for instance, the monitoring of rapidly rising carbon dioxide levels and their effect on climate, doubtless recognizing the consequences of any action to reduce its levels globally for industrialized nations.[68] From 1976, more specific programs were set up concerning the collection of data on the air (GEMS/Air), water (GEMS/Water), and the food chain (GEMS/Food). The data collected were processed under the responsibility of one or several organizations, depending on the program concerned.[69]

In 1975 UNEP made available to the GEMS a Programme Activity Centre (PAC) in Nairobi and appointed the former secretary of UNSCEAR, the Italian geneticist and epidemiologist Francesco Sella, as GEMS director.[70] The PAC, however, was woefully understaffed—a decision that necessarily limited GEMS's potential ambitions, to the relief of other UN agencies. GEMS would primarily be collecting information from existing UN and non-UN monitoring agencies rather than creating new ones. In particular, GEMS would routinely interact and exchange information and data with the International Atomic Energy Agency, World Health Organization (WHO), WMO, UNESCO, and FAO.[71] Through the Information Referral System, GEMS could also examine data available to ICSU's World Data centers.[72]

Existing US monitoring and science-policy organizations thus played prominent roles in the GEMS activities. For instance, the chief source of data on pollution in the GEMS climate-related monitoring program was the WMO Background Air Pollution Monitoring Network (BAPMoN) housed in the US EPA. This brought the US National Oceanic and Atmospheric Organization (NOAA) into the project, aside from UNEP.[73] While NOAA would contribute substantially to GEMS's monitoring program, other US agencies would took responsibility for reviewing it. For instance at the request of the US State Department, the International Environmental Programs Committee of the US National Academy of Sciences organized a conference on October 9–11, 1975, to assess the GEMS project.[74] The proceedings, financed by the Rockefeller Foundation, were published in the report *Design Philosophy for GEMS*.[75] The following year SCOPE published a second report on GEMS.[76]

By then however, criticism was about to surface on GEMS activities because of this overreliance on UN and US organizations that effectively reduced the space for contributions from other countries. Although the GEMS projects did not face lack of funding as such, by the end of the 1970s there were clear concerns that global economic problems would pose the biggest risk to monitoring of global environmental problems—even if the ostensible rationale was to aid human development. The surge in oil prices that followed the 1973 Arab-Israeli War catalyzed an economic crisis in the West that in turn made spending on environmental diplomacy and monitoring less politically appealing. By the end of the 1970s the United States had pushed for a reduction in direct contributions to international organizations and the targeting of specific, carefully negotiated, programs. US scientists involved in the GEMS now urged those nations sponsoring UNEP to make sure that "it maintains a catalytic capability by not exhausting available funds to support programs."[77]

The threat of further funding cuts, in addition to the existing squeeze on resources and pressure to focus on existing systems, reinforced the need for GEMS administrators to make use of existing surveillance networks that had been established during the Cold War. Monitoring of atmospheric pollution was carried out through the WMO's World Weather Watch (WWW) service. Dating back to 1963, the WWW had envisioned the concept of a global weather monitoring system comprising stations across the globe and satellites. As Clark A. Miller has argued, WWW was both a means of knowing the earth and a means of keeping up to date with the enemy by enabling meteorologists from the superpowers to meet and gauge of their respective capacities.[78] Likewise, a planned GEMS assessment of natural disasters (earthquakes and tsunami) aimed to make available risk maps for national and local authorities dealing with these disasters, but used a typical Cold War monitoring device for this purpose. The project entailed collecting information from the Worldwide Standardized Seismic Network (WWSSN), the global system of seismic stations that was conceived to monitor Soviet nuclear tests (and secretly sponsored by the Vela Uniform project that Frosch directed).[79]

When GEMS did not turn to traditional Cold War monitoring systems, its data collection strategy showed some naiveté, and from the late 1970s it came under closer scrutiny. By 1977 70 governments has registered to the IRS, thus gaining access to data and agreeing to supply information to partners. Training workshops had also taken place in 154 countries.[80] The proponents of the monitoring program hoped to achieve a global reach, but openly stated that "monitoring implies participation of many nations *without regard to their stage of economic development* [emphasis mine]," believing it possible for countries that were fragile economically and scientifically to take part. They also hoped to access local data regardless of the existence of competing local, national, or even international programs.[81] This approach was soon revealed to have critical limitations. For instance the section of the program GEMS/Air devoted to monitoring industrial pollution, which was also housed at the EPA, failed to take off due to the lack of data on the industrial sites selected. The EPA ended up using other data put together at the WHO office in Geneva instead, but, as one UNEP critical review later pointed out, the sites often bore "no relation to environmental monitoring networks within a country."[82] In 1977, law expert John W. Head thus emphasized that the GEMS's assessment function consisted of "a categorization of its responsibilities and the completion of a few specialized studies" and concluded that GEMS "apparently achieved very few of its stated goals."[83]

This criticism was responsible for the significant redefinition of GEMS's activities that began in the 1980s. In 1978 Michael D. Gwynne was appointed in the PAC and he would rise from the ranks to soon become deputy director and then, from 1989, GEMS director. Less familiar than Sella with the UN bureaucracy (he had worked for FAO from 1973) and much closer to the affairs of African countries, Gwynne helped GEMS to focus much more on ecological matters. Gwynne, a British graduate of Oxford University, had worked in Kenya from 1959 and conducted extensive research into crop production and land utilization.[84] Almost from the start Gwynne's task was to manage GEMS through a time of potential crisis, since widespread criticism of the GEMS activities eventually urged its backers to think about its future. At its seventh session, the UNEP governing council requested that Sella convene a group of government-designated experts to consult on the GEMS "mechanisms and procedures for conducting environmental assessments."[85] At a meeting in Geneva in November 1979 this group drew up a detailed plan of actions that effectively led to a new phase in the history of global environmental monitoring—one conditioned not only by the emergence of ecological science and environmental consciousness, but by the politics of international organizations and the powerful legacy of Cold War geophysical surveillance mechanisms.

Conclusions

Once the GEMS was made operational, it exemplified a science-based approach to environmental problems that drew heavily on preexisting monitoring systems laid out during the early Cold War years. The implementation of GEMS was a further step in the development of the technical systems for environmental monitoring that constructed the entire earth as a subject of research and investigation. In over 30 years of functioning, the system has collected an impressive volume of data on the state of the earth's atmosphere, water, air, and forests. It has been mobilized under five large programs corresponding to types of pollution or themes, whose importance is recognized by the United Nations Environment Programme.

GEMS is thus a constituent part of the global environmental regime that was set up from the late 1960s. By examining it we see that globalization, based partly on this regime, is at once scientific, institutional, and political. It has been marked and shaped, in the GEMS case in particular, by technology and geopolitics, which continue to determine all environmental activities at international level. The case of GEMS shows that, even after the end of the Cold War, the observation of environmental pollution and the definition of environmental policy are still shaped not only by observation techniques and techniques for collecting and processing data, but also by the geopolitical stakes involved in monitoring and controlling territories, and in the construction of political hegemony.

While the principle of GEMS was laudable, the manner in was initially executed was problematic. Its origins lay in the environmental concerns within the rich world—middle-class American readers of *Silent Spring* rather than impoverished residents of the developing world. The asymmetries in power between the rich world and the rest were inescapably inscribed upon GEMS. Such a link was particularly clear given the association between national wealth and ideological authority, especially in Western capitalism but also the industrialization-worshipping Soviet Bloc. Given that so much Cold War monitoring had its origins in the race to secure an advantage over enemy states, who could blame representatives of the states on the receiving end of surveillance conducted by others for being skeptical?

This chapter has shown how critics of GEMS emphasized its limitations and at times commented harshly on its shortcomings. But at the core of the problems lay the inequality of scientific and financial means between the world's states, which prevented GEMS from truly serving the global population it was intended to aid. This disparity was a legacy of events such as the IGY, which established international collaboration in the geosciences as global in terms of the spaces placed under surveillance, but less so in terms of the individuals and institutions that conducted the research. Specialized communities, technical networks, and data processing capabilities where concentrated within a small number of states, almost all outside the Non-Aligned world. Even the collection of data within their own territories was beyond the means of many states.

This inequality was difficult to reconcile with the stated aims of UNEP and the promoters of GEMS, who wanted to obtain the participation of as many participants as possible in environmental monitoring activities. Ownership of the means of surveillance would open the door to the conflict of interest problems mentioned above, but also to a potential strengthening of the grip developed nations held on the developing world by using their lack of scientific and technical capacity as a reason to entrench a postcolonial authority over those states could develop.

One might ask whether the initial curse of GEMS, its inability to do much more than simply bring together projects that already existed, turned into a blessing when the independence of those projects ensured that a decline in funding for GEMS specifically did not kill the network of monitoring projects already underway, from the atmosphere to the oceans and the food cycle. And yet this chapter shows that the solution was hodgepodge: reliant upon existing monitoring programs from the Cold War and insufficiently resourced to pursue new and ambitious research with a distinctive environmental sentiment.

GEMS is reliant upon observation and monitoring techniques, as well as the availability of a volume of data unprecedented in history, that would not be possible without the significant investments of states engaged in strategic surveillance during the Cold War. The sense of environmental consciousness that produced the Stockholm Conference, UNEP, Earthwatch, and eventually GEMS was a necessary but not a sufficient condition for the development of modern global environmental monitoring, which ought to be considered as a political in addition to a technical infrastructural phenomenon. GEMS continues to thrive and the organizations that have been pivotal to its establishment and development—the WMO, the UNEP, the ICSU, and the SCOPE—have more recently embarked on the far more ambitious project of tackling climate change, also establishing a monitoring and assessment mechanism to provide recommendations to policymakers: the Intergovernmental Panel on Climate Change. The fortunes and perils of global environmental technocracy and management have not been consigned to history, but are due to stay with us for the foreseeable future. The infrastructure persists.

Notes

1. Geoffrey Bowker and Susan Leigh Star, *Sorting Things Out: Classification and Its Consequences* (Cambridge, MA: MIT Press, 1999).
2. Thomas P. Hughes, *Networks of power: Electrification in western society, 1880–1930* (Baltimore, MD: John Hopkins University Press, 1983); Erik van der Vleuten, "Toward a transnational history of technology: meanings, promises, pitfalls," *Technology and Culture* 49 (2008): 974–994; Thomas Misa and Johan Schot, "Inventing Europe: technology and the hidden integration of Europe," *History and Technology* 21 (2005): 1–20; Erik van der Vleuten and Arne Kaijser, "Networking Europe," *History and Technology* 21 (2005): 21–48; E. van der Vleuten and A. Kaijser (eds.), *Networking Europe: Transnational Infrastructures and the Shaping of Europe, 1850–2000* (Sagamore Beach, MA: Science History Publishing, 2006); Helmut Trischler and Hans Weinberger, "Engineering Europe: big technologies and military systems in the making of twentieth-century Europe," *History and Technology* 21 (2005): 49–84.
3. Martin Hewson, "Did global governance create informational globalism?" in Martin Hewson and Timothy J. Sinclair (eds.), *Approaches to Global Governance Theory* (Albany, NY: State University of New York Press, 1999): 97–113.
4. See Herran and Turchetti in this volume.
5. Paul Edwards, "Meteorology as infrastructural globalism," *Osiris* 21 (2006): 229–250.

6. See Poole in this volume.

7. Jacob D. Hamblin, "Environmentalism for the Atlantic Alliance," *Environmental History* 15 (2010): 54–75.

8. Hewson, "Did global governance create informational globalism?" 97–116. See also Paul Edwards, *A Vast Machine: Computer Models, Climate Data, and the Politics of Global Warming* (Cambridge, MA: MIT Press, 2010), Chapter 1.

9. It was during the 1970s that international relations experts Robert Keohane and Joseph Nye outlined new approached focussing on transnational relations and globalization. See in particular: J. Nye, *Transnational Relations and World Politics* (Cambridge, MA: Harvard University Press, 1973); J. Nye and R. Keohane, *Power and Interdependence: World Politics in Transition* (Boston: Little, Brown, 1977). See also Niall Ferguson, Charles S. Maier, Erez Manela, and Daniel J. Sargent (eds.), *The Shock of the Global: the 1970s in Perspective* (Cambridge, MA: Harvard University Press, 2011). See also: Robbie Robertson, *The Three Waves of Globalization: A History of a Developing Global Consciousness* (London, Zed Books, 2003).

10. Clark A. Miller, "Resisting empire: globalism, relocalization, and the politics of knowledge," in Sheila Jasanoff and Marybeth Long-Martello (eds.), *Earthly Politics: Local and Global in Environmental Governance* (Cambridge, MA: MIT Press, 2004), 81–102.

11. On the history of the United Nations, see among others Robert C. Hilderbrand, *Dumbarton Oaks: The Origins of the United Nations and the Search for Postwar Security* (Chapel Hill: University of North Carolina Press, 2001).

12. See for instance Ethan B. Pollock, *Stalin and the Soviet Science Wars* (Princeton, NJ: Princeton University Press, 2006).

13. On the origins of the IGY see Fae Korsmo, "The genesis of the International Geophysical Year," *Physics Today* 60 (2007): 40–43.

14. Alan Needell, *Science, Cold War and the American State: Lloyd V. Berkner and the Balance of Professional Ideals* (Washington, DC: Smithsonian/Harwood, 2000).

15. On the IGY see: Walter Sullivan, *Assault on the Unknown: The International Geophysical Year* (New York: McGraw Hill, 1961). On ICSU see: Frank Greenaway, *Science International. A History of ICSU* (Cambridge: Cambridge University Press, 1996), 147.

16. Ron Doel, "Constituting the postwar earth sciences: the military's influence on the environmental sciences in the USA after 1945," *Social Studies of Science* 33 (2003): 635–666.

17. See for instance Joseph Masco, "Bad weather: On planetary crisis." *Social Studies of Science* 40:1 (2010): 7–40.

18. Lawrence S. Wittner, *The Struggle against the Bomb* (Stanford, CA: Stanford University Press, 1997), 2 Vols. Paul Boyer, *By the Bomb's Early Light: American Thought and Culture at the Dawn of the Atomic Age* (Chapel Hill: The University of North Carolina Press, 1994); Soraya Boudia, "Global regulation: Controlling and accepting radioactivity risks," *History and Technology* 23 (2007): 389–406.

19. Jacob Darwin Hamblin, *Poison in the Well. Radioactive Waste in the Oceans at the Dawn of the Nuclear Age* (New Brunswick, NJ: Rutgers University Press, 2008).

20. Rachel Carson, *Silent Spring* (London: Penguin, 1999 [1962]). See also: Samuel Hays, *Beauty, Health, and Permanence: Environmental Politics in the United States, 1955–1985* (Cambridge: Cambridge University Press, 1989); Karl Boyd Brooks, *Before Earth Day: The Origins of American Environmental Law, 1945–1970* (Lawrence: University of Kansas Press, 2009).

21. For an overview se the WWF website: wwf.panda.org/who_we_are/history (accessed 13.2.2014). For a recent reappraisal see: Kate Kellaway, "How the Observer brought the WWF into being," *The Observer*, 7 November 2010.

22. See J. Hamblin, "Environmentalism for the Atlantic Alliance," 54–75.
23. Sheila Jasanoff, *The Fifth Branch: Science Advisers as Policymakers* (Cambridge, MA: Harvard University Press, 1990); S. Jasanoff, "Science, politics and the renegotiation of expertise at EPA," *Osiris* 7 (1992): 194–217; Carl Cranor, *Regulating Toxic Substances: A Philosophy of Science and the Law* (New York and Oxford: Oxford University Press, 1993); James T. Patterson, *The Dread Disease: Cancer and Modern American Culture* (Cambridge, MA: Harvard University Press, 1989); Samuel S. Epstein, *The Politics of Cancer Revisited* (New York: East Ridge Press, 1998).
24. Greenaway, *Science International*, 147.
25. Soraya Boudia and Nathalie Jas (eds.), *Toxicants, Health and Regulation since 1945* (London: Pickering and Chatto, 2013). On the role of different scientific communities in the emergence and constitution of a global environmental regime, see: Peter M. Haas, *Saving the Mediterranean: The Politics of International Environmental Cooperation, Political Economy of International Change* (New York: Columbia University Press, 1990); Peter M. Haas (ed.), *Knowledge, Power, and International Policy Coordination* (Columbia, SC: University of South Carolina Press, 1997); Gareth Porter and Janet Welsh Brown, *Global Environmental Politics* (Boulder, CO: Westview Press, 1991); Peter Dauvergne, *Handbook of Global Environmental Politics* (Cheltenham and Northampton: Edward Elgar Publishing, 2006); Chunglin Kwa, "Representations of nature mediating between ecology and science policy: The case of the International Biological Programme," *Social Studies of Science* 17 (1987): 413–442; Yannick Mahrane, Marianna Fenzi, Céline Pessis and Christophe Bonneuil, "De la nature à la biosphère. L'invention politique de l'environnement global, 1945–1972," *Vingtième Siècle* 1:113 (2012): 127–141; Lynton Keith Caldwell and Paul Stanley Weiland, *International Environmental Policy, from the Twentieth to the Twenty-first century* (Durham, NC: Duke University Press, 1996).
26. Soraya Boudia and Nathalie Jas (eds.), *Powerless Science: Science and Politics in a Toxic World* (London: Berghan Books, 2014).
27. 1967: Bombs rain down on Torrey Canyon,' BBC (available at: http://news.bbc .co.uk/onthisday/hi/dates/stories/march/29/newsid_2819000/2819369.stm).
28. Jane M. Hightower, *Diagnosis: Mercury* (Washington, DC: Island Press, 2008).
29. UN General Assembly, 23rd session, December 3, 1968. http://daccess-dds -ny.un.org/doc/RESOLUTION/GEN/NR0/243/58/IMG/NR024358 .pdf?OpenElement
30. UN General Assembly Resolution on Prevention of Marine Pollution, January 12, 1970 in *International Legal Material* 9 (1970): 424–426.
31. Barbara Ward and René Dubos, *Only One Earth. The care and maintenance of a small planet* (New York: W.W. Norton & Co., 1972).
32. The notion emerged within the circle of experts advising the US president and led Ward to publish her book in 1966: B. Ward, *Spaceship Earth* (New York: Columbia University Press, 1966).
33. Amongst the leading figures in the Club of Rome were the economist Aurelio Peccei, who was member of the board of directors of the car manufacturer Fiat, and Alexander King, a British chemist and diplomat who had been scientific attaché at the embassy of Great Britain in Washington DC during World War II. From 1961 King was the chief scientific officer of the Organisation for Economic Co-operation and Development (OECD). The Club of Rome sought to analyze worldwide issues such as overpopulation, environmental degradation, poverty and misuse of technology. In 1971 the Club published *The Limits of Growth* as a contribution to the Stockholm conference. Donella H. Meadows, Dennis

Meadows, Jorgen Randers, and William W. Behrens III, *The Limits of Growth. Report of The Club of Roma* (New York, Universe Books, 1972).

34. SCEP, ed., *Man's Impact on the Global Environment Assessment and Recommendations for Action* (Cambridge, MA: MIT Press, 1970). See also: Élodie Vielle Blanchard, "Les limites à la croissance dans un monde global. Modélisations, prospectives, refutations," PhD dissertation (Paris: École des hautes études en sciences sociales, 2011).

35. SCOPE, *Global Environment Monitoring* (Stockholm: SCOPE/ICSU, 1971), 16.

36. Jacob Darwin Hamblin, *Oceanographers and the Cold War: Disciples of Marine Science* (Seattle: University of Washington Press, 2005).

37. SCOPE, *Global Environmental Monitoring.* See also G. F. White, "SCOPE. The first sixteen years," *Environmental Conservation* 14 (1987): 7–13. See also Greenaway, *Science International,* 176–177.

38. SCOPE, *Global Environmental Monitoring,* 23.

39. Ibidem, 48.

40. On the genesis of the UNEP see Maria Ivanova, "Looking forward by looking back: Learning from UNEP's history," in Lydia Swart and Estelle Perry (eds.), *Global Environmental Governance: Perspectives on the Current Decade* (New York: Center for UN Reform Education, 2007), 26–47; Mostafa Kamel Tolba and Iwona Rummel-Bulska, *Global Environmental Diplomacy: Negotiating Environmental Agreements for the World, 1973–1992* (Cambridge, MA: MIT Press, 1998); Lynton K. Caldwell, *International Environmental Policy* (Durham, NC: Duke University Press, 1984); James Gustave Speth and Peter M. Haas, *Global Environmental Governance* (Washington, DC: Island Press, 2006).

41. "Global environmental action plan proposed for Stockholm Conference", press release, March 15, 1972. UN archives, New York, series 0913–0016, box 01.

42. Clayton E. Jensen, Dail W. Brown and John A. Morabito, "Earthwatch. Guidelines for implementing this global assessement program are presented," *Science* 190 (1975): 432–438.

43. Brian Martin and F. Sella, "Earthwatching on a macroscale," *Environmental Science and Technology* 10 (1976), 230–233.

44. Jensen, Brown and Morabito, "Earthwatch," 433.

45. Wade Rowland, *The plot to save the world, the life and times of the Stockholm Conference on the Human Environment* (Toronto: Clarke, Irwin & Co., 1973), 2–3.

46. Joshua P. Howe, "Making Global Warming Green: Climate Change and American Environmentalism," PhD dissertation, Stanford University, 152.

47. On the Antarctic Treaty System see Philip W. Quigg, *A Pole Apart. The Emerging Issue of Antarctica* (New York: McGraw-Hill, 1983). See also Simone Turchetti, Simon Naylor, Katrina Dean, and Martin Siegert, "On thick ice: Scientific internationalism in Antarctic affairs, 1957–1980," *History and Technology* 24:4 (2008): 351–376.

48. Howe, "Making global warming green," 153.

49. Doel, "Constituting the Postwar Earth Sciences," 635–666. See also James R. Fleming, *Fixing the Sky: The Checkered History of Weather and Climate Control* (New York: Columbia University Press, 2010).

50. See for instance Mahrane et al., "De la nature à la biosphère"; Craige, Beatty Jean. *Eugene Odum: Ecosystem ecologist and environmentalist* (Athens, GA: University Georgia Press, 2001); Jacob Darwin Hamblin, *Arming Mother Nature: The Birth of Catastrophic Environmentalism* (Oxford: Oxford University Press, 2013).

51. Thomas Robertson, "This is the American Earth: American empire, the Cold War, and American environmentalism," *Diplomatic History* 32 (2008): 561–584.

52. J. Brooks Flippen, "Richard Nixon, Russell Train, and the birth of modern American environmental diplomacy," *Diplomatic History* 32 (2008): 613–638; Stephen Macekura, "The limits of community: the Nixon administration and global environmental politics," *Cold War History* 11 (2011): 489–518.

53. Matthew Evangelista, *Unarmed Forces: The Transnational Movement to End the Cold War* (Ithaca, NY: Cornell University Press, 1999); John McCormick, *Reclaiming Paradise: The Global Environmental Movement* (London: Belhaven Press, 1989).

54. On the fundamentally pragmatic nature of Nixon's environmental policies, see J. Brooks Flippen. *Nixon and the Environment.* (Albuquerque: University of New Mexico Press, 2000).

55. Howe, "Making Global Warming Green," 158.

56. Hamblin, "Environmentalism for the Atlantic alliance," 54–56.

57. Ibidem. See also Andrew Hurrell and Benedict Kingsbury (eds.), *The International Politics of the Environment* (Oxford: Clarendon Press, 1992); Alan K. Henrikson, ed., *Negotiating World Order: The Artisanship and Architecture of Global Diplomacy* (Wilmington, NC: Scholarly Resources, Inc., 1986); Kai Hünemörder, "Environmental crisis and soft politics: détente and the global environment, 1968–1975," in John R. McNeill and Corinna R. Unger (eds.), *Environmental Histories of the Cold War* (Cambridge: Cambridge University Press, 2010), 257–276.

58. Jay H. Blowers, "The United States' position on the environment," *American Journal of Public Health* 62 (1972): 634–638.

59. Hamblin shows that the use of NATO as a vehicle for environmental diplomacy struck its allies as counterproductive. NATO would not take part to the UN conference in Stockholm. Hamblin, "Environmentalism for the Atlantic Alliance," 54–56.

60. These objectives were clearly set out in the various meetings at the State Department. The content of these talks has recently become available in Susan K. Holly and William B. McAllister (eds.), *Foreign Relations of the United States, Foreign Relations 1969–1976, Documents on Global Issues 1969–1972, Volume E-1, Chapter V, International Environment Policy* (Washington, DC: US Department of State [Office of the Historian], 2005). See also Flippen, "Richard Nixon, Russell Train, and the birth of modern American environmental diplomacy," 613–638.

61. Louis B. Sohn, "The Stockholm Declaration on the human environment," *Harvard International Law Journal* 14 (1973): 423–515, on 431.

62. Ibidem, 454.

63. Ibidem, 509.

64. Jensen, Brown and Morabito, "Earthwatch," 434.

65. 40% of total contributions would come from the USA and 95 percent of the overall funding from 15 countries. Ibidem, 433.

66. Michael D. Gwynne, "The Global Environment Monitoring System (GEMS) of UNEP," *Environmental Conservation* 9 (1982): 35–41, on 35.

67. In this order from 1 to 8: Sulfur Dioxide and Radionuclides; Ozone and DDT; Cadmium and Nitrates; Mercury, Lead and Carbon Dioxide; Carbon Monoxide and Petroleum Hydrocarbons; Fluorides; Asbestos and Arsenic; Mycotoxins and Microbial Contaminants. See Jensen, Brown and Morabito, "Earthwatch," 433.

68. SCOPE, *Global Environmental Monitoring*, 23.

69. Gwynne, "The Global Environment Monitoring System (GEMS) of UNEP," 37–40.

70. Brian Martin and F. Sella, "Earthwatching on a Macroscale," 230–233. Michael D. Gwynne, "The Global Environment Monitoring System (GEMS) of UNEP," *Environmental Conservation* 9 (1982): 35–41. On the UNSCEAR, see Herran in this volume.

71. Martin and Sella, "Earthwatching on a macroscale," 231.
72. Jensen, Brown and Morabito, "Earthwatch," 437.
73. Gwynne, "The Global Environment Monitoring System (GEMS) of UNEP," 36
74. International Environmental Programs Committee, *Institutional Arrangements of International Environmental Cooperation* (Washington, DC: NAS, 1972); International Environmental Programs Committee, *Early Action on the Global Environmental Monitoring Systems* (Washington, DC: NAS, 1976).
75. Committee on International Environmental Affairs, *Design Philosophy for the Global Environmental Monitoring System* (Washington, DC: Department of State, 1976).
76. SCOPE, *Global Environmental Monitoring System. Action Plan for Phase I*, Report n. 3, 1973.
77. C. E. Jensen and D. W. Brown, "Earthwatch—Global environmental assessment," *Environmental Management* 5 (1981): 225–232, on 228.
78. Clark A. Miller, "Scientific internationalism in American foreign policy: the case of meteorology," in Clark A. Miller and Paul N. Edwards (eds.), *Changing the Atmosphere: Expert Knowledge and Environmental Governance* (Cambridge, MA: MIT Press, 2001), 167–217.
79. Jensen, Brown and Morabito, "Earthwatch," 435. On the WWSSN see J. Oliver and L. Murphy, "WWNSS: Seismology's global network of observing stations," *Science* 174 (1971): 254–261 and Turchetti in this volume.
80. John W. Head, "Challenges in international environmental management: A critique of the united nations environment programme," *Virginia Journal of International Law* 18 (1977): 269–288, on 273.
81. "The Global Environmental Monitoring System of the United Nations Environmental Programme," Doc. A/CONF.62/C.3?l/23, March 17, 1975, 207–209 (available at: http://legal.un.org/diplomaticconferences/lawofthesea-1982/docs/vol_IV/a_conf-62_c-3_l-23.pdf, accessed 4.3.2014).
82. United Nations System-Wide Earthwatch, Environmental Observing and Assessing Strategy, 1992, 4 (available at: www.un.org/earthwatch/about/docs/unepstrx.htm, accessed March 4, 2014).
83. Head, "Challenges in international environmental management," 280.
84. Christine McCulloch, "Obituary: Michael Douglas Gwynne," *The Geographical Journal* 178 (2012): 383–384.
85. Jensen and D. W. Brown, "Earthwatch," 228.

Chapter 10

What Was Whole about the Whole Earth?
Cold War and Scientific Revolution

Robert Poole

In 1948 the astronomer Fred Hoyle speculated what the Earth would look like from space, and predicted: "once a photograph of the Earth, taken from outside, is available, we shall, in an emotional sense, acquire an additional dimension... once let the sheer isolation of the Earth become plain to every man whatever his nationality or creed, and a new idea as powerful as any in history will be let loose."[1] In 1970 he found himself in a position to reflect upon his prophecy.

> Well, now we have such a photograph, and I've been wondering how this old prediction stands up. Has any new idea in fact been let loose? It certainly has. You will have noticed how quite suddenly everybody has become seriously concerned to protect the natural environment. Where has this idea come from? You could say from biologists, conservationists and ecologists. But they have been saying the same things now as they have been saying for many years. Previously they never got on base. Something new has happened to create a world-wide awareness of our planet as a unique and precious place. It seems to me more than a coincidence that this awareness should have happened at exactly the moment man took his first step into space.[2]

Hoyle placed first the whole Earth photographs firmly in a technological rather than an environmental context. The space program, despite its orientation defined by the surveillance imperative, allowed ordinary citizens to share an objective view of the Earth as an object in the solar system previously attained only by scientific thinkers: the revolution was technological, not ecological. The *Apollo* Earth photographs have often been seen as a technological windfall for an environmental movement that was unprepared for it but rapidly recognized its significance.[3] In Thomas Kuhn's model, a scientific revolution is a "change of world view," which follows a period of intellectual struggle within an existing model that no longer fits the observations. Here, it seems, the new world view arrived first and the rethink followed.[4]

Hoyle's judgment that the whole Earth pictures had a powerful impact has been widely shared. The *Apollo* "Blue Marble" Earth has been called both "the most influential scientific photograph ever taken" and "the most influential environmental photograph ever taken" (see Figure 10.1).[5] Donald Worster writes that

Figure 10.1 The famous "Blue Marble" picture taken from the crew of *Apollo 17* on 7 December 1972

Source: visibleearth.nasa.gov.

the image of Earth from space came as "a stunning revelation" that nourished the young discipline of ecology.[6] Betty Jean Craige, biographer of the ecologist Eugene Odum, finds that the surge of interest in ecology at the end of the 1960s was stimulated by the first distant views of the Earth from space: "From the perspective of the Moon, human beings were indistinguishable components of the indivisible biosphere. The sight of the blue planet spinning in space alerted its inhabitants to its vulnerability and reminded us of our dependence on its stability...Americans turned to ecology." J. R. McNeill and Corinna Unger have argued that satellite photography of the Earth "fostered a rediscovery of organic thinking and the emergence of deep ecology,"[7] and Erik Conway states that the *Apollo* Earth photos "became the root of a global environmental consciousness."[8] Several articles and even whole books have been written attempting to analyze and explain this phenomenon.[9] The *Apollo* years of 1968–72 coincided with rise of the modern environmental movement and also with the run-up to the UN Conference on the Human Environment in Stockholm, the first "Earth Summit," providing a context in which the two were likely to be linked.

At the same time it has to be recognized that the first whole Earth photographs did not, as it were, drop out of a clear blue sky. There had been a period when number of thinkers had been dissatisfied with the divided understanding of the Earth's dynamic processes produced by the separate scientific disciplines. As Worster writes:

> The view of the Earth as organism was an old one, going back into prehistoric cultures, but it was reborn in the modern age, and ironically the image of an ailing but ancient organic planet came from the highly polished lens of a mechanical camera carried aloft in a mechanical spaceship.[10]

In the nineteenth and twentieth centuries, individual investigators with unusual powers of vision had conceptualized the Earth as an integrated whole. "The field of the Geologist's inquiry is the Globe itself," declared the British geologist William Buckland (1784–1856).[11] Reading Alexander von Humboldt's encyclopedic work of natural history *Cosmos* (1845–58), Laura Dassow Wild comments: "In mind's eye, Humboldt saw Earth as Sagan's generation learned to see it: a blue globe above, alone, an astonishment in the black abyss of space." Humboldt's original title had been *Gaia*.[12] In the late nineteenth century the Swiss biologist Eduard Suess, in coining the term "biosphere," imagined gazing from space at "the face of the Earth." Alexander Vernadsky, popularizing the term in the 1920s, also imagined studying the Earth from space as "a harmonious integration of parts that must be studied as an indivisible mechanism."[13] The work of Suess and Vernadsky helped bring into being a compound field of science known for a time as "biogeochemistry," which resurfaced in the work of James Lovelock in the 1960s and 1970s. Lovelock formulated the Gaia hypothesis, that the Earth as a whole behaved as a self-regulating entity, after conceptualizing the Earth from the outside, and felt that the *Apollo* pictures when they arrived confirmed and deepened his view.[14] While they might rhetorically have resembled long-standing organic philosophies, all of these interpretations of the dynamic workings of the planet were based on interdisciplinary investigations that challenged the distinction between the life and the non–life sciences.

In the postwar decades, there appeared for the first time planetary-scale research to match these planetary-scale hypotheses, thanks to the military research programs of the Cold War. The Pentagon had declared in 1961: "[the] environment in which the Army, Navy, Air Force, and Marine Corps will operate covers the entire globe and extends from the depths of the ocean to the far reaches of interplanetary space."[15] These programs supported not only surveillance-driven space exploration programs but also a huge growth in what have since become known as the Earth sciences. The environmental sciences and even environmentalism were the beneficiaries of these programs long before the apparent windfall of the whole Earth photographs, which were the product of the Cold War space race. As Michael Aaron Dennis puts it: "going about the task of understanding how to destroy the enemy, the Earth sciences produced a new picture of the Earth and its complexities."[16] Joseph Masco, enlarging on the work of Paul Edwards, writes that: "the Cold War nuclear project enabled a new vision of the planet as an integrated biosphere [...] a new vision of the globe as an integrated political, technological and environmental space."[17] This begins to sound like a change of world view, which anticipated the images of the Earth from space. So, were the first whole Earth images just incidental pictures, afterward conscripted into the service of various versions of globalism and environmentalism? Or were they themselves products of scientific globalism, historically connected with the themes and discoveries that they were held to represent? Was there anything in the Cold War Earth sciences that corresponded to the holistic claims made about the Earth in the aftermath of those first photographs from space? In short, we have to ask: What was whole about the whole Earth?

This chapter attempts a kind of high-altitude survey of planetary concepts and models in the Cold War Earth sciences, broadly defined, in four sections. Any account of the global Earth sciences has to begin with the International

Geophysical Year (IGY) of 1957–8. This is followed by an overview of what seems (at least to this nonspecialist) to amount to an Earth sciences revolution, singling out geodesy, plate tectonics, and atmospheric science. Thirdly, attention shifts to the related fields of cybernetics, systems theory, and ecology. Here, it is argued, there occurred the key development in scientific whole Earth thinking: the convergence of biological and nonbiological models. This leads into a fourth section on James Lovelock's Gaia hypothesis, which related to and anticipated both orthodox planetary science and the first pictures of the Earth from space. Like a space-age version of Newton's windfall apple, the image of the whole Earth fell ripe from orbit in full view of a scientific public ready to receive it.

The International Geophysical Year

Geophysics has been described as "the area of science in which the whole Earth is the laboratory and nature conducts the experiments," and the IGY was presented as "the world studying itself."[18] Experiments were conducted on a global scale to explore the electromagnetic radiation in the atmosphere, the solar storms through which the planet occasionally passed, the cosmic rays reaching the surface, the temperature, pressure, and chemical composition of the atmosphere, the global circulation of both atmospheric and ocean currents, the dynamics of the "energy balance" as the Earth simultaneously absorbed and radiated solar heat, the topography and seismology of the sea bed, and the extent and nature of the polar ice caps. The now-global phenomenon of nuclear fallout was studied from a planet-wide network of monitoring stations. Through its sheer scale, the IGY fostered an understanding of the Earth as a set of integrated systems.[19] As yet there was no camera stationed beyond the atmosphere but several "world days" of simultaneous observation offered what amounted to "a snapshot of the Earth."[20]

It would be a mistake to project back onto the IGY a whole Earth concept which was developed later. It took place at a period when understanding of the Earth was most commonly associated with surveillance, exploitation, and control, and when the despoliation of the global environment was decisively accelerating, a phenomenon that has been diagnosed as "1950s syndrome." There were proposals to use atomic explosions to dig a new Panama Canal, melt the arctic icecap, and destroy the newly discovered van Allen belts.[21] When a stratospheric nuclear test did seriously disrupt the Earth's electromagnetic field, the *New York Times* science correspondent welcomed it as "an intellectual triumph...an experiment that enveloped almost the entire planet."[22] The IGY project had a contentious Cold War history, which belied its idealistic aspirations. Even its global icon, which incorporated zones of both day and night, seemed to mirror the divided world in which it took place.[23]

Yet, as so often during the Cold War, divisive forces generated unifying visions which acquired a life of their own. President Eisenhower's promotion of the IGY as "a striking example of the opportunities which exist for cooperative action among the peoples of the world" may have been a maneuver in the Cold War but it drew upon a widespread ideal that science could provide "the common language of mankind." One of the IGY's most important consequences was the 1961 Antarctic Treaty, which (albeit for geopolitical reasons) suspended national claims to sovereignty and declared the continent an international reservation for science.[24] The Antarctic Treaty in its turn became a model for the 1963 nuclear test ban treaty, which has been described as "the first global environmental treaty," and for the 1967 Outer Space Treaty, which declared outer space to be "the province of all mankind."[25] Although clandestine Cold War ambitions often lay behind such treaties, their adoption of legislation extending to the whole Earth entailed an enhanced understanding of human stewardship of the natural world. The systems for global scientific monitoring which were established were of great long-term significance. The CO_2 measuring station in Hawaii and the polar research bases set up during the IGY were eventually to provide conclusive evidence for climate change on a planetary scale. The IGY gave a decisive push to the convergence of the Earth sciences, which yielded important insights into the interdependence of the Earth's natural systems.

One thing that the IGY lacked was an actual image of the Earth. The US National Academy of Sciences issued a lavishly produced booklet entitled *Planet Earth: the Mystery with 10,000 Questions* complete with six specially commissioned color posters representing the different scientific fields, each incorporating an image of the Earth. Before the space age, however, all of these images were necessarily schematic.[26] The most naturalistic of the posters incorporated a painting of what appeared to be the whole Earth commissioned by the chief US meteorologist Harry Wexler. On closer inspection, it showed weather systems over North America converted into a globe through a fish eye lens effect, but its depiction of land, water, and clouds, without any of the traditional geographical grids and boundaries, was innovative. Wexler had been inspired by the earlier V-2 pictures of the curving planet, and perhaps too by the experimental color photographs of North America taken in 1954 by the *Aerobee* sounding rocket. His own concern was with the details rather than the whole: "by a bird's-eye view of a good portion of the Earth's surface and the cloud structure," he wrote, "it should be possible by inference to identify, locate, and track storm areas and other meteorological features."[27] When *Life* magazine published an issue titled "A New Portrait of our Planet," on the IGY's findings, its cover featured an image of a cloudless geographic globe.[28] It is an interesting counterfactual exercise to consider what the impact of the IGY's survey of the Earth would have been had the *Apollo* whole Earth pictures been available a decade earlier.

The coming of the space age provided the technology to continue the IGY's program Earth sciences at a new level, but at the same time it shifted attention from the exploration of the Earth to the exploration of space. The launch of *Sputnik* in October 1957, although presented as part of the Soviet IGY effort, created an association between space and national security that dominated the 15 years of the first space age (1957–72) and hampered the kind of cooperation upon which the IGY had been built. Cold War priorities affected not only space technology but less obviously contentious areas such as oceanography, where genuinely international activity was replaced by (at best) intergovernmental cooperation with a secondary brief of "easing tensions."[29] NASA itself had been founded during the IGY as a civilian agency (albeit one sustained by extensive "black" programs funded by the Department of Defense), with a brief that included study of the Earth, but the order in 1961 to race the Soviet Union to the Moon ensured that the US space program faced away from the Earth for most of the 1960s and 1970s.

The IGY model of a synoptic project to study the Earth was taken up by biologists and ecologists in the International Biological Program (1964–74). The diversity of the biological sciences however made a single focus impossible to achieve, and when a theme was settled upon – 'the biological basis of human productivity and welfare' – it provided a focus only for disagreement. Oceanography and the emerging field of ecosystems ecology had unifying ambitions but these foundered on resistance from more traditional biologists who regarded big science as a 'contagion' and a 'disease.' Half-way through the program, however, a group of ecologists set up a Global Network of Environmental Monitoring, which was adopted by the International Council of Scientific Unions SCOPE Commission, and thence by the 1972 UN Conference on the Human Environment—the first Earth summit.[30] Notwithstanding the reservations of many biologists, it was to be the coming together of the physical and natural sciences that would generate a new understanding of the Earth as whole as the climactic years of the space age coincided with the environmental renaissance of the late 1960s and early 1970s.

Thus it was that during a formative period for the Earth sciences NASA suffered from institutional Earth blindness, only occasionally disturbed by second thoughts. This helps to account for the space agency's notable lack of preparation for the first views of Earth from space, and the sense of incongruity and surprise that accompanied their arrival. But while NASA turned its corporate back on the Earth, advances in the Earth sciences in several fields were constructing models of the planet which meant that, ironically, other parts of the scientific community were better prepared than the space agency for the sight of the whole Earth.

The Earth Sciences Revolution

During the 1950s and 1960s the physical Earth sciences were expanding their observations and models to a global scale, putting together large-scale observations and measurements to develop an understanding of the Earth's systems on a planetary level on the back of military programs. This section will survey three such fields: geodesy, plate tectonics, and meteorology and climate.

In the 1960s one lesser-known discipline provided an unseen image of the Earth: geodesy, the exact measurement of the shape of the planet, or geoid. This was, according to its historian John Cloud, a planetary enterprise that provided "one of the most important intellectual achievements of the Cold War."[31]

Geodesy had become a pressing practical problem with the advent of long-range ballistic missile. The hoard of maps seized from Germany at then end of the war had revealed discrepancies of hundreds of meters between national maps prepared from different reference points—enough to make a decisive difference in the targeting of long-range missiles, as the V-2 program had discovered to its cost. The problem was that the Earth's shape was neither a globe nor even regular, as assumed by cartographers, owing to the combination of the flattened shape caused by the planet's rotation and the irregular distribution of land masses. The exact shape was difficult to measure since conventional methods relied upon gravity, whose force varied with the radius of the Earth. The relationship between gravity and radius, however, was not constant, varying in its turn according to the mass and density of the Earth at the point of measurement. Ingenious attempts to measure the shape of the Earth independently of gravity by taking highly accurate photographs of the stars in relation to the Moon and the Earth had not quite come off.[32] The coming of the satellite made it possible to measure the geoid independently of gravity.

The image of the geoid remained invisible partly because it was constructed from a variety of nonvisual data and partly because it was obtained through the US Department of Defense's satellite surveillance programs, which remained a military secret until after the end of the Cold War. Between 1960 and 1972 the CORONA satellite network took high-altitude photographs of the Earth, parachuting the cameras back to Earth in reentry capsules which were caught in mid-air by cargo planes equipped with nets. The pictures were reconciled with German and Soviet geodetic charts and correlated against other satellite observations from the Department of Defense's World Geodetic System. One Department of Defense satellite, named DODGE, produced the first color picture of the whole Earth as early as August 1967, a low-resolution television image taken through colored glass filters. Although it prompted one of the first ever color printings of a major newspaper and made its way into *National Geographic*, the DODGE Earth photo made little impact compared with the more naturalistic Earth images that were soon to follow.[33] The accurate reconstruction of the geoid was significant for the whole Earth in another way for, writes Cloud, it involved "a great re-convergence of the now disparate disciplines of astronomy, geodesy, geography, geology, cartography, photogrammetry, and geophysics." The processes behind this clandestine development of an invisible image of the Earth thus paralleled the convergence of the Earth sciences happening elsewhere.[34] Geodetic measurements were also important for the manned space program. As Cloud puts it, nicely reversing the more familiar Earthrise story, "reaching the Moon required first discerning the Earth."[35]

For geologists the Earth came to life in the 1960s as a synthesis of work in geology, seismology, oceanography, vulcanology, and studies of the Earth's magnetic field came together in the discipline of plate tectonics. Ever since Lyell, the orthodoxy had been that geological processes were extremely gradual. The continents were essentially static, modified incrementally over eons by slow processes such as upheaval, sedimentation, and erosion, with limited local assistance from earthquakes and volcanoes. Lyell's views in turn conditioned Darwin's model of evolution as a steady accumulation of small variations, although it is worth noting that Darwin, having experienced earthquakes, found himself "impressed with the never-ceasing mutability of the crust of this our world."[36] In the late nineteenth century the Swiss geologist Eduard Suess, impressed by the evidence for rapid geological upheavals, had challenged but not dented the static Earth orthodoxy. The early twentieth-century German meteorologist Alfred Wegener had put forward a theory of continental drift, but in the absence of a plausible mechanism or even a coherent set of measurements his ideas were widely rejected.[37]

After World War II the US Navy became the major patron of oceanography, transporting scientists around the world's oceans to develop new technologies

of measurement. Deep-sea topography mapped the boundaries of the continental shelves, which revealed a much better fit between continents than the
visible coastlines. Investigations of the ocean floor revealed a "world-girdling"
system of ocean ridges and rifts ripe for further exploration during the IGY.[38]
Surveys of thermoclines, prompted by the need to understand how they altered
sonar signals, yielded evidence of high heat flow in geologically significant patterns: at the mid-ocean ridges new rock was emerging as magma.[39] Meanwhile
the World-Wide Seismography project, designed to detect underground nuclear
tests and to distinguish them from earthquakes, provided a kind of x-ray of the
Earth. It revealed that earthquakes were clustered along the boundaries where
continental plates slowly moved under or past each other.[40] The final piece in the
jigsaw was provided by studies of the magnetism of the ocean floor, arising from
the military need for accurate magnetic navigation. This revealed barcode-style
patterns of magnetic stripes imprinted on the emerging magma as it solidified,
evidence of successive reversals of the Earth's magnetic poles. This calibrated
the spreading sea floor over time and enabled mobile plate boundaries to be
matched and mapped. Through a series of international conferences and high-
profile discoveries in the years 1962–66 there emerged a unified account of plate
tectonics, amounting, in the words of one participant, to a "revolution in Earth
science."[41] In a related development, the US Navy's investigations into deep-sea
listening posts led to the discovery of deep ocean vents and of new forms of life
based on chemosynthesis rather than photosynthesis; undersea geology was connecting with the life sciences.[42]

   The dynamic view of the Earth's geology was associated with visual thinking.
Eduard Suess in his 1885 book *The Face of the Earth* had imagined the Earth as
it appeared to a visitor from space, "pushing aside the belts of red-brown clouds
which obscure our atmosphere, to gaze for a whole day on the surface of the
Earth as it rotates beneath him".[43] Richard Fortey comments that with seismic
mapping of the ocean floor "it was possible to look at the whole Earth for the
first time". In October 1967, just as the first color satellite photographs of the
whole Earth were appearing, *National Geographic* began publishing a series of
color maps of the ocean floors, crafted to show rifts and mountain ranges, continental shelves, and mid-ocean ridges. Widely used in schools and colleges, such
maps conveyed a sense of the planet as a single geological entity.[44] When in the
early 1970s the cell biologist Lewis Thomas put into words his response to the
first whole Earth photos, he had plate tectonics very much in mind: "If you had
been looking for a very long, geologic time, you could have seen the continents
themselves in motion, drifting apart on their crustal plates, held aloft by the
fire beneath. It has the organized, self-contained look of a live creature, full of
information, marvelously skilled in handling the sun".[45]

   Even more than studies of continental plates and oceans, study of the atmosphere involved global model-building. In the late 1940s the head of the US
Weather Bureau Harry Wexler had given a contract to the Lowell Observatory
to try and understand the general circulation of the atmospheres of Mars and
Venus, but astronomers were not able to see well enough.[46] The IGY of 1957–8,
wrote Walter Sullivan, had brought an awareness that the planet was surrounded
by a single "ocean of air…one great, mobile reservoir covering two-thirds of
the globe and carrying, within its deep, slow currents, the seeds of latent climate

change that might destroy existing civilizations and make possible new ones." The comment now appears prophetic, but at the time studies of the atmosphere were driven primarily by meteorology and the desire for better weather forecasting. As Sebastian Grevsmühl shows in this volume (chapter 8), it took some time for meteorologists—even the globally minded Harry Wexler—to see satellites as more than just a better method of observing existing weather systems, although the satellite perspective of the Earth's atmosphere from the outside did lead to new classifications and insights. Big-picture thinking about the dynamics of the planetary climate as a whole, over time scales much longer than those of ordinary weather forecasting, emerged more gradually.

From the beginning, the World Meteorological Organization, conceived by the UN in 1947 and founded in 1951, aimed to study the Earth as a single physical system.[47] Advances in computing led to the first general circulation models of the atmosphere in the mid-1950s, supplemented by visual monitoring from the TIROS satellite series from 1960 onward and the first satellite TV weather pictures from NIMBUS in 1964. At first it was hoped that the sight of weather systems from orbit would lead to much longer-range forecasts, but the see-and-predict model produced disappointing results. Television pictures proved intractable and were soon abandoned, and even when a global network of seven satellites was set up in the 1970s they could not improve upon the existing five-day forecast horizon. This in turn prompted the development of the mathematics of complex systems, which gave rise to chaos theory. The key insight here was that while small changes in one part of the atmosphere could give rise to large changes in another part, this did not happen in any consistent way: what could be modeled in principle could not be predicted in practice. As Edwards explains: "conceiving weather and climate as global phenomena helped promote an understanding of the world as a single physical system." This, however, was a complex process mediated by layer upon layer of data processing and modeling procedures; there was no sudden rise in awareness.[48] While the work of meteorologists involved some of the first truly global datasets, they were using global tools for local purposes; even when instrumental in securing photographs of the whole Earth from space they were unable to see the Earth for the clouds.[49]

For all the impulse that meteorology gave to global atmospheric modeling, fully integrated study of the global atmosphere was stimulated by environmental concerns. An early instance of this was provided by the international network of monitoring stations to measure the levels of radioactive carbon in the atmosphere from nuclear bomb tests. This made possible the 1963 Partial Test Ban Treaty, which Edwards describes as "not only [...] the first global environmental treaty, but also [...] the first to recognize atmosphere as a circulating global commons that could be directly affected on the planetary scale by human activities."[50] The next major push came with the four-year program of preparations for the 1972 United Nations Conference on the Human Environment in Stockholm: the first Earth summit. While the *Apollo* photographs of 1968–72 are the most famous, distant images of the whole Earth began to appear in the late summer of 1966 while stunning orbital photographs taken from outside the capsule by spacewalking Gemini astronauts had begun to appear in mid-1965. These helped build public support for the creation in 1966 of the Earth Resources Observation Satellite program (EROS), which eventually developed into the Landsat program.[51]

Concerns raised in both the UN and the WMO about the effect of CO_2 and chlorofluourocarbon (CFC) emissions on the climate created a need for global data sets in order to filter out long-term "signals" of climate change from short-term "noise" of natural variation. In 1970, at the instigation of the UN, Massachusetts Institute of Technology (MIT) produced its *Study of Critical Environmental Problems,* with follow-up reports in 1971 and 1972. The long-delayed Global Atmospheric Research Program was developed during the 1970s, with NASA at last adopting the program; the last Nimbus weather satellite (1978–84) was modified to detect atmospheric pollution, yielding data for the first maps of the global biosphere. In 1980 NASA put together 20 months of data on the distribution of marine phytoplankton in the oceans collected by the Nimbus-7 satellite's Coastal Zone Color Scanner with three years of observations of land surface vegetation from the National Oceanographic and Atmospheric Agency (NOAA-7) satellite to produce what it called "the first composite image of the global biosphere."

These programs culminated in 1979 with a massive global atmospheric observation project reminiscent of the IGY. The WMO held its first global climate conference in the same year, launching the World Climate Program of the 1980s, which in turn led to the establishment in 1988 of the Intergovernmental Panel on Climate Change and the vast programs of scientific and political activity which followed.[52] In the end, as Edwards puts it: "meteorology was only one part of a larger project in constructing a global panopticon."[53] Thus, idealized models of the Earth first developed for meteorology soon became bound up with the emergence of concerns about the planetary environment as a whole. These concerns were inspired in part by the first views of the Earth from space, and they fostered an interdisciplinary understanding of the planetary climate.

Ecology and Ecosystems

An image of Earthrise from the Moon formed the frontispiece the 1971 edition of Eugene P. Odum's foundational textbook *Fundamentals of Ecology.* It was described in the caption as a photograph of Earth at "the biosphere level." Odum, described by Joel Hagen as "the philosophical leader of modern ecosystem ecology," liked to compare the Earth to a space capsule, in that the inhabitants of both were part of a closed ecosystem, mutually dependent upon each other and upon their environment in order to survive. The parallel had occurred to him when the *Apollo 13* accident, which left three astronauts struggling for survival as they gazed down upon their own receding planet, occurred around the time of the first Earth Day in 1970. As the astronauts urgently tried to understand what had gone wrong with the space capsule in order to save it, Odum mused that the situation was not so different on Spaceship Earth: "Our global life-support system that provides air, water, food and power is being stressed by pollution, poor management, and population pressure." He kept a poster of the *Apollo 8* Earthrise on his study wall.[54] For Odum in 1971 ecology entailed both the study of the interacting forces at work within and between species in nature and a philosophical commitment to the principles of group selection and "coevolution" (or "reciprocal selection"). As Odum emphasized in his preface, "the holistic approach and ecosystems theory [...] are now matters of world-wide concern," applicable to human survival and environmental stability as well as to understanding of the natural world.[55]

There are so many overlapping ideas here, jostling for position around the still-fresh image of the whole Earth, that it is difficult to know where to begin unraveling them. They are perhaps most familiar from ecological and counter-cultural activism but for that reason the links with science are perhaps less clearly appreciated, at least outside the specialist literature. This section will look at two areas where interaction between living and nonliving systems formed part of orthodox science from the 1940s: systems theory and ecology. When pictures of the Earth from space arrived in the late 1960s much of the talk was about the Earth as a set of systems, of which humankind was (visibly) a part. The picture was novel but the mode of thinking was well-established in two related fields that had both been established in the mid-1940s: cybernetics (or systems theory) and ecosystems ecology. Both, in different ways, arose out of military problems in the war and early Cold War.

The founding text of systems theory was Norbert Wiener's *Cybernetics, or Control and Communication in the Animal and the Machine* (1948). As its title suggests, Wiener ranged across the disciplines, developing his principles through work on problems as apparently diverse as antiaircraft fire, the physiology of the heart, and computing, to arrive at a general science of control. Wiener's key concept was "feedback," the means by which a movement in one variable triggers compensating movements in other variables and even in other linked systems. Applicable to systems of any kind, mechanical, biological, or social, Wiener's work generated insights in just about every area of scientific and intellectual endeavor. The dust jacket advertised the book as "a study of vital importance to psychologists, physiologists, electrical engineers, radio engineers, sociologists, philosophers, mathematicians, anthropologists, psychiatrists and physicists," and so it proved to be.[56] *Cybernetics* was the most prominent product of a series of 10 conferences on the subject held in New York between 1946 and 1951, which attracted many leading thinkers in the natural and social sciences, eager to be involved with what was proclaimed as "one of the major transitions or upheavals in the history of ideas." They ranged from associates of the RAND Corporation seeking in systems theory a "complete science of warfare," such as John von Neumann, who was in the process of developing game theory into the mathematics of Armageddon, to the anthropologist Gregory Bateson who was seeking to put together a social science equivalent of the Manhattan project in order to discern the deeper causes of conflict and so avert atomic warfare.[57] The processing and transmission of signals, which was the concern of cybernetics, was also fundamental in the development of Cold War surveillance networks.

The science of systems was first scaled up to global level through the development of world modeling (or world dynamics) in the late 1960s and early 1970s. Buckminster Fuller's "World Game," originally proposed as an exhibit for Expo '67, was played across university campuses in the United States, Canada, and Britain in the summer of 1969 and was the subject of a supplement to the *Whole Earth Catalog* in 1970. Fuller, originator of the term "spaceship Earth" and author of *Operating Manual for Spaceship Earth*, envisaged that "The young may take over and operate 'Spaceship Earth.'"[58] The 1972 Club of Rome report *Limits to Growth*—conceived and researched during the Apollo years of 1968–72—brought world modeling to a mass audience, its central argument being that economic growth was already coming up against environmental limits. Jay Forrester of MIT, the founder of system dynamics, later recalled a conversation on a plane returning from an international economic conference in Switzerland conference. "We haven't tackled the rally hard problem," he said to a colleague. "What's that?" "The world." Forrester sketched a flow diagram of the forces operating in the planet with feedback loops, which gave the same results every time: excessive population growth, collapse of population and living standards, and slow recovery.[59] By later standards the techniques now appear crude and simplistic, modeling human activity with the environment appearing simply as a resource constraint. The significance of these exercises in world modeling was that they sought to demonstrate the interaction of science and technology with politics, society, economy, and the environment, and that they popularized an integrated mode of thinking about global developments. They were developed and publicized in parallel with the development of similar modes of thinking in the Earth sciences, and with the appearance of images of the Earth from space which showed, as no model could, that the Earth was indeed both whole and limited.

Systems theory was also a resource for ecologists, whose discipline had run into trouble. Ecology had grown up in the 1920s, 1930s, and 1940s, led by the Chicago school of animal ecology, which argued for the role of coexistence and cooperation in evolution. For ecologists, the natural world was only fully intelligible at the level of the group or (to use a term coined in 1935) the ecosystem. This vision of harmony in nature had a powerful appeal and an affinity with organic models of human society. Such "organicist" thinking, however, had become tainted by its ideological associations with Nazism and by its scientific associations with vitalism, the idea that natural processes were driven by intangible inner forces. Systems theory promised a new integral approach to understanding the world, based not on intangible forces but on the measurable interactions of a myriad individuals. It indicated a way forward for ecology that was compatible with the evolutionary "modern synthesis" established in the 1940s.[60]

Among the early enthusiasts for cybernetics was the ecologist G. Evelyn Hutchinson, whose 1946 paper "Circular Causal Systems in Ecology" argued that groups of organisms used feedback loops to maintain their state (for example in the way that populations tend over time to maintain a viable balance within their environments) and could be considered as self-regulating systems. He was an early practitioner of the integrated discipline of biogeochemistry and the champion in the west of the work of the Russian Vladimir Vernadsky and his 1926 book *The Biosphere*. Hutchinson was one of the first to suggest that the carbon balance in the atmosphere might be regulated biologically—a suggestion that would later find full expression in James Lovelock's Gaia hypothesis.[61] Among Hutchinson's pupils was Howard T. Odum, younger brother of Eugene, who made early use of cybernetics in a 1950 PhD which argued that 'energy flows' had kept the chemical balance of the oceans constant for over millions of years.[62] Howard Odum developed a distinctively technocratic approach to ecology—he would later write of "ecological engineering"—which appealed beyond environmentalists to policy analysts and, in time, even economists.[63]

In the 1950s Eugene and Howard Odum established systems ecology as a distinct subdiscipline. Like its cousins in the postwar Earth sciences, systems ecology (initially known as "radiation ecology") piggybacked on military and atomic programs to generate insights at a global level. Eugene Odum gained grants from the Atomic Energy Commission to do ecological research at nuclear sites, first at the Savannah River atomic plant in Georgia and subsequently at the hydrogen bomb test site of Eniwetok atoll in the Pacific Ocean and at nuclear test sites in Nevada. With Howard Odum he traced concentrations of radiation through the food chain and the environment, founding an Institute of Radiation Ecology. At Eniwetok they were able to show that the coral reefs were stable because of the mutual relationship of coral and algae. At the 1955 Atoms for Peace conference Odum urged that atomic programs of all kinds should proceed cautiously until the total effects of radiation on ecosystems were known. The Odums also pioneered the study of energy flow in ecosystems, demonstrating that the ecosystem of the coral at Eniwetok was not only self-sustaining but actually generated energy.[64]

When the "year of ecology" was proclaimed by *Time* on August 15, 1969, and the "age of ecology" by *Newsweek* on January 26, 1970, Eugene Odum was featured on the covers of both along with his dictum that "all nature is interconnected." His 1953 textbook *Fundamentals of Ecology*, with its second edition in 1959 and its third on the way in 1971 (complete with Earthrise frontispiece), had taught generations of ecologists and environmentalists to see the natural world as an interconnected web of systems and prepared them to interpret the visual revelation of the whole Earth in similar terms.[65] The fundamental compatibility of cybernetics and systems on the one hand and ecology and ecosystems on the other lay in the way that they treated living and nonliving phenomena in similar terms, fostering study of the links between them. Cybernetic concepts scaled up easily to planetary level, ready to inform scientific as well as public responses to the *Apollo* pictures. In the 1960s some philosophers of biology, inspired by Gregory Bateson, proposed that living organisms could be understood not as physical entities but as systems, whose enduring core feature was self-organization and which in turn acted as elements of higher order ecosystems.[66] Thus for the biologist and philosopher Rene Dubos, who in 1969 was among the first publicly to compare the Earth as seen from space to a living organism, "Earth and man are thus two complementary components of a system, which might be called cybernetic, since each shapes the other in a continuous act of creation."[67] Successive editions of the *Whole Earth Catalog*, published from 1968 to 1972, printed the *Apollo 8* Earthrise picture, the last of them with a quotation which reflected its editor, Stuart Brand's, interest in cybernetics: "The flow of energy through a system acts to organize that system."[68] When Lewis Thomas reflected upon the image of the Earth from space, he too used the language of systems:

> The most beautiful object I have ever seen in a photograph, in all my life, is the planet Earth seen from the distance of the moon, hanging there in space, obviously alive. Although it seems at first glance to be made up of innumerable separate species of living things, on closer examination every one of its working parts, including us, is interdependently connected to all the other working parts. It is, to put it one way, the only truly closed ecosystem any of us know about. To put it another way, it is an organism.[69]

All these observations came from biologists. The suggestion that the Earth's complex systems were analogous to a living thing was, however, most fully set out by James Lovelock, an engineer and physical scientist, and it is to this larger vision that we now turn.

The Gaia Hypothesis

"Can there have been any more inspiring vision this century than that of the Earth from space?" exclaimed James Lovelock in his autobiography *Homage to Gaia*. Yet while the Apollo Earth images seemed to embody Lovelock's understanding of the Earth, he made clear that his own revelation of Earth as a living planet had already been formed through orthodox scientific endeavor. "Moments of intuition do not come from an empty mind; they require the gathering together of many apparently unconnected facts. The intuition that the Earth controls its surface and atmosphere to keep the environment always benign for life came to me one afternoon in September 1965 at the Jet Propulsion Laboratory (JPL) in California and it was here that most of the facts were gathered."[70] Lovelock's Gaia hypothesis, of an Earth where living and nonliving systems interact to regulate the global environment, offers an apt case study for Kuhn's model of scientific revolution as a change of world view.

Before Lovelock was given the term "Gaia" in 1970 by his friend and neighbor the novelist William Golding his working description of the Earth was "a cybernetic system with homeostatic tendencies." The whole Earth rhetoric in which the Gaia hypothesis was packaged for a wider public came later.[71] In late 1964 or early 1965, while working for NASA, Lovelock was asked to advise a team at JPL working on life detection experiments intended for a Mars lander.[72] After listening for hours to a gathering of biologists discussing ways to directly detect Earth-like life-forms in the Martian soil, Lovelock turned the problem round. Instead of looking for specific types of life, NASA should look for the generic signature effects of life in the Martian atmosphere through "a general experiment...that looked for entropy reduction." His remarks "seemed to annoy many of those present," who complained to management. Challenged to come up with an experiment in a matter of days to test his ideas he turned to Erwin Schroedinger's 1944 book *What is Life?* which discussed the subject from the point of view of a physical scientist, making use of the concept of entropy reduction. Lovelock reasoned that the atmosphere of a planet that harbored life would exhibit "effects which cannot be accounted for by abiological processes," such as a strong presence of oxygen or other combustible gases, a complex structure in a state of disequilibrium, or other anomalously orderly features – perhaps even regular sounds. In short, "knowledge of the composition of the Martian atmosphere may...reveal the presence of life."[73] NASA was impressed enough to make him acting chief scientist for the life detection program in March 1965, but within six months, thanks in part to lobbying from indignant biologists, the Voyager Mars lander program was cancelled.

Lovelock developed his ideas further in another visit to JPL in September 1965. In the meantime, images from the *Mariner* spacecraft had shown that Mars was "all rock or desert." This time he was present when results of an infrared spectrographic analyses of the Martian and Venusian atmospheres from ground-based radio telescope came through, showing that both were overwhelmingly dominated by carbon dioxide (see Figure 10.2).

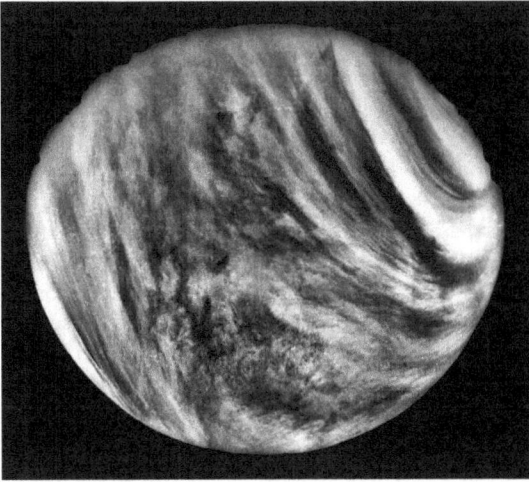

Figure 10.2   Picture of Venus and its atmosphere from the Mariner 10 spacecraft taken on February 5, 1974

Source: //solarsystem.nasa.gov.

"I knew instantly that Mars was lifeless," recalled Lovelock. "It was an equilibrium atmosphere." He immediately switched viewpoints to ask himself how Earth's complex atmosphere could also remain stable.

> It came to me suddenly, just like a flash of enlightenment, that to persist and keep stable, something must be regulating the atmosphere and so keeping it at its constant composition. Moreover, if most of the gases came from living organisms, then life at the surface must be doing the regulation.

Afterwards Carl Sagan told him that the Sun was thought to have been some 30 percent less luminous early in the life of the Earth than it was now, yet Lovelock also knew that there had been no corresponding long-term rise in the temperature of the Earth.

> Suddenly the image of the Earth as a living organism able to regulate its temperature and chemistry at a comfortable steady state emerged in my mind.[74]

Around the same time, Sagan was working on an American edition of a 1962 book called *Universe, Life, Mind* by the Soviet astrophysicist Iosif S. Shklovskii. It came out in 1966 as *Intelligent Life in the Universe* by Sagan and Shklovskii and sold very well. Shklovskii, doubtless aware of Vernadsky's earlier work on the biosphere, had written: "Such a vast amount of oxygen as is present in the Earth's atmosphere can be explained only in terms of extensive biological activity." Sagan was doubtful, but his contribution on this point seems to bear the influence of his conversations with Lovelock: "I wonder whether an intelligent anaerobic organism, who finds oxygen a poison gas, would conclude very readily that an extensive oxygen atmosphere can only be the product of biological activity."[75]

Lovelock went on to attend the second "Origins of Life" conference at Princeton in May 1968, where the reception he received showed how far from the mainstream his ideas were at that stage. His attempts to suggest that the Earth's atmosphere had a partly biological origin met with blank incomprehension; natural scientists rejected him as a physicist, while physical scientists marked him down as a biologist.[76] Lyn Margulis, editor of the conference proceedings and his future coauthor, was also present and later wondered aloud to Lovelock why they had not met. "He said, because the first time I opened my mouth, Preston Cloud yelled at me and was so intimidating and rude that I didn't speak for the rest of the conference." Margulis's innovative work on cell biology provided Lovelock with the missing link, a biological mecahnism to account for the presence of methane in the Earth's atmosphere. Another space scientist interested in extraterrestrial life proved receptive: this time it was Alistair Cameron, editor of an early volume of essays on the subject and now chairman of the National Academy of Sciences Space Science Board. "He saw a couple of paragraphs that Lovelock and I had written about the effect of life as a planetary phenomenon," recalled Margulis. "He totally and immediately understood. He told me he never understood anything biologists talked about at all. It doesn't make any sense to me at all; this is the first time I've seen sense."[77]

By 1972 US and Soviet probes had established that both Mars and Venus had simply structured atmospheres hostile to life, Mars at extremely low temperatures and pressures and Venus at extremely high ones. This highlighted the question of how Earth alone had remained hospitable to life over billions of years.[78] In a series of articles in 1972–4 Lovelock and Margulis discussed the explanation for the "anomalous nature" of the Earth's atmosphere and presented the Gaia hypothesis, "the concept that the Earth's atmosphere is actively maintained and regulated by life on the surface, that is by the biosphere." They explained: "We have written the paper to be comprehensible to a wide scientific audience, recognizing that an understanding of the Earth's atmosphere will come only from the cooperation of many scientists: planetary astronomers, geologists, meteorologists, chemists, physicists, and biologists."[79] The wider philosophical claims of the Gaia hypothesis proved controversial but, as Conway points out, it provided "a view of Earth that could be grasped by systems engineers."[80] Indeed, Margulis and Lovelock presented a speculative graph of oscillations of planetary temperature over the past 100,000 years and suggested that a "hypothetical planetary engineer would probably recognize this as a chart of the behavior of an unstable control system in which instability had developed leading to oscillation yet control had not failed altogether."[81]

The Gaia hypothesis broke scientific ground in several ways. First, it fostered a convergence of the physical and biological sciences. Second, the Gaia hypothesis represented the ultimate application of system theory and cybernetics. Gaia was not at first the living planet but the homeostatic planet. At the formative period Lovelock was not aware of Vladimir Vernadsky's 1926 organicist work *The Biosphere,* with its claim that "life is a geological force." The Russian practice of integrating the study of geology, chemistry and biology – 'biogeochemistry' – was rooted in the study of particular environments and lacked the capacity to travel which the language of systems would have allowed it. When Lovelock was eventually introduced to this work he commented that Vernadsky "did not seem to have a feeling for system science."[82] Gaia was particularly strongly attacked by evolutionists, led by Richard Dawkins, for the concept of a single collective organism managing its own evolution appeared to violate the basic principles of natural selection.[83] Among life scientists, felt Lovelock, only Eugene Odum "understood that an ecosystem is a deterministic feedback system," reflecting Lovelock's understanding that "Gaia is the ecosystem of the Earth." Thirdly, Lovelock's work married the perspectives of ecology with those of planetary science, allied to a shift of time scale from the historical to the astronomical. Gaia, in the words of Donald Worster, was "how things look to the cosmic eyeball"— that is, in time as well as space.[84] The serendipitous appearance of the first photographs of the Earth from space helped to propagate this mode of whole Earth thinking to a global public just as the environmentalist renaissance took off. But Lovelock (a visual thinker affected by dyslexia) had already achieved his insights imaginatively, without the aid of pictures from space.

By the 1980s, notwithstanding considerable scientific hostility to the full-blown Gaia hypothesis, the view that life played a role in forming the physical Earth had become orthodox.[85] As Erik Conway has shown, the scientists who worked with the planetary probes of the 1960s and 1970s—Mariner, Viking, Pioneer, and Voyager—which searched for evidence of dynamic change and life elsewhere in the solar system, became used to combining physical, chemical, geological, atmospheric, and biological investigations in pursuit of planetary questions, much as the freelance Lovelock had sought to do for the Earth. The budgetary crisis that afflicted planetary science in the late 1970s and early 1980s brought many of these NASA planetary scientists to the study of the Earth, at a time when the Earth sciences were acquiring global data sets as a consequence of their Cold War expansion. This in turn generated a wave of research and observation of the planetary dynamics of the Earth, and a swell of concern over environmental issues such as ozone depletion and climate change. Acceptance of a world view that integrated life and non–life sciences was fostered by the name chosen for the new field in 1986: distancing itself from organicist views (such as Vernadsky's "biogeochemistry") NASA opted for "Earth systems science."[86]

Conclusion

In September 1970 *Scientific American* produced a special issue on "The Biosphere," later published as a book (see Figure 10.3). It opened with the observation that "photographs of the Earth show it has a blue-green color" and continued with an introductory essay by G. Evelyn Hutchinson, invoking Vernadsky as the father of the concept of the biosphere and rewriting the entire history of life on Earth in its light. Successive essays explained the various cycles operating at global level: the energy cycles of both planet and biosphere; the water, carbon, oxygen, nitrogen, and mineral cycles; and the human cycles of food, energy, and metal production, each identified as "a cycle in the biosphere." Similar flow diagrams in each chapter signified that all these cycles could be understood in systems terms. Hutchinson's essay featured the master diagram of the biosphere, showing physical, biological, and human cycles interacting. It demonstrated how far holistic thinking about the Earth had come in the year of the first Earth Day, even before the *Apollo 17* Blue Marble appeared.[87]

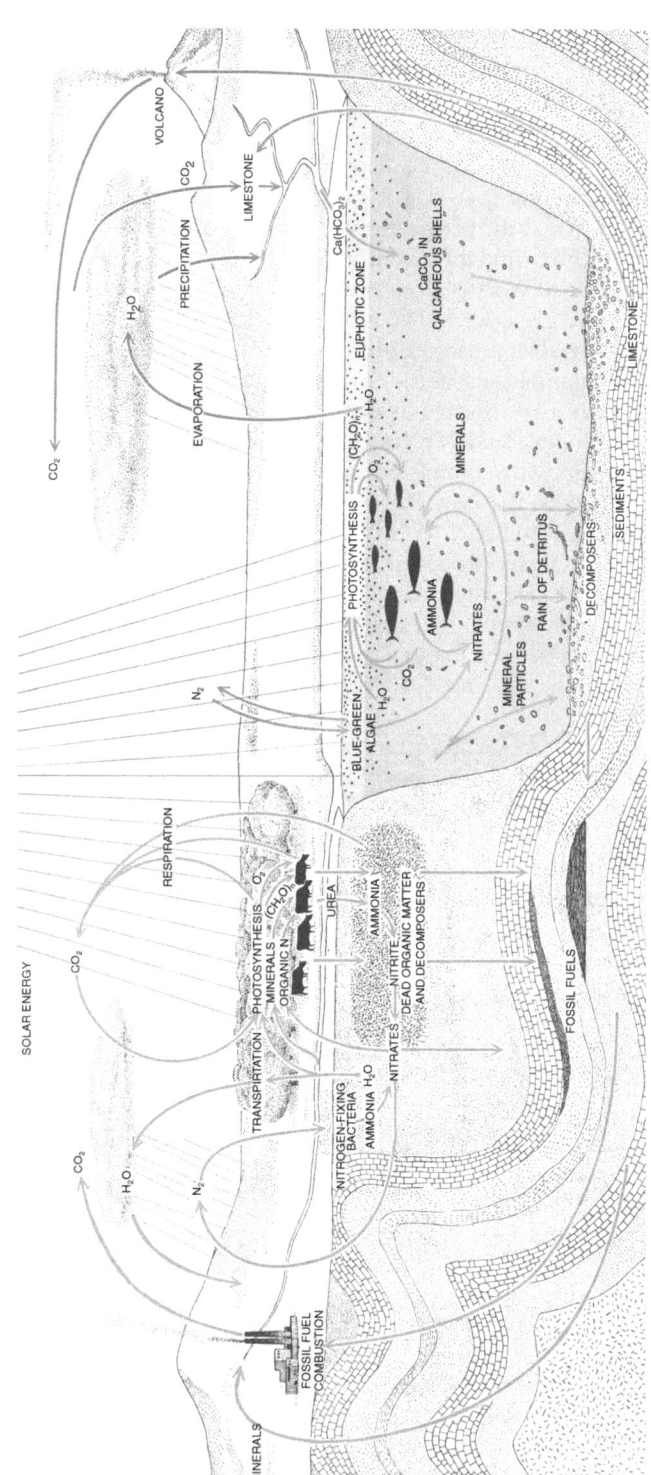

Figure 10.3 The biosphere

Source: Scientific American, September 1970 [reprinted in *The Biosphere* (San Francisco: W. H. Freeman & Co, 1970), 8–9].

Ideas of the whole Earth preceded the pictures, but the pictures had a powerful impact because there already existed ideas and models of planetary processes which had been developed in several different fields, conceptualized through dynamic models of the Earth as seen from the outside. The early images of the home planet in turn accelerated and propagated a whole Earth style of thinking whose defining mark was the understanding of biological and physical mechanisms as interdependent. The Cold War expansion of the Earth sciences in association with the space program generated both the research data and the Earth images upon which the new understanding was founded. The combined insights of a dozen separate disciplines had shown that the Earth was whole after all. Both literally and metaphorically, it was a change of world view. Arguably, it was a scientific revolution.

Notes

1. Fred Hoyle, *The Nature of the Universe* (Oxford, 1950), on pp. iii, 9–10. Hoyle's comments were first aired in a radio broadcast in 1948: Robert Poole, *Earthrise: How Man First Saw the Earth* (New Haven, CT: Yale University Press, 2008), 37–9.

2. From Hoyle's after-dinner speech at the *Apollo 11* lunar science conference, 6 January 1970, quoted in Donald D. Clayton, *The Dark Night Sky* (New York: Quadrangle, 1975).

3. For discussions of this line of argument, see: Frank White, *The Overview Effect* (Reston, VA: 1998 [1987]); De Witt Douglas Kilgore, *Astrofuturism* (Philadelphia: University of Pennsylvania Press, 2003); Charles S. Cockell, *Space on Earth: Saving Our World by Seeking Others* (London: Macmillan, 2007); Fred W. Spier, *Big History and the Future of Humanity* (London: Wiley-Blackwell, 2010).

4. Thomas S. Kuhn, *The Structure of Scientific Revolutions* (Chicago: Chicago University Press, 1996 [1962]), chap. 10.

5. Jon Darius, *Beyond Vision: One Hundred Historic Scientific Photographs* (Oxford: Oxford University Press, 1984), 142; Galen Rowell, in *Sierra*, September 1995, 73, quoted in Robert Zimmerman, *Genesis: the Story of Apollo 8* (New York, 1998), 284.

6. Donald Worster, *Nature's Economy: A History of Ecological Ideas* (Cambridge: CUP, 1994), 358–9, 387.

7. J. R. McNeill and Corinna Unger (eds.), *Environmental Histories of the Cold War* (Cambridge: Cambridge University Press, 2010), 17.

8. Erik M. Conway, *Atmospheric Science at NASA* (Baltimore, MD: Johns Hopkins University Press, 2008), 122–123.

9. Daniel C. Noel, "Re-Entry: Earth Images in Post-Apollo Culture," *Michigan Quarterly Review* 18:2 (1979): 155–176; Dennis Cosgrove, "Contested Global Visions: *One-World, Whole-Earth*, and the Apollo Space Photographs," *Annals of the Association of American Geographers* 84:2 (1994): 270–294; Sheila Jasanoff, "Image and Imagination: the Formation of Global Environmental Consciousness," in Clark A. Miller and Paul N. Edwards (eds.), *Changing the Atmosphere: Expert Knowledge and Environmental Governance* (Cambridge, MA: MIT Press, 2001), 309–37; Neil Maher, "Shooting the Moon," *Environmental History* 9:3 (2004): 526–531. Poole, *Earthrise* (the article in this volume supplements *Earthrise*).

10. Worster, *Nature's Economy*, 387. See also D. Worster, "The Vulnerable Earth: Towards a Planetary History," in D. Worster (ed.), *The Ends of the Earth: Perspectives on Modern Environmental History* (Cambridge: Cambridge University Press, 1988).

11. Charles C. Gillespie and Nicolaas Rupke, *Genesis and Geology* (Cambridge, MA: Harvard University Press, 1951), 104.
12. Laura Dassow Wild, *The Passage to Cosmos: Alexander von Humboldt and the Shaping of America* (Chicago: Chicago University Press, 2009), 217, 241.
13. Vladimir I. Vernadsky, *The Biosphere* (New York: Copernicus, 1998 [1926]), 41, 43–5, and Lyn Margulis et. al., "Foreword," 14–19.
14. Poole, *Earthrise*, Chapter 8.
15. Ronald E. Doel, "Constituting the Postwar Earth Sciences: The Military's Influence on the Environmental Sciences in the USA after 1945," *Social Studies of Science* 33:5 (2003): 635–666, on 656–7.
16. Michael Aaron Dennis, "Earthly Matters: on the Cold War and the Earth Sciences," *Social Studies of Science* 33:5 (Oct. 2003): 809–819, on 817.
17. Joseph Masco, "Bad Weather: On Planetary Crisis," *Social Studies of Science* 40:1 (2010): 7–40, on 9.
18. Walter Sullivan, *Assault on the Unknown* (London: Hodder and Stoughton, 1962), 150, 48.
19. Dian Olson Belanger, *Deep Freeze* (Boulder, CO: University Press of Colorado, 2006); Roger T. Launius, James R. Fleming & David H. Devorkin (eds.), *Globalising Polar Science* (New York and London: Palgrave Macmillan, 2010).
20. Jacob Hamblin, *Oceanographers and the Cold War* (Seattle, WA: University of Washington Press, 2005), 66 and Chapter 3.
21. Chunglin Kwa, "Radiation Ecology, Systems Ecology and the Management of the Environment," in Michael Shortland (ed.), *Science and Nature: Essays in the History of Environmental Sciences*, BSHS Monograph 8 (1993): 213–251; Joachim Radkau, *Nature and Power: a Global History of the Environment* (Cambridge: Cambridge University Press, 2008 [2002]), 250–60.
22. Sullivan, *Assault on the Unknown*, 136–137, 163.
23. Michael Aaron Dennis, "A Polar Perspective," in Launius, Fleming and Devorkin, *Globalising Polar Science*, 13–22.
24. Simone Turchetti et al., "On Thick Ice: Scientific Internationalism in Antarctic Affairs, 1957–1980," *History and Technology* 24:4 (2008): 351–376.
25. Miller and Edwards, "Introduction," in *Changing the Atmosphere*, 1; Paul Edwards, *A Vast Machine: Computer Models, Climate Data, and the Politics of Global Warming* (Cambridge, MA: MIT Press, 2010), 226–227; Poole, *Earthrise*.
26. The booklet is available at the NAS website [http://www7.nationalacademies.org/archives/IGYPlanetEarthPosters.html].
27. Harry Wexler, "Observing the Weather from a Satellite Vehicle," *Journal of the British Interplanetary Society* 13:5 (1954): 269–76; James R. Fleming, "A 1954 Color Painting of Weather Systems as Viewed from a Future Satellite," *Journal of the American Meteorological Society* 88 (2007): 1525–1527.
28. *Life*, November 7, 1960.
29. Hamblin, *Oceanographers and the Cold War*, 91–94, 99–100, 144–145, 177–178.
30. E. Aronova, K. Baker, and N. Oreskes, "Big Science and Big Data in Biology: From the International Geophysical Year through the International Biological Program to the Long-Term Ecological Research Program, 1957–present," *Historical Studies in the Natural Sciences* 40:2 (2010): 183–224, on 203; E. Aronova, "Environmental Monitoring in the Making: From Surveying Nature's Resources to Monitoring Nature's Change," *Historical Social Research*, special issue "Climate and Beyond," eds. Andrea Westerman and Christian Rohr (forthcoming).
31. John Cloud, "Imagining the World in a Barrel: CORONA and the Clandestine Convergence of the Earth sciences," *Social Studies of Science* 31:2 (2001): 240,

244–246; John Cloud, "Crossing the Olentangy River," *Studies in the History and Philosophy of Modern Physics* 31:3 (2000): 371–404; Doel, "Constituting the Post-war Earth sciences," 638–641.

32. Sullivan, *Assault on the Unknown*, 392–9; Steven J. Dick, "Geodesy, Time, and the Markowitz Moon Camera Program: An Interwoven International Geophysical Year Story," in Launius, Fleming, and Devorkin (eds), *Globalizing Polar Science*, 307–326.

33. Poole, *Earthrise*, 82–84.

34. Cloud, "Crossing the Olentangy River," 376–379; Conway, "The IGY and Planetary Science," in Launius et. al. (eds.), *Globalizing Polar Science*, 331–342.

35. Cloud, "Imagining the World in a Barrel," 244–246.

36. Charles Darwin, *Voyage of the Beagle* (London: Penguin, 1989 [1839]), 356.

37. James Lawrence Powell, *Mysteries of Terra Firma: the Age and Evolution of the Earth* (New York: Simon and Schuster, 2001); Richard Fortey, *The Earth* (London: Harper Collins, 2004), 26.

38. Hamblin, *Oceanographers and the Cold War*, 77–78.

39. Ibidem, 41 and Chapter 2, *passim*.

40. See Turchetti's chapter in this volume.

41. J. Tuzo Wilson in 1968, cited in Hamblin, *Oceanographers and the Cold War*, 198.

42. Naomi Oreskes, "A Context of Motivation: US Navy Oceanographic Research and the Discovery of Sea-Floor Hydrothermal Vents," *Social Studies of Science* 33:5 (2003): 697–742.

43. Quoted in Lyn Margulis et al., "Foreword," to Vernadsky, *The Biosphere*, 14–19.

44. Fortey, *The Earth*, 84–7; Cathy Barton, "Marie Tharp, Oceanographic Cartographer, and Her Contributions to the Revolution in the Earth Sciences," *Geological Society of London Special Publications* 192 (2002): 215–228.

45. Thomas, *Lives of a Cell*, 147–148.

46. James R Fleming, "A 1954 Color Painting of Weather Systems as Viewed from a Future Satellite," 1525–7.

47. Edwards, *A Vast Machine*, 204.

48. Ibidem, xix.

49. Amy Dahan Dalmedico and Helene Guillemot, "Climate Change: Scientific Dynamics, Expertise, and Geopolitical Challenges," in Gregoire Mallard, Catherine Paradeise, and Ashveen Peerbaye (eds.), *Global Science and National Sovereignty* (London: Routledge, 2009), 196–9; Conway, *Atmospheric Science at NASA*, 41–44, 62–63, 70–77; Poole, *Earthrise*, 59–65; Edwards, *A Vast Machine*, 234.

50. Ibidem, 207–15, 226–227.

51. Andrew K. Johnston, "Exploring Planet Earth: the Development of Satellite Remote Sensing for Earth Science," in Roger D. Launius (ed.), *Exploring the Solar System: the History and Science of Planetary Exploration* (New York, London: Palgrave Macmillan, 2013), 205.

52. Conway, *Atmospheric Science at NASA*, 62–93; Paul Edwards, "Representing the Global Atmosphere: Computer Models, Data and Knowledge about Climate Change," in Miller and Edwards (eds), *Changing the Atmosphere*, 31–65.

53. Edwards, *A Vast Machine*, 253, 189.

54. Eugene P. Odum, *Fundamentals of Ecology* (Philadelphia: W. B. Sounders, 1971); Worster, *Nature's Economy*, 369–370; E. P. Odum, *Ecology and our Endangered Life-Support Systems* (Sunderland, MA: 1989), 1–7; Joel B. Hagen, *The Entangled Bank: The Origins of Ecosystem Ecology* (New Brunswick, NJ: Rutgers University Press, 1992), 189–197.

55. Odum, *Fundamentals of Ecology*, 271–272.

56. Norbert Wiener, *Cybernetics, or Control and Communication in the Animal and the Machine* (New York: Wiley, 1948); Jon Agar, *Science in the Twentieth Century and Beyond* (Cambridge: Polity Press, 2012), 373–5; David Hounshell, "The Cold War, RAND and the Generation of Knowledge 1946–62," *Historical Studies in the Physical Sciences* 27:2 (1997): 237–267.

57. Ibidem, 244. See also Agatha C. Hughes and Thomas P. Hughes (eds.), *Systems, Experts, and Computers: The Systems Approach in Management and Engineering, World War II and After* (Cambridge, MA: MIT Press, 2000); Gregory Bateson, "Physical Thinking and Social Problems," *Science* 103 (1946): 717–718.

58. Buckminster Fuller, "Vertical is to Live—Horizontal is to Die," *American Scholar* 39 (1969). See also Hal Aigner, "Buckminster Fuller's World Game," *Mother Earth News*, Dec. 1970, 62–68; Diedrich Diedrichsen and Anselm Franke (eds.), *The Whole Earth* [catalogue of exhibition at *Haus de Kulturen der Welt*, Berlin, June 2013]; Poole, *Earthrise*, 146–147.

59. Richard K. Ashley, "The Eye of Power: the Politics of World Modelling," *International Organization* 37:3 (1983): 495–535; Edwards, *A Vast Machine*, 366–74; video interview with Jay Forrester in *The Whole Earth* exhibition, Berlin, June 2013.

60. Peter Bowler, *The Fontana History of The Environmental Sciences* (London: Fontana Press, 1992), 527–34; Greg Mitman, *The State of Nature: Ecology, Community and American Social Thought, 1900–1950* (Chicago: Chicago University Press, 1992), 201–209; Peter J. Taylor, "Technocratic Optimism, H. T. Odum and the Partial Transformation of Ecological Metaphor after WWII," *Journal of the History of Biology* 21 (1988): 213–244; George Gaylord Simpson, *The Meaning of Evolution* (New Haven, CT: Yale University Press, 1949).

61. Taylor, "Technocratic Optimism," 215–23; Bowler, *History of the Environmental Sciences*, 537–546; Thomas E. Lovejoy, "George Evelyn Hutchinson, 1903–1991," *Biographical Memoirs of Fellows of the Royal Society* 57 (2011): 167–177. See also Jonathan D. Oldfield and Denis Shaw, "V. I. Vernadsky and the Noosphere Concept: Russian Understanding of Society-Nature Interaction," *Geoforum* 37 (2006): 145–154.

62. Bowler, *History of the Environmental Sciences*, 537–546.

63. Taylor, "Technocratic Optimism," 233–244.

64. Beatty Jean Craige, *Eugene Odum: Ecosystem Ecologist and Environmentalist* (Athens: GA: University Georgia Press, 2001), chap. 9; Chunglin Kwa, "Radiation Ecology, Systems Ecology and the Management of the Environment," 213–251.

65. Craige, *Eugene Odum*, xi–xii and chaps. 4–5; Mittman, *State of Nature*, 209–210.

66. Fritjof Capra, *The Web of Life* (New York: Random House, 1996).

67. Rene Dubos, "A Theology of the Earth," in R. Dubos (ed.), *The God Within* (New York: Charles Scribner, 1972), 38–39.

68. *The Last Whole Earth Catalogue* (California: Portola Institute, and Harmondsworth: Penguin, 1971), inside front cover. The source given in the quotation is: Harold J. Morowitz, *Energy Flow in Biology* (Woodbridge, CT: Ox Bow Press, 1979).

69. Lewis Thomas, "Foreword," in Paul R. Ehrlich, Carl Sagan, Donald Kennedy, and Walter Orr Roberts, *The Cold and the Dark: the World after Nuclear War* (New York: W. W. Norton, 1984), xxi–xxiv. See also Ann Woodlief, "Lewis Thomas," in *Dictionary of Literary Biography*, vol. 275 [available at: http://www.vcu.edu/engweb/LewisThomas.htm]

70. James Lovelock, *Homage to Gaia* (Oxford: Oxford University Press, 2001), 241. See also Jon Turney, *Lovelock and Gaia* (New York: Columbia University Press, 2004); James Lovelock, interview for British Library Oral History of British Science, sections 12–13, at http://sounds.bl.uk/Oral-history/Science.

71. Lynn Margulis, *The Symbiotic Planet: a New Look at Evolution* (New York: Basic Books, 1998), 147–148.
72. Lovelock, *Homage to Gaia*, 241–255.
73. Dian Hitchcock and James Lovelock, "Life Detection by Atmospheric Analysis," *Icarus* 7(1967): 149–159; J. Lovelock, "A Physical Basis for Life Detection Experiments," *Nature* 207 (1965): 568–570.
74. Lovelock, *Homage to Gaia*, 251–254.
75. Carl Sagan and Iosif S. Shklovskii, *Intelligent Life in the Universe* (San Francisco: Holden-Day, 1966), 221, 254.
76. Lyn Margulis (ed.), *Proceedings of the Second Conference on the Origin of Life* (Washington, DC: Interdisciplinary Communication Associates, 1971).
77. Interview with Lyn Margulis, 1998, NASA History Office oral history collection, Washington, DC.
78. Conway, *Atmospheric Science at NASA,* 96–103; Erik M. Conway, "Planetary Science and the 'Discovery' of Global Warming," in Roger D. Launius (ed.), *Exploring the Solar System: the History and Science of Planetary Exploration* (London: Palgrave Macmillan, 2013), 183–202; Roger D. Launius, "Venus-Earth-Mars: Comparative Climatology and the Search for Life in the Solar System," in Launius (ed.), *Exploring the Solar System,* 223–247.
79. James Lovelock, "Gaia as Seen through the Atmosphere," *Atmospheric Environment* 6:8 (1972): 579–580; James Lovelock and Lynn Margulis, "Atmospheric Homeostasis by and for the Biosphere: the Gaia Hypothesis," *Tellus* 26 (1974): 1–10; L. Margulis and J. Lovelock, "Biological Modulation of the Earth's Atmosphere," *Icarus* 21 (1974): 471–489 (quotations on 471).
80. Conway, *Atmospheric Science at NASA,* 112–121.
81. Margulis and Lovelock, "Biological Modulation", 487.
82. James Lovelock, *The Ages of Gaia* (Oxford: Oxford University Press, 1995), 8–10; Giulia Rispoli, "Between Biosphere and Gaia: Earth as a Living Organism in Soviet Geo-Ecology," unpublished paper delivered at International Congress for the History of Science, Technology and Medicine, University of Manchester, July 2013; Jonathan D. Oldfield and Denis J. B. Shaw, 'V. I. Vernadsky and the Noosphere Concept: Russian Understandings of Society-Nature Interaction', *Geoforum* 37 (2006), 145–154.
83. Richard Dawkins, *The Extended Phenotype* second edition (Oxford: Oxford University Press, 1982).
84. D. Worster, "Doing Environmental History," in D. Worster (ed.), *The Ends of the Earth: Perspectives on Modern Environmental History* (Cambridge: Cambridge University Press, 1988).
85. Conway, "Planetary Science", 83–84.
86. Kim McQuaid, "Selling the Space Age: NASA and Earth's Environment, 1958–1990," *Environment and History* 12:2 (2006): 127–163.
87. Scientific American, *The Biosphere* (San Francisco: W. H. Freeman, 1970).

Bibliography

ACDA. *Arms Control and Disarmament Agreements: Texts and History of Negotiations.* Washington, DC: US Government Printing Office, 1977.

Adams, Mark B. "Networks in Action: The Khrushchev Era, the Cold War, and the Transformation of Soviet Science. In *Science, History and Social Activism: A Tribute to Everett Mendelsohn*, edited by G. E. Allen and R. M. MacLeod, 255–276. Dordrecht and Boston: Kluwer, 2001.

AEC. *Worldwide Effects of Atomic Weapons. Project Sunshine, August 6, 1953.* Santa Monica: Rand Corporation, 1956.

Agar, Jon, and Brian Balmer. "British Scientists and the Cold War: the Defence Research Policy Committee and Information Networks, 1947–1963." *Historical Studies in the Physical Sciences* 28:2 (1998): 209–252.

Agar, Jon. *Science in the Twentieth Century and Beyond.* Cambridge: Polity Press, 2012.

Agrawala, Shardul. "Context and Early Origins of the Intergovernmental Panel on Climate Change". *Climatic Change* 39:4 (1998): 605–620.

Aigner, Hal. "Buckminster Fuller's World Game." *Mother Earth News*, Dec. 1970, 62–68.

Aitor Anduaga. *Geofísica, economía y sociedad en la España contemporánea.* Madrid: CSIC, 2009.

Aldrich, Richard. "British Intelligence and the Anglo-American 'Special Relationship' during the Cold War." *Review of International Studies* 24:3 (1998): 331–351.

Aldrich, Richard. *Espionage, Security, and Intelligence in Britain, 1945–1970.* Manchester: Manchester University Press, 1998.

Aldrich, Richard. *GCHQ: Britain's Most Secret Intelligence Agency.* London: HarperCollins, 2011.

Allaud, Louis A., and Maurice H. Martin. *Schlumberger: The History of a Technique.* New York: Wiley & Sons, 1978.

Althoff, William F. *Drift Station: Arctic Outposts of Superpower Science.* Washington, DC: Potomac Books, 2007.

Anderson, Irvine H. *ARAMCO. The United States and Saudi Arabia. A Study of the Dynamics of Foreign Oil Policy, 1933–1950.* Princeton: Princeton University Press, 1981.

Anker, Peder. *Imperial Ecology: Environmental Order and the British Empire, 1895–1945.* Cambridge, MA: Harvard University Press, 2001.

Anon. "A 100 Mile High Portrait of Earth." *Life Magazine* 39 (1955), 10–11.

Anson, Peter, and Dennis Cummings. "The First Space War: The Contribution of Satellites to the Gulf War." *Royal United Services Institute Journal* 136 (1991): 45–53.

Appleyard, Ray K. "The Birth of UNSCEAR—the Midwife's Tale." *Journal of Radiological Protection* 30 (2010): 621–626.

Armstrong, Terence. "Soviet Work on Sea Ice Forecasting." *Polar Record* 7:49 (1955): 302–311.

Armstrong, Terence. "The Voyage of the *Komet* Along the Northern Sea Route, 1940." *Polar Record* 5:37–38 (1949): 291.

Armstrong, Terence. *Sea Ice North of the USSR*. London: Admiralty Hydrographic Department, 1958.

Armstrong, Terence. *The Northern Sea Route*. Cambridge: Cambridge University Press, 1952.

Armstrong, Terence. *The Russians in the North: Aspects of Soviet Exploration and Exploitation of the Far North, 1937–57*. London: Methuen, 1960.

Arnold, David C. "Corona." In *Space Exploration and Humanity: A Historical Encyclopedia*, edited by Stephen B. Johnson, 873–875. Santa Barbara, CA: ABC-CLIO, 2010.

Arnold, Lorna (with Katherine Pine). *Britain and the H-Bomb*. London: Palgrave Macmillan, 2001.

Arnold, Lorna. *A Very Special Relationship: British Weapon Trials in Australia*. London, HMSO, 1987.

Aronova, Elena, Karen Baker, and Naomi Oreskes. "Big Science and Big Data in Biology: From the International Geophysical Year through the International Biological Program to the Long Term Ecological Research Program, 1957 to present." *Historical Studies in the Natural Sciences* 40:2 (2012): 183–224.

Aronova, Elena. "Environmental Monitoring in the Making: From Surveying Nature's Resources to Monitoring Nature's Change." *Historical Social Research* (Special Issue "Climate and Beyond" edited by Andrea Westerman and Christian Rohr -forthcoming).

Ashley, Richard K. "The eye of power: the politics of world modelling." *International Organization* 37:3 (1983): 495–535.

AWRE. *The Detection and Recognition of Underground Explosions. A Special UKAEA Report*. London: AWRE/HMSO, 1965.

Baggio, P., F. Ippolito,, S. Lorenzoni,, G. Marinelli,, and F. Silvestro. "Occurrence of uranium in Italy." *Energia Nucleare* 4:3 (1957): 196–198.

Baines, Phillip J. "Prospects for "Non-Offensive" Defenses in Space." *Center for Nonproliferation Studies Occasional Paper* 12 (2004).

Baldwin, F. "Patrolling the Empire: Reflections on the USS Pueblo," *Bulletin of Concerned Asian Scholars,* Summer 1972: 54–75.

Baliles, Maurice D., and Herbert Neiss, eds. *Conference on Satellite Ice Studies* [Meteorological Satellite Laboratory report no. 20]. Washington, DC: US Weather Bureau, 1963.

Ball, Kirstie and Kevin Haggerty. "Editorial: Doing Surveillance Studies." *Surveillance & Society* 3 (2005): 129–138.

Ball, Kirstie, Kevin Haggerty, and David Lyon, eds. *Routledge Handbook of Surveillance Studies*. Abingdon: Routledge, 2012

Bamberg, James H. *The History of the British Petroleum Company, 2 Vols.* Cambridge: Cambridge University Press, 1994.

Barca, Francesc X. "Nuclear Power for Catalonia: the Role of the Official Chamber of Industry of Barcelona, 1953–1962." *Minerva, 43* (2005): 163–181.

Barth, Kai-Henrik. "Catalysts of Change: Scientists as Transnational Arms Control Advocates in the 1980s." In *Global Power Knowledge: Science and Technology in International Affairs (Osiris* 21), edited by J. Krige and Kai-Henrik Barth, 161–181. Chicago: University of Chicago Press, 2006.

Barth, Kai-Henrik. "Science and Politics in Early Nuclear Arms Control Negotiations." *Physics Today* 51 (1998): 34–39.

Barth, Kai-Henrik. "The Politics of Seismology." *Social Studies of Science* 33 (2003): 743–781.

Barton, Cathy. "Marie Tharp, Oceanographic Cartographer, and Her Contributions to the Revolution in the Earth Sciences." *Geological Society of London Special Publications* 192 (2002): 215–228.

Bartzen Culver, Kathleen. "From Battlefield to Newsroom: Ethical Implications of Drone Technology in Journalism." *Journal of Mass Media Ethics: Exploring Media Morality* 29 (1) 2014: 52–56.

Bates, Charles C., Thomas F. Gaskell, and Robert B. Rice. *Geophysics in the Affairs of Man: A Personalized History of Exploration Geophysics and its Allied Sciences of Seismology and Oceanography.* Oxford and New York: Pergamon, 1982.

Bates, Charles. "Vela Uniform. The Nation's quest for better detection of underground nuclear explosions." *Geophysics* 26:4 (1961): 499–507.

Bateson, Gregory. "Physical Thinking and Social Problems." *Science* 103 (1946): 717–18.

Battimelli, Gianni, ed. *L'Istituto Nazionale di Fisica Nucleare.* Bari: Laterza, 2001.

Baucom, Donald R. "The Rise and Fall of Brilliant Pebbles." *Journal of Social, Political and Economic Studies* 29 (2004): 145–190.

Baucom, Donald R. *The Origins of SDI: 1944–1983.* Lawrence, KA: University Press of Kansas, 1992.

Bayly, Christopher A., Sven Beckert, Matthew Connelly, Isabel Hofmeyr,, Wendy Kozol, and Patricia Seed. "AHR Conversation: On Transnational History." *American Historical Review* 111 (2006): 1441–1464.

Bazan, Elizabeth B. "The Foreign Intelligence Surveillance Act: An Overview of the Statutory Framework and Recent Judicial Decisions." *Congressional Research Service,* RL30465, April 2005.

Belanger, Dian Olson. *Deep Freeze.* Boulder, CO: University Press of Colorado, 2006.

Beltran, Alain, and Sophie Chauveau, *Elf Aquitaine, des origines a 1989.* Paris: Fayard, 1998.

Bergstralh, Thor. "Photography from the V-2 at altitudes ranging up to 160 kilometers." In *Upper Atmosphere Research Report no. IV* [NRL report R-3171], edited by H. E. Newell and J. W. Siry, 119–130. Washington, DC: Naval Research Laboratory, 1947.

Berton, Pierre. *The Arctic Grail: The Quest for the North West Passage and the North Pole, 1818–1909.* New York: Viking, 1988.

Beschloss, Michael R. *Mayday: Eisenhower, Khrushchev and the U-2 Affair.* New York: Harper & Row, 1986.

Bijker, Wiebe, Roland Bal, and Ruud Hendriks, *The Paradox of Scientific Authority: The Role of Scientific Advice in Democracies.* Cambridge MA: MIT Press, 2009.

Bird, J. Brian, and A. Morrison. "Space Photography and Its Geographical Applications," *Geographical Review* 54 (1964): 463–486.

Black, Jeremy. *The British Seaborne Empire.* New Haven, CT: Yale University Press, 2004.

Blanchard, Élodie Vielle. "Les limites à la croissance dans un monde global. Modélisations, prospectives, refutations." Ph.D. dissertation, Paris: École des hautes études en sciences sociales, 2011.

Blowers, Jay H. "The United State's Position on the Environment." *American Journal of Public Health* 62 (1972): 634–638.

Bodu, Robert. *Les secrets des cuves d'attaque: 40 ans de traitement des minerais d'uranium.* Cogéma, 1994.

Bolt, Bruce. *Nuclear Explosions and Earthquakes. The Parted Veil.* San Francisco: W. H. Freeman & Co., 1976.

Bossuat, Gérard. *La France, l'aide américaine et la construction européenne, 1944–1954.* Paris: Comité pour l'histoire économique et financière de la France, 1997.

Boudia, Soraya, and Nathalie Jas, eds. *Powerless Science: Science and Politics in a Toxic World.* London: Berghan Books, 2014.

Boudia, Soraya, and Nathalie Jas, eds., *Toxicants, Health and Regulation since 1945.* London: Pickering and Chatto, 2013.

Boudia, Soraya, Néstor Herran, and Simone Turchetti, eds., Special Issue on the Transnational History of Science, *British Journal for the History of Science* 45: 3 (2012).

Boudia, Soraya. "Global Regulation: Controlling and Accepting Radioactivity Risks." *History and Technology* 23:4 (2007): 389–406.

Bowker, Geoffrey, and Susan Leigh Star. *Sorting Things Out: Classification and Its Consequences.* Cambridge, MA: MIT Press, 1999.

Bowler, Peter. *The Fontana History of The Environmental Sciences.* London: Fontana Press, 1992.

Boyer, Paul. *By the Bomb's Early Light: American Thought and Culture at the Dawn of the Atomic Age.* Chapel Hill: The University of North Carolina Press, 1994.

Brigham, Lawson. "Armstrong, Terence." In *Encyclopedia of the Arctic: A-F*, edited by Mark Nuttall, 154–155. New York: Routledge, 2005.

Brink Jr., Frank. *Detlev Wulf Bronk 1897–1975: a Biographical Memoir.* Washington, DC: National Academy of Sciences, 1978.

Brooks, Karl Boyd. *Before Earth Day: The Origins of American Environmental Law, 1945–1970.* Lawrence: University of Kansas Press, 2009.

Brown, J. H., and E.E. Howick. "Physical Measurements of Sea Ice" (Technical Report 825). San Diego, CA: US Navy Electronics Laboratory, 1958).

Brugioni, Dino A. *Eyes in the Sky: Eisenhower, the CIA and Cold War Aerial Espionage.* Annapolis, MD: Naval Institute Press, 2010.

Buccianti, Giovanni. *Enrico Mattei: Assalto al potere petrolifero mondiale.* Milan: Giuffrè, 2005.

Bulkeley, Rip. *Sputniks Crisis and Early United States Space Policy.* Bloomington: Indiana University Press, 1991.

Burrows, William E. *By Any Means Necessary: America's Secret Air War in the Cold War.* New York: Farrar, Straus and Giroux, 2001.

Burrows, William E. *Deep Black: Space Espionage and National Security.* New York: Random House, 1988.

Calder, Kent E. *Embattled Garrisons: Comparative Base Politics and American Globalism.* Princeton: Princeton University Press, 2010.

Caldwell, Lynton K. *International Environmental Policy.* Durham, NC: Duke University Press, 1984.

Caldwell, Lynton Keith, and Paul Stanley Weiland. *International Environmental Policy, From the Twentieth to the Twenty-first Century.* Durham, NC: Duke University Press, 1996.

Callaway, Elliott B. *An Analysis of Environmental Factors Affecting Ice Growth.* Washington, DC: US Navy Hydrographic Office, September 1954.

Canizares-Esguerra, Jorge. *Nature, Empire, and Nation: Explorations of the History of Science in the Iberian World.* Stanford: Stanford University Press, 2006.

Cantoni, Roberto. "Oily Deals. Exploration, Diplomacy and Security in early Cold War France and Italy," Ph.D. thesis, Manchester: University of Manchester, 2014.

Capra, Fritjof. *The Web of Life.* New York: Random House, 1996.

Carbonell, Antonio. "Nota sobre los minerales de uranio." *Revista Ejército* 72 (1946).

Cargill Hall, R. "Earth Satellites: A Few Look by the United States Navy." In *Essays on the History of Rocketry and Astronautics: Proceedings of the Third through the Sixth History Symposia of the International Academy of Astronautics,* edited by R. Cargill Hall, 253–78. San Diego, CA: Univelt, Inc., 1986.

Cargill Hall, R. "From Concept to National Policy: Strategic Reconnaissance in the Cold War." *Prologue: The Journal of the National Archives* 28:3 (1996): 106–23.

Cargill Hall, R. "Origins of U.S. Space Policy: Eisenhower, Open Skies, and Freedom of Space." In *Exploring the Unknown: Selected Documents in the History of the U.S. Civil Space Program, Vol. 1,* edited by John M. Logsdon, 213–229. Washington, DC: NASA, 1995 [SP-4407].

Cargill Hall, R. "The Eisenhower Administration and the Cold War: Framing American Aeronautics to Serve National Security." *Prologue: The Journal of the National Archives* 27 (1995): 61–70.

Cargill Hall, R. "The NRO in the 21st Century: Ensuring Global Information Supremacy." *Quest: The History of Spaceflight Quarterly* 11:3 (2004): 4–11.

Cargill Hall, R. *A History of the Military Polar Orbiting Meteorological Satellite Program.* Chantilly, VA: NRO Office of the Historian, 2001.

Cargill Hall, R. *Samos to the Moon: The Clandestine Transfer of Reconnaissance Technology Between Federal Agencies.* Washington, DC: National Reconnaissance Office, 2001.

Cargill Hall, R., "Postwar Strategic Reconnaissance and the Genesis of CORONA." In *Eye in the Sky: The Story of the Corona Spy Satellite,* edited by Dwayne A. Day, John M. Logsdon, and Brian Latell. Washington, DC: Smithsonian Institution Press, 1998.

Cargill Hall, R., and Clayton Laurie, eds. *Early Cold War Overflights: Symposium Proceedings.* Washington, DC: Office of the Historian, National Reconnaissance Office, 2003.

Cargill Hall, R., and Robert Butterworth. *Military Space and National Policy: Record and Interpretation.* Washington, DC: George C. Marshall Institute, 2006.

Caro, Rafael, et al., eds. *Historia Nuclear de España.* Madrid: Sociedad Nuclear Española, 1955.

Carpenter, Eric. "An Historical Review of Seismometer Array Development." *IEEE Proceedings* 53 (1965): 1816–1821.

Carson, Rachel. *Silent Spring.* London: Penguin, 1999 [1962].

Catta, Emmanuel. *Victor De Metz. De la CFP au Groupe TOTAL.* Paris: Total Edition Presse, 1990.

Chafer, Tony. *The End of Empire in French West Africa: France's Successful Decolonization?* Oxford: Berg, 2002.

Chapman, Gary. "Rep. Brown Left a Legacy for Science." *Los Angeles Times* August 2, 1999.

Chapman, Richard Leroy. "A Case Study of the United States Weather Satellite Program: The Interaction of Science and Politics." Ph.D. diss., Syracuse: Syracuse University, 1967.

Charnok, Herny. "John Crossley Swallow." *Biographical Memoirs of the Fellows of the Royal Society* 43 (1997): 505–519.

Clayton, Donald D. *The Dark Night Sky.* New York: Quadrangle, 1975.

Cloud, John G. "American Cartographic Transformations During the Cold War." *Cartography and Geographic Information Science* 29 (2002): 261–282.

Cloud, John, and Keith C. Clarke. "Through a Shutter Darkly: The Tangled Relationships Between Civilian, Military, and Intelligence Remote Sensing in the Early U.S. Space Program." In *Secrecy and Knowledge Production,* edited by Judith Repps, 36–56. Ithaca: Cornell University, 1999.

Cloud, John. "Crossing the Olentangy River." *Studies in the History and Philosophy of Modern Physics* 31:3 (2000): 371–404.

Cloud, John. *Hidden in Plain Sight: CORONA and the Clandestine Geography of the Cold War.* Ph.D. thesis, Santa Barbara: University of California, 1999.

Cockell, Charles S. *Space on Earth: Saving our World by Seeking Others.* London: Macmillan, 2007.

Collins, Alan, ed. *Contemporary Security Studies.* Oxford: Oxford University Press, 2007.

Collis, Christy, and Klaus Dodds. "Assault on the Unknown: The Historical and Political Geographies of the International Geophysical Year (1957–8)." *Journal of Historical Geography* 34:4 (2008): 555–573.

Committee on International Environmental Affairs. *Design Philosophy for the Global Environmental Monitoring System.* Washington, DC: Department of State, 1976.

Connelly, Matthew. "Rethinking the Cold War and Decolonization: the Grand Strategy of the Algerian War for Independence." *International Journal of the Middle East* 33:2 (2001): 221–245.

Conway, Erik M. "Drowning in Data: Satellite Oceanography and Information Overload in the Earth Sciences." *Historical Studies in the Physical and Biological Sciences* 37 (2006): 127–151.

Conway, Erik M. "Planetary Science and the 'Discovery' of Global Warming." In *Exploring the Solar System: the History and Science of Planetary Exploration*, edited by Roger D. Launius, 183–202. London: Palgrave Macmillan, 2013.

Conway, Erik M. *Atmospheric Science at NASA: A History.* Baltimore: Johns Hopkins University Press, 2008.

Cook, F. A., "Ice Studies of the Canadian Geographical Branch." *Polar Record* 10:65 (1960): 123–125.

Cookman, Aubrey O. "Top of the World Weather Run." *Popular Mechanics* (November 1948): 97–264.

Corton, Edward L. *Climatology of the Ice Potential as Applied to the Beaufort Sea and Adjacent Waters* (Technical Report TR-30). Washington, DC: US Navy Hydrographic Office, 1955.

Corton, Edward L. *The Ice Budget of the Arctic Pack and its Application to Ice Forecasting* Washington, DC: US Navy Hydrographic Office, September 1954.

Cortright, Edgar M. *Exploring Space with a Camera.* Washington, DC: NASA, 1968.

Cosgrove, Dennis. "Contested Global Visions: *One-World, Whole-Earth*, and the Apollo Space Photographs." *Annals of the Association of American Geographers* 84:2 (1994): 270–294.

Craige, Beatty Jean. *Eugene Odum: Ecosystem Ecologist and Environmentalist.* Athens, GA: University Georgia Press, 2001.

Cranor, Carl. *Regulating Toxic Substances: A Philosophy of Science and the Law.* New York and Oxford: Oxford University Press, 1993.

Craven, John. *The Silent War: The Cold War Battle Beneath the Sea.* New York: Touchstone, 2002.

Critchley, Harriet. "Polar Deployment of Soviet Submarines." *International Journal* 39:4 (1984), 836–7.

Crouch, Tom D. *Lighter Than Air: An Illustrated History of Balloons and Airships.* Baltimore, MD: Johns Hopkins University Press, 2009.

Crouch, Tom D. *The Eagle Aloft: Two Centuries of the Baloon in America.* Washington, DC: Smithsonian Institution Press, 1983.

Crowson, D. L. "Cloud Observations From Rockets," *Bulletin of the American Meteorological Society* (1949): 17–22.

D'Anieri, Paul. *International Politics: Power and Purpose in Global Affairs.* Boston: Wadsworth, 2011.

Dahan, Amy. "Putting the Earth System in a Numerical Box? The Evolution from Climate Modeling toward Global Change." *Studies in History and Philosophy of Modern Physics* 41 (2010): 282–292.

Dallin, Alexander. *Black Box: KAL 007 and the Superpowers.* Berkeley: University of California Press, 1985.

Dalmedico, Amy Dahan, and Helene Guillemot. "Climate Change: Scientific Dynamics, Expertise, and Geopolitical Challenges." In *Global Science and National Sovereignty: Studies in Historical Sociology of Science*, edited by Grégoire Mallard, Catherine Paradeise, and Ashveen Peerbaye, 196–9. London: Routledge, 2009.

Dansk Udenrigspolitisk Institut. *Grønland under den kolde krig: dansk og amerikansk sikkerhedspolitik 1945–68.* Copenhagen: Dansk Udenrigspolitiks Institut, 1997.

Darius, Jon. *Beyond Vision: One Hundred Historic Scientific Photographs.* Oxford: Oxford University Press, 1984.

Darwin Charles. *Voyage of the Beagle.* London: Penguin, 1989 [1839].

Dassow Wild, Laura. *The Passage to Cosmos: Alexander von Humboldt and the Shaping of America.* Chicago: Chicago University Press, 2009.

Daugherty, Charles Michael. *City Under the Ice: The Story of Camp Century.* New York: Macmillan, 1963.

Dauvergne, Peter. *Handbook of Global Environmental Politics.* Cheltenham and Northampton: Edward Elgar Publishing, 2006.

Davies, David. "Hal Thirlaway (obituary)." *The Guardian* January 19, 2010.

Davies, Merton E., and William R. Harris, *RAND's Role in the Evolution of Balloon and Satellite Observation Systems and Related U.S. Space Technology* [RAND report R-3692-RC]. Santa Monica: RAND Corporation, 1988.

Davies, Mike. *City of Quartz: Excavating The Future in Los Angeles.* New York: Verso, 1992.

Day, Dwayne A "The Development and Improvement of the CORONA Satellite." In *Eye in the Sky: The Story of the Corona Spy Satellites,* edited by Dwayne A. Day, John M. Logsdon, and Brian Latell, 48–85. Washington, DC: Smithsonian Institution Press, 1998.

Day, Dwayne A. "Blunt Arrows: The Limited Utility of ASATs." *Space Review.* Available on-line at http://www.thespacereview.com/article/388/1 [accessed 10/11/2006].

Day, Dwayne A. "Corona: America's First Spy Satellite Program." *Quest* 4 (1995): 4–21.

Day, Dwayne A. "Invitation to Struggle: The History of Civilian-Military Relations in Space." In *Exploring the Unknown: Selected Documents in the History of the U.S. Civilian Space Program. Vol.II: External Relationships,* edited by John M. Logsdon [Report NASA SP-4407], 233–70. Washington, DC: NASA, 1996.

Day, Dwayne A. "Mapping the Dark Side of the World, Part 1: The KH-5 ARGON Geodetic Satellite." *Spaceflight* 40 (1998): 264–269.

Day, Dwayne A. "Mapping the Dark Side of the World, Part 2: Secret Geodetic Programmes After ARGON." *Spaceflight* 40 (1998): 303–310.

Day, Dwayne A. "New Revelations About the American Satellite Programme Before Sputnik." *Spaceflight* 36 (1994): 372–73.

Day, Dwayne A., Logsdon, John M., and Brian Latell, eds. *Eye in the Sky: The Story of the Corona Spy Satellite.* Washington, DC: Smithsonian Institution Press, 1998.

De Maria, Michelangelo, and Lucia Orlando, eds. *Italy in Space: In Search of a Strategy, 1957–1975.* Paris: Beauchesne, 2008.

Deblois, Bruce M. "Space Sanctuary: A Viable National Strategy." *Aerospace Power Journal* 12 (1998): 41–57.

Dennis, Michael Aaron. "Earthly Matters: on the Cold War and the Earth Sciences." *Social Studies of Science* 33:5 (2003): 809–819.

Dennis, Michael Aaron. "Secrecy and Science Revisited: From Politics to Historical Practice and Back." In Ronald E. Doel and Thomas Söderqvist, eds., *The Historiography of Contemporary Science, Tecnology and Medicine: Writing Recent Science.* London: Routledge, 2006, 172–184.

DeVorkin, David. *Science with a Vengeance: How the Military Created the US Space Sciences After World War II.* New York: Springer, 1992.

Di Nolfo, Ennio. "The Cold War and the transformation of the Mediterranean." In *The Cambridge History of the Cold War,* edited by M. P. Leffler and O. A. Westad, Vol. 2, 238–57. Cambridge: Cambridge University Press, 2010.

Dick, Steven J. "Geodesy, Time, and the Markowitz Moon Camera Program: An Interwoven International Geophysical Year Story." In *Globalizing Polar Science: Reconsidering the International Polar and Geophysical Years,* edited by James R. Fleming, Roger D. Launius, and David H. DeVorkin, 307–326. New York: Palgrave, 2010

Diedrichsen, Diedrich, and Anselm Franke, eds. *The Whole Earth* [catalogue of exhibition at *Haus de Kulturen der Welt,* Berlin, June 2013].

Dirnwoeber, M., Machan, R., and J. Herler. "Coral Reef Surveillance: Infrared-Sensitive Video Surveillance Technology as a New Tool for Diurnal and Nocturnal Long-Term Field Observations." *Remote Sensing* 4:11 (2012): 3346–3362.

Divine, Robert A. *The Sputnik Challenge*. New York and Oxford: Oxford University Press, 1993.

Divine, Robert. *Blowing on the Wind/ The Nuclear Test Ban Debate*. New York: Oxford University Press, 1978.

Dobson, Alan. *Anglo-American Relations in the Twentieth Century: Of Friendship, Conflict and the Rise and Decline of Superpowers*. London: Routledge, 1995.

Dockrill, Michael. *British Defence since 1945*. Oxford: Wiley-Blackwell, 1988.

Doel, Ronald E., and Allan A. Needell, "Science, Scientists and the CIA: Balancing International Ideals, National Needs, and Professional Opportunities." *Intelligence and National Security* 12 (1997): 59–81.

Doel, Ronald E., "Does Scientific Intelligence Matter?" *Centaurus* 52 (2010): 311–322.

Doel, Ronald E. "Constituting the Postwar Earth Sciences: The Military's Influence on the Environmental Sciences in the USA after 1945." *Social Studies of Science* 33:5 (2003): 635–666.

Doel, Ronald E. "Quelle place pour les sciences de l'environnement physique dans l'histoire environnementale." *Revue d'histoire moderne et contemporaine* 56 (2009): 137–164.

Doel, Ronald E. "The Earth Sciences and Geophysics." In *Companion to Science in the 20th Century*, edited by John Krige and Dominique Pestre, 391–417. London: Routledge, 2003.

Doel, Ronald E. "Why Value History?" *Eos, Transactions of the American Geophysical Union* 83:47 (2002): 544–45.

Doel, Ronald E., Levin, Tanya J., and Mason K. Marker. "Extending Modern Cartography to the Ocean Depths: Military Patronage, Cold War Priorities, and the Heezen–Tharp Mapping Project 1952–1959." *Journal of Historical Geography* 32 (2006): 605–626.

Doel, Ronald E., and Naomi Oreskes. "The Physics and Chemistry of the Earth." In *The Cambridge History of Science*, Vol. 5, edited by Mary Jo Nye, 538–557. Cambridge: Cambridge University Press, 2012.

Doel, Ronald E. "Scientists as Policymakers, Advisors and Intelligence Agents: Linking Contemporary Diplomatic History with the History of Contemporary Science." *The Historiography of Contemporary Science and Technology*, edited by Thomas Söderqvist 215–244. Amsterdam: Harwood, 1997.

Dolman, Everett. "U.S. Military Transformation and Weapons in Space," *SAIS Review* 26 (2006): 163–174.

Dolman, Everett. *Astropolitik: Classical Geopolitics in the Space Age*. London: Frank Cass, 2002.

Douglas Aircraft Company, *Preliminary Design of an Experimental World-Circling Spaceship* [Report no. SM-11827]. Santa Monica: Douglas Aircraft Company, 1946.

Drayton, Richard. *Nature's Government: Science, Imperial Britain, and the "Improvement" of the World*. New Haven, CT: Yale University Press, 2000.

Dubos, Rene. "A Theology of the Earth." In *The God Within*, edited by R. Dubos. New York: Charles Scribner, 1972.

Duckworth, W. E. "Ieuan Maddock, 1917–1988." *Biographic Memoir of the Fellows of the Royal Society* 37 (1991): 323–340.

Dupas, Alain. *La nouvelle conquête spatial*. Paris: Odile Jacob, 2010.

Dylan, Huw. "The Joint Intelligence Bureau: (Not So) Secret Intelligence for the Post-War World." *Intelligence and National Security* 27:1 (2012): 27–45.

Edgington, Ryan. "An 'All-seeing Flying Eye': V-2 Rockets and the Promises of Earth Photography." *History and Technology* 28 (2012): 363–371.

Edwards, Paul N. "Meteorology as infrastructural globalism." *Osiris* 21 (2006): 229–250.

Edwards, Paul. *A Vast Machine: Computer Models, Climate Data, and the Politics of Global Warming*. Cambridge, MA: MIT Press, 2010.

Ellingsen, Gunnar. "Instrumentutvikling med NATO-bistand." In *I vinden: Geofysisk Institutt 90 år*, edited by Edgar Hovland, 112–115. Bergen: Fagbokforlaget, 2007.

Ellingsen, Gunnar. "Varme havstrømmer og kald krig: 'Bergensstrømmåleren' og vitenskapen om havstrømmer fra 1870-årene til 1960-åreme [Warm ocean currents and the cold war: the Bergen current meter and the science of ocean currents from the 1870s to the 1960s]." Ph.D. thesis, Bergen: Universitetet i Bergen, 2012.

Elliott, Derek W. "Finding an Appropriate Commitment: Space Policy Under Eisenhower and Kennedy." PhD Diss., George Washington University, 1992.

Engel, Jeffrey A. *Cold War at 30,000 Feet: The Anglo-American Fight for Aviation Supremacy*. Cambridge, MA: Harvard University Press, 2007.

Epstein, Samuel S. *The Politics of Cancer Revisited*. New York: East Ridge Press, 1998.

Erlich, Paul R., Sagan, Carl, Kennedy, Donald, and Walter Orr Roberts, *The Cold and the Dark: the World After Nuclear War*. New York: W. W. Norton, 1984.

Evangelista, Matthew. *Unarmed Forces: The Transnational Movement to End the Cold War*. Ithaca: Cornell University Press, 1999.

Ewing, Gifford C., ed. *Oceanography from Space, Proceedings of the Conference on the Feasibility of Conducting Oceanographic Explorations from Aircraft, Manned Orbital and Lunar Laboratories. Held August 24–28 at Woods Hole, Massachusetts, USA* [ref. no.65–10]. Woods Hole: Woods Hole Oceanographic Institution, 1965.

Faul, Henry, ed. *Nuclear Geology. A Symposium on Nuclear Phenomena in The Earth Sciences*. New York: John Wiley, 1954.

Finn, Rachel L., and David Wright. "Unmanned Aircraft Systems: Surveillance, Ethics and Privacy in Civil Applications." *Computer Law & Security Review* 28 (2) 2912: 184–194.

Fisher, William A., and Charles J. Robinove. "A Rationale for a General Purpose Earth Resources Observation Satellite." In *Proceedings of University of Washington Remote Sensing Symposium*. Washington, DC: Washington University, 1968.

Fitzgerald, Frances. *Way Out There in the Blue: Reagan, Star Wars, and the End of the Cold War*. New York: Simon and Schuster, 2000.

Fleming, James R. "A 1954 Color Painting of Weather Systems as Viewed From a Future Satellite." *Bulletin of the American Meteorological Society* 88 (2007): 1525–1527.

Fleming, James R. "Earth Observations from Space: Achievements, Challenges, and Realities." In *NASA's First Fifty Years: Historical Perspectives*, edited by Steven J. Dick, 543–62. Washington, DC: NASA History Division, 2010.

Fleming, James R. "Polar and Global Meteorology in the Career of Harry Wexler," In *Globalizing Polar Science: Reconsidering the International Polar and Geophysical Years*, edited by James R. Fleming, Roger D. Launius, and David H. DeVorkin, 225–241. New York: Palgrave, 2010.

Fleming, James R. *Fixing the Sky. The Checkered History of Weather and Climate Control*. New York: Columbia University Press, 2010.

Fletcher, Roy J. "Military Radar Defence Lines of Northern North America: An Historical Geography." *Polar Record* 26:159 (1990): 265–76.

Flippen, J. Brooks Flippen. *Nixon and the Environment*. Albuquerque: University of New Mexico Press, 2000.

Flippen, J. Brooks. "Richard Nixon, Russell Train, and the Birth of Modern American Environmental Diplomacy." *Diplomatic History* 32 (2008): 613–638.

Fontaine, Pierre. *La morte étrange de Conrad Kilian, inventeur du pétrole saharien*. Paris: Les Sept Couleurs, 1959.

Forman, Paul. "Behind Quantum Electronics: National Security as Basis for Physical Research in the US, 1940–1960." *Historical Studies in the Physical Sciences* 18 (1985): 149–229.

Fortey, Richard. *The Earth*. London: HarperCollins, 2004.

Foucault, Michael. *Discipline and Punish*. New York: Pantheon, 1977.

Fraser, J. Keith. "Activities of the Geographical Branch in Northern Canada, 1947–1957." *Arctic* 10:4 (1957): 246–250.

Freedman, Lawrence. *US Intelligence and the Soviet Strategic Threat.* London: Macmillan, 1977.

Friedman, Norman. *Seapower as Strategy: Navies and National Interests.* Annapolis: Naval Institute Press, 2001.

Friedman, Robert Marc. "Background to the Establishment of Norsk Polarinstitutt: Postwar Scientific and Political Agendas." The Northern Space: The International Research Network on the History of Polar Science, Working Paper 2 (1995).

Fuller, Buckminster. "Vertical is to Live—Horizontal is to Die." *American Scholar* 39 (1969).

Fursenko, Aleksandr, and Timothy Naftali, *Kruschev's Cold War.* New York: W. W. Norton & Co., 2006.

Gaddis, John L. *Strategies of Containment: A Critical Appraisal of American National Security Policy during the Cold War.* Oxford: Oxford University Press, 2005.

Gaddis, John Lewis. *The Long Peace: Inquiries Into the History of the Cold War.* New York: Oxford University Press, 1987.

Gaddis, John Lewis. *We Now Know. Rethinking Cold War History.* Oxford: Oxford University Press, 1997.

Galison, Peter. "Judgment against Objectivity." In *Picturing Science, Producing Art,* edited by Caroline Jones and Peter Galison, 327–359. New York and London: Routledge, 1998.

Geffroy, Jacques, and J. A. Sarcia, "La notion de 'gite épithermal uranifère' et les problèmes qu'elle pose." In *Compte-rendu du Colloque de Géologie des Gisements de Minerais d'Uranium et Méthodes de Prospection, 9–11 May 1957,* edited by EAES, 159–179. Madrid: JEN, 1957.

Gillespie, Charles C., and Nicolaas Rupke. *Genesis and Geology.* Cambridge, MA: Harvard University Press, 1951.

Glaser, Charles L. "The Security Dilemma Revisited." *World Politics* 50 (1997): 171–201.

Gold, Peter. *A Stone in Spain's Shoe: The Search for a Solution to the Problem of Gibraltar.* Liverpool: Liverpool University Press, 1994.

Goldschmidt, Bertrand. *The Atomic Complex: A Worldwide Political History of Nuclear Energy.* La Grange Park, IL: American Nuclear Society, 1982.

Goncharov, G. A. "Thermonuclear milestones." *Physics Today* 49:11 (1996): 44–61.

Goodman, Michael. "With a Little Help from My Friends: The Anglo-American Intelligence Partnership, 1945–1958." *Diplomacy and Statecraft* 18:1 (2007): 155–153.

Goodman, Michael. *Spying on the Polar Bear.* Stanford: Stanford University Press, 2007.

Gordienko, P. A., "Arctic Sea Ice Research," *Priroda* 9 (1958): 68–71.

Gould, W. John. "From Swallow floats to Argo—the development of neutrally buoyant floats." *Deep-Sea Research* 52:3 (2005): 529–543.

Gowing, Margaret. *Independence and Deterrence.* London: Palgrave Macmillan, 1974, 2 Vols.

Grandhouser, Larry K. "Sentinels Rising: Commercial High-Resolution Satellite Imagery and Its Implications for US National Security." *Aerospace Power Journal* 12 (1998): 61–80.

Greenaway, Frank. *Science International: History of the International Council of Scientific Unions.* Cambridge: Cambridge University Press, 1996.

Greenfield, Stanley M., and William W. Kellogg. *Inquiry Into the Feasibility of Weather Reconnaissance From a Satellite Vehicle* [RAND report no. R-365]. Santa Monica, CA: RAND Corporation, 1951.

Grevsmühl, Sebastian Vincent. "Epistemische Topografien. Fotografische und radartechnische Wahrnehmungsräume." In *Verwandte Bilder. Die Fragen der Bildwissenschaft,* edited by Ingeborg Reichle, Steffen Siegel, and Achim Spelten. Berlin: Kadmos, 2007.

Grevsmühl, Sebastian. "A la recherche de l'environnement global: De l'Antarctique à l'Espace et retour." Ph.D. dissertation, Paris: University of Paris—Ecole des Hautes Etudes en Sciences Sociales, 2012.

Grevsmühl, Sebastian. *La Terre vue d'en haut: l'invention de l'environnement global.* Paris: Seuil, 2014.

Griggs, D. T., and F. Press. "Probing the Earth with Nuclear Explosions." *Journal of Geophysical Research* 66:1 (1961): 237–258.

Grove, Eric. *Vanguard to Trident: British Naval Policy since World War II.* Annapolis: Naval Institute Press, 1987.

Gwynne, Michael D. "The Global Environment Monitoring System (GEMS) of UNEP." *Environmental Conservation* 9 (1982): 35–41.

Haas, P. M., ed., *Knowledge, Power, and International Policy Coordination.* Columbia, SC: University of South Carolina Press, 1997.

Haas, Peter M. *Saving the Mediterranean: The Politics of International Environmental Cooperation, Political Economy of International Change.* New York: Columbia University Press, 1990.

Hagen, Joel B. *The Entangled Bank: The Origins of Ecosystem Ecology.* New Brunswick, NJ: Rutgers University Press, 1992.

Hahn, Peter. *The United States, Great Britain, Egypt, 1945–1956: Strategy and Diplomacy in the Early Cold War.* Chapel Hill: University of North Carolina Press, 1991.

Haines, Gerald K., and Robert E. Leggett, eds. *Watching the Bear: Essays on CIA's Analysis of the Soviet Union.* Langley, VA: Center for the Study of Intelligence, 2003.

Hallion, Richard P. *Rise of the Fighter Aircraft, 1914–1918.* Baltimore, MD: The Nautical and Aviation Press, 1984.

Hamblin, Jacob D. "'A Dispassionate and Objective Effort': Negotiating the First Study on the Biological Effects of Atomic Radiation." *Journal of the History of Biology* 40:1 (2007): 147–177.

Hamblin, Jacob D. "Exorcising Ghosts in the Age of Automation: United Nations Experts and Atoms for Peace." *Technology and Culture* 47:4 (2006): 734–756.

Hamblin, Jacob D. *Arming Mother Nature: The Birth of Catastrophic Environmentalism.* Oxford: Oxford University Press, 2013.

Hamblin, Jacob D. *Oceanographers and The Cold War. Disciples of Marine Science.* Seattle: University of Washington Press, 2005.

Hamblin, Jacob D. *Poison in the Well. Radioactive Waste in the Oceans at the Dawn of the Nuclear Age.* New Brunswick: Rutgers University Press, 2008.

Hamblin, Jacob D. "Environmentalism for the Atlantic Alliance." *Environmental History* 15 (2010): 54–75.

Hamblyn, Richard. *The Invention of Clouds: How An Amateur Meteorologist Forged the Language of the Sky.* London: Picador, 2001.

Hamilton, Clive. *Earthmasters: The Dawn of the Age of Climate Engineering.* New Haven, CT: Yale University Press, 2013.

Hansen, Chuck. "Open Secrets, Closed Minds." *Bulletin of the Atomic Scientists* 51:4 (1995): 16–17.

Harkavy, Robert E. *Bases Abroad: The Global Foreign Military Presence.* Stockholm: SIPRI, 1989.

Harkavy, Robert E., *Strategic Basing and the Great Powers, 1200–2000.* London: Routledge, 2007.

Harper, Kristine C., "Climate Control: United States Weather Modification in the Cold War and Beyond." *Endeavour* 32,1 (2008): 20–26.

Harper, Kristine C., *Weather by the Numbers. The Genesis of Modern Meteorology.* Cambridge (MA): 2008.

Harley, John H., Hallden, Naomi A., and Long D. Y. Ong, *Summary of Gummed Film Results through December, 1959*. New York: U.S. Atomic Energy Commision's Health and Safety Laboratory, 1960.

Harris, Shane. *The Watchers: The Rise of the Surveillance State*. New York: Penguin Books, 2001.

Harvey, Fiona. "Doha Climate Change Deal Clears Way for 'Damage Aid' to Poor Nations," *The Observer* December 8, 2012.

Hastedt, Glenn. "Reconnaissance Satellites, Intelligence, and National Security." In *Societal Impact of Spaceflight*, edited by Steven J. Dick and Roger D. Launius, eds., 369–385. Washington, DC: NASA, 2007 [SP-4811].

Haydon, Stansbury F., and Tom D. Crouch. *Military Ballooning during the Early Civil War*. Baltimore, MD: Johns Hopkins University Press, 2000.

Hays, Samuel. *Beauty, Health, and Permanence: Environmental Politics in the United States, 1955–1985*. Cambridge: Cambridge University Press, 1989.

Head, John W. "Challenges in International Environmental Management: A Critique of the United Nations Environment Programme." *Virginia Journal of International Law* 18 (1977): 269–288.

Hecht, Gabrielle, ed. *Entangled Geographies. Empire and Technopolitics in the Global Cold War*. Cambridge, MA: MIT Press, 2011.

Hecht, Gabrielle. *Being Nuclear. Africans and the Uranium Trade*. Cambridge, MA: MIT Press, 2012.

Hecht, Gabrielle. *The Radiance of France*. Cambridge, MA: MIT Press, 1998.

Helmreich, J. E. "The United States and the Formation of EURATOM." *Diplomatic History* 15:3 (1991): 387–410.

Helmreich, Jonathan E. *Gathering Rare Ores: The Diplomacy of Uranium Acquisition, 1943–1954*. Princeton: Princeton University Press, 1986.

Henrikson, Alan K., ed. *Negotiating World Order: The Artisanship and Architecture of Global Diplomacy*. Wilmington: Scholarly Resources, Inc., 1986.

Hernando, José L., and Rafael Hernando. "Descubrimiento, explotación y tratamiento de los minerales de uranio en Sierra de Albarrana, El Cabril (Córdoba)." *Boletín de la Real Academia de Córdoba* 143 (2002): 161–178.

Herran, Néstor. *Aguas, semillas y radiaciones. El Laboratorio de Radiactividad de la Universidad de Madrid, 1904–1929*. Madrid: CSIC, 2008

Hersh, Seymour M. *The Target is Destroyed*. New York: Random House, 1986.

Hewlett, Richard, and Francis Duncan. *Atomic Shield: A History of the AEC, 1947–1952*. Berkeley, CA: University of California Press, 1990.

Hewson, Martin. "Did Global Governance Create Informational Globalism?" In *Approaches to Global Governance Theory*, edited by M. Hewson and Timothy J. Sinclair, 97–113. Albany, NY: State University of New York Press, 1999.

Hightower, Jane M. *Diagnosis: Mercury*. Washington, DC: Island Press, 2008.

Higuchi, Toshihiro. "An Environmental Origin of Antinuclear Activism in Japan, 1954–1963: The Politics of Risk, the Government, and the Grassroots Movement." *Peace & Change* 33:3 (2008): 333–366

Higuchi, Toshihiro. "Atmospheric Nuclear Weapons Testing and the Debate on Risk Knowledge in Cold War America, 1945–1963." *Environmental Histories of the Cold War*, edited by J. R. McNeill and Corinna R. Unger, 301–322. Cambridge: Cambridge University Press, 2010.

Higuchi, Toshihiro. "Tipping the Scale of Justice: the Fallout Suit of 1958 and the Environmental Legal Dimension of Nuclear Pacifism." *Peace & Change* 38:1 (2013): 33–55.

Hildebrand, Robert C. *Dumbarton Oaks: The Origins of the United Nations and the Search for Postwar Security*. Chapel Hill: University of North Carolina Press, 2001.

Hitchcock, Dian, and James Lovelock, "Life Detection by Atmospheric Analysis." *Icarus* 7 (1967): 149–159.

Hitchcock, William I. *France Restored: Cold War Diplomacy and the Quest for Leadership in Europe, 1944–1954.* Chapel Hill and London: University of North Carolina Press, 1998.

Holly, Susan K., and William B. McAllister, eds. *Foreign Relations of the United States, Foreign Relations 1969–1976, Documents on Global Issues 1969–1972, Volume E-1, Chapter V, International Environment Policy.* Washington, DC: US Department of State [Office of the Historian], 2005.

Hounshell, David. "The Cold War, RAND and the Generation of Knowledge 1946–62." *Historical Studies in the Physical Sciences* 27:2 (1997): 237–267.

Houtermans, Frederich G., and E. Picciotto. *Primo Convegno sulla Geologia Nucleare.* Rome: CNRN, 1955.

Howe, Joshua P. "Making Global Warming Green: Climate Change and American Environmentalism." PhD Dissertation, Stanford: Stanford University, 2010.

Hoyle, Fred. *The Nature of the Universe.* Oxford: Blackwell, 1950.

Hubert, Lester F., and Otto Berg. "A Rocket Portrait of a Tropical Storm." *Monthly Weather Review* 83 (1955): 119–124.

Huchthausen, Peter, and Alexander Sheldon-Duplaix. *Hide and Seek: The Untold Story of Cold War Naval Espionage.* London: Wiley, 2009.

Hug, Peter. "La Génèse de la technologie nucléaire en Suisse." *Relations internationales* 68 (1991): 325–344.

Hughes, Agatha C., and Thomas P. Hughes, eds. *Systems, Experts, and Computers: The Systems Approach in Management and Engineering, World War II and After.* Cambridge, MA: MIT Press, 2000.

Hughes, Thomas P. *Networks of Power: Electrification in Western Society, 1880–1930.* Baltimore, MD: John Hopkins University Press, 1983.

Humphreys, F. E., "The Wright Flyer and its Possible Uses in War." *Journal of the United States Artillery* 33:2 (1910).

Hünemörder, Kai. "Environmental Crisis and Soft Politics: Détente and the Global Environment, 1968–1975." In *Environmental Histories of the Cold War*, edited by John R. McNeill and Corinna R. Unger, 257–276. Cambridge: Cambridge University Press, 2010.

Hurrell, Andrew, and Benedict Kingsbury, eds. *The International Politics of the Environment.* Oxford: Clarendon Press, 1992.

Huskisson, Darren. "Protecting the Space Network and the Future of Self-Defense." *Astropolitics: The International Journal of Space Politics & Policy* 5:2 (2007): 123–143.

Inhofe, James. *The Greatest Hoax: How the Global Warming Conspiracy Threatens Your Future.* Washington, DC: WND Books, 2012.

Ippolito, Felice. "Dieci Anni di Ricerca Uranifera in Italia." *Notiziario CNRN* 9:7 (1963): 22–33.

Ippolito, Felice. "Stato presente delle ricerche di uranio e torio in Italia." *Energia Nucleare* 17 (1955): 479–489.

Ippolito, Felice. *Intervista sulla Ricerca Scientifica.* Bari: Laterza, 1978.

Iriye, Akira. *Global Community: The Role of International Organizations in the Making of the Contemporary World.* Berkeley: University of California Press, 2002.

Ivanov, Konstantin. "Science after Stalin: Forging a New Image of Soviet Science." *Science in Context* 15:2 (2002): 317–338.

Ivanova, Maria. "Looking Forward by Looking Back: Learning from UNEP's History." In *Global Environmental Governance: Perspectives on the Current Decade*, edited by Lydia Swart and Estelle Perry, 26–47. New York: Center for UN Reform Education, 2007.

Jacobson, Harold K., and Eric Stein. *Diplomats, Scientists, and Politicians: The United States and the Nuclear Test Ban Negotiations.* Ann Arbor: University of Michigan Press, 1966.

James F. Keeley, *A List of Bilateral Civilian Nuclear Co-Operation Agreements.* http://dspace.ucalgary.ca/bitstream/1880/47373/10/Treaty_List_Volume_04.pdf (accessed May 8, 2012).

Jasanoff, Sheila. "Image and Imagination: the Formation of Global Environmental Consciousness." In *Changing the Atmosphere: Expert Knowledge and Environmental Governance,* edited by Clark A. Miller and Paul N. Edwards, 309–337. Cambridge, MA: MIT Press, 2001.

Jasanoff, Sheila. "Science, Politics and the Renegotiation of Expertise at EPA." *Osiris* 7 (1992): 194–217.

Jasanoff, Sheila. *The Fifth Branch: Science Advisers as Policymakers.* Cambridge, MA: Harvard University Press, 1990.

Jenkins, Dennis R. *Lockheed U-2 Dragon Lady.* North Branch MN: Specialty Press Publishers and Wholesalers, 1998.

Jense, Kurt F. *Cautious Beginnings: Canadian Foreign Intelligence 1939–51.* Vancouver: University of British Columbia Press, 2009.

Jensen, C. E., and D. W. Brown. "Earthwatch—Global Environmental Assessment." *Environmental Management* 5 (1981): 225–232.

Jensen, Clayton E., Brown, Dail W., and John A. Morabito. "Earthwatch. Guidelines for Implementing This Global Assessement Program Are Presented." *Science* 190 (1975): 432–438.

Jervis, Robert. "Cooperation Under the Security Dilemma." *World Politics* 30:2 (1978): 167–214.

Jervis, Robert. "Was the Cold War a Security Dilemma?" *Journal of Cold War Studies* 3 (2001): 36–60.

Johnson, G. A. L. "Sir Kingsley Charles Dunham, 1910–2001." *Biographical Memoirs of Fellows of the Royal Society* 49 (2003): 147–162.

Johnson, Nicholas Johnson. "Space Traffic Management: Concepts and Practices." *Space Policy* 20 (2004): 79–85.

Johnson, Richard William. *Shootdown: Flight 007 and the American Connection.* New York: Viking, 1986.

Johnson, Stephen B. "The Political Economy of Spaceflight." in *Societal Impact of Spaceflight,* edited by Steven J. Dick and Roger D. Launius, 141–91. Washington, DC: NASA History Division, 2007.

Johnston, Andrew K. "Exploring Planet Earth: the Development of Satellite Remote Sensing for Earth Science." In *Exploring the Solar System: the History and Science of Planetary Exploration,* edited by Roger D. Launius, 203–222. New York and London: Palgrave Macmillan, 2013.

Jones, Howard. *Crucible of Power: A History of American Foreign Relations from 1945.* Lanham, MD: Rowman & Littlefield, 2009.

Jones, Matthew. *After Hiroshima: The United States, Race and Nuclear Weapons in Asia, 1945–1965.* Cambridge: Cambridge University Press, 2010.

Jordan, David Alan. "Decrypting the Fourth Amendment: Warrantless NSA Surveillance and the Enhanced Expectation of Privacy Provided by Encrypted Voice over Internet Protocol." *Boston College Law Review* 47:1 (2006): 1–42.

Joyner, Christopher C. *Antarctica and the Law of the Sea.* Dordrecht: Martinus Nijhoff, 1992.

Kaminski, Henry S. *Distribution of Ice in Davis Strait and Baffin Bay.* Washington, DC: US Navy Hydrographic Office, 1955.

Kaufman, I. B. *The Oil Cartel Case: A Documentary Study of Antitrust Activity in the Cold War Era.* Westport: Greenwood Press, 1978.

Kellway, Kate. "How the Observer brought the WWF into being." *The Observer* November 7, 2010.

Kennett, Lee. *The First Air War: 1914–1918.* New York: Free Press, 1991.

Kent, P. E. "North Sea Exploration—A Case History." *The Geographical Journal* 133:3 (1967): 289–301.

Kent, P. E. "The North Sea—Evolution of a Major Oil and Gas Play." *Facts and Principles of World Petroleum Occurrence* 6 (1980), 633–652.

Keskitalo, Carina. *Negotiating the Arctic: The Construction of an International Region.* New York: Routledge, 2004.

Ketov, Ryurik A. "The Cuban Missile Crisis as Seen through a Periscope." *The Journal of Strategic Studies* 28:2 (2005): 217–231.

Kevles, Daniel J. "Cold War and Hot Physics: Science, Security, and the American State, 1945–1956." *Historical Studies in the Physical Sciences* 20 (1990): 239–264.

Khalturin, Vitaly, et al. "A Review of Nuclear Testing by the Soviet Union at Novaya Zemlya, 1955–1990." *Science and Global Security* 13:1 (2005): 1–42.

Kilgore, De Witt Douglas. *Astrofuturism.* Philadelphia: University of Pennsylvania Press, 2003.

Kistiakowsky, George Bogdan. *A Scientist at the White House: The Private Diary of President Eisenhower's Special Assistant for Science and Technology.* Cambridge MA: Harvard University Press, 1976.

Kleinberg, Howard. "On War in Space." *Astropolitics: The International Journal of Space Politics & Policy* 5:1 (2007): 1–27.

Kletter, Leopold. "Die praktische Auswertung der Bildsendungen der Wettersatelliten." *Schriften des Vereins zur Verbreitung naturwissenschaftlicher Kenntnisse* 110 (1970): 23–35.

Kletter, Leopold. "Meteorologische Satelliten erforschen das Weltwetter," *Schriften des Vereins zur Verbreitung naturwissenschaftlicher Kenntnisse* 103 (1963): 1–17.

Kondorskaya, N. V., and Z. I. Aronovich. "The Uniform System of Seismic Observations of the U.S.S.R. and Prospects of Its Development." *Physics of The Earth and Planetary Interiors* 18:2 (1979): 78–86.

Korsmo, Fae. "The Genesis of the International Geophysical Year." *Physics Today* 60 (2007): 40–43.

Kranzberg, Melvin. "Technology and History: 'Kranzberg's Laws'." *Technology and Culture* 27 (3): 544–560.

Krementsov, Nikolai. *Stalinist Science.* Princeton: Princeton University Press, 1997.

Krepon, Michael, and Michael Katz-Hyman. "The Responsibilities of Space Faring Nations." *Defense News* October 16, 2006.

Krige, John and Arturo Russo. *A History of the European Space Agency,* 2 Vols. Noordwijk: ESA, 2000.

Krige, John, and Dominique Pestre."Some Thoughts on the History of CERN in the 50s and 60s." In *Big Science: The Growth of Large Scale Research,* edited by P. Galison and B. Hevly, 78–99. Stanford: Stanford University Press, 1992.

Krige, John. "Atoms for Peace, Scientific Internationalism, and Scientific Intelligence." In *Global Power Knowledge: Science and Technology in International Affairs (Osiris 21),* edited by J. Krige and Kai-Henrik Barth, 161–181. Chicago: University of Chicago Press, 2006.

Krige, John. "Hybrid Knowledge: the Transnational Co-Production of the Gas Centrifuge for Uranium Enrichment in the 1960s." *British Journal for the History of Science* 45:3 (2012): 337–358.

Krige, John. "The Peaceful Atom as Political Weapon: Euratom and American Foreign Policy in the Late 1950s." *Historical Studies in the Natural Sciences* 38:1 (2008): 5–44.

Krige, John. *American Hegemony and the Postwar Reconstruction of Science in Europe.* Cambridge, MA: MIT Press, 2006.

Krypton, Constantine. *The Northern Sea Route and the Economy of the Soviet North.* London: Methuen, 1956.

Kwa, Chunglin. "Radiation Ecology, Systems Ecology and the Management of the Environment." In *Science and Nature: Essays in the History of Environmental Sciences,* edited by Michael Shortland. BSHS Monograph 8 (1993): 213–251.

Kwa, Chunglin. "Representations of Nature Mediating between Ecology and Science Policy: The Case of the International Biological Programme." *Social Studies of Science* 17 (1987): 413–442.

Kwa, Chunglin. "The Rise and Fall of Weather Modification: Changes in American Attitudes toward Technology, Nature, and Society." in *Changing the Atmosphere: Expert Knowledge and Environmental Governance,* edited by Clark A. Miller and Paul N. Edwards, 135–165. Cambridge, MA: MIT Press, 2001.

Lackenbauer, P. Whitney, Farish, Matthew J., and Jennifer Arthur-Lackenbauer. *The Distant Early Warning (DEW) Line: A Bibliography and Documentary Research List.* Calgary: Arctic Institute of North America, 2005.

Lajus, Julia, and Sverker Sörlin. "Melting the Glacial Curtain: The Politics of Scandinavian-Soviet Networks in the Geophysical Field Sciences between Two Polar Years, 1932/33–1957/58." *Journal of Historical Geography* 2 (in press—2014).

Lambakis, Steven. "The World's First Space War." *Orbis* 39 (1995): 417–3.

Lambright, W. Henry. *NASA and the Environment: The Case of Ozone Depletion.* Washington, DC: NASA, 2005.

Latter, A. L., Martinelli, E. A., and E. Teller. "Seismic Scaling Law for Underground Explosions." *Physics of Fluids* 280 (1959): 280–282.

Laughton, Anthony S. "The Future of Oceanographic Research in the Light of the UN Convention." In *The UN Convention on the Law of the Sea: Impact and Implementation,* edited by E. D. Brown and R. R. Churchill. Honolulu: University of Hawaii, 1987.

Launius, Roger D. "American Memory, Culture Wars, and the Challenge of Presenting Science and Technology in a National Museum." *The Public Historian* 29:1 (2007): 13–30.

Launius, Roger D. "Eisenhower, Sputnik, and the Creation of NASA: Technological Elites and the Public Policy Agenda." *Prologue: Quarterly of the National Archives and Records Administration* 28 (1996): 127–143.

Launius, Roger D. "Venus-Earth-Mars: Comparative Climatology and the Search for Life in the Solar System." In *Exploring the Solar System: the History and Science of Planetary Exploration,* edited by Roger D. Launius, 223–247. London: Palgrave Macmillan, 2013.

Launius, Roger D. "What Are Turning Points in History, and What Were They for the Space Age?" In *Societal Impact of Spaceflight,* edited by Steven J. Dick and Roger D. Launius, 19–39. Washington, DC: NASA History Division, 2007.

Launius, Roger D. et al., "Spaceflight: The Development of Science, Surveillance, and Commerce in Space." *Proceedings of the IEEE* 100 (2012): 1785–1818.

Launius, Roger D., ed. *Organizing for the Use of Space: Historical Perspectives on a Persistent Issue.* San Diego, CA: Univelt, 1995 [AAS History Series, Vol. 18].

Launius, Roger D., Fleming, James Rodger, and David H. DeVorkin, eds. *Globalizing Polar Science: Reconsidering the International Polar and Geophysical Years.* New York: Palgrave Macmillan, 2008.

Launius, Roger D., Smith, Robert W., and John M. Logsdon, eds. *Reconsidering Sputnik: Forty Years Since the Soviet Satellite.* Amsterdam: Harwood, 2000.

Layat, C., A. Clement, G. Pommier, and A. Buffet. "Some Technical Aspects of Refraction Seismic Prospecting in the Sahara." *Geophysics* 26:4 (1961): 437–446.

Lazareva, A., and A. Sytinskiy, "Seismic Observations at Mirny in 1960." In *Pyataya Kontinentalnaya Expeditsia 1959–1961. Nauchnye rezultaty*, edited by E. Korotkevich, 280–284. Leningrad, 1967.

Leary, William M. *Under Ice: Waldo Lyon and the Development of the Arctic Submarine.* College Station: Texas A&M University Press, 1999.

Lee, Owen S., and Lloyd S. Simpson. *A Practical Method of Predicting Sea Ice Formation and Growth.* Washington, DC: United States Navy Hydrographic Office, 1954.

Leffler, Melvyn P. *A Preponderance of Power: National Security, the Truman Administrations, and the Cold War.* Stanford, CA: Stanford University Press, 1992.

Lenoble, André, and Jacques Geffroy. "Province uranifère en Europe: place occupée par la France." In *Compte-rendu du Colloque de Géologie des Gisements de Minerais d'Uranium et Méthodes de Prospection, 9–11 May 1957*, edited by EAES, 273–298. Madrid: JEN, 1957.

Lerner, Mitchell B. *The Pueblo Incident. A Spy Ship and the Failure of American Foreign Policy.* Lawrence: University Press of Kansas, 2002.

Leslie, Stuart W. *The Cold War and American Science: The Military-Industrial-Academic Complex at MIT and Stanford.* New York: Columbia University Press, 1993.

Library of Congress, *Meteorological Satellites, Staff Report Prepared for the Use of the Committee on Aeronautical and Space Sciences United States Senate, Eighty-seventh Congress, Second Session.* Washington, DC: Government Printing Office, 1962.

Lieber, Robert J. "Energy, Economics and Security in Alliance Perspective," *International Security* 4 (1980): 139–163.

Lindee, Susan. *Suffering Made Real, American Science and the Survivors of Hiroshima.* Chicago: University of Chicago Press, 1994.

Lindsay, D. G., "Sea Ice in the Canadian Archipelago." MSc thesis, Toronto: McGill University, 1968.

Linke, Alexander, and Dominique Rudin. "The Earth as Seen from Apollo 8 in Space." *Rheinsprung 11: Zeitschrift für Bildkritik* 1 (2011): 147–156.

Locardi, E., "Uranium and Thorium in the Volcanic Processes," *Bulletin Volcanologique* 31:1 (1967): 235–260.

Lord, Lance W. "Space Superiority." *High Frontier* 1 (2005): 4–5.

Lovejoy, Thomas E. "George Evelyn Hutchinson, 1903–1991." *Biographical Memoirs of Fellows of the Royal Society* 57 (2011): 167–177.

Lovelock, James, and Lynn Margulis. "Atmospheric Homeostasis by and for the Biosphere: the Gaia Hypothesis." *Tellus* 26 (1974): 1–10.

Lovelock, James. "A Physical Basis for Life Detection Experiments." *Nature* 207 (1965), 568–570.

Lovelock, James. "Gaia as Seen through the Atmosphere." *Atmospheric Environment* 6:8 (1972): 579–580.

Lovelock, James. *The Ages of Gaia.* New York: Norton, 1995 [1988].

Lovelock, James. *Homage to Gaia.* Oxford: Oxford University Press, 2000.

Lutts, Ralph H. "Chemical Fallout: Rachel Carson's Silent Spring, Radioactive Fallout, and the Environmental Movement." *Environmental Review* 9:3 (1985): 210–225.

Macekura, Stephen. "The Limits of Community: The Nixon Administration and Global Environmental Politics." *Cold War History* 11 (2011): 489–518.

Mack, Pamela E. *Viewing the Earth: The Social Construction of the Landsat Satellite System.* Cambridge, MA: MIT Press, 1990.

Mack, Pamela E., and Ray A. Williamson. "Observing the Earth from Space." In *Exploring the Unknown: Selected Documents in the History of the U.S. Civil Space Program, Vol.III: Using Space*, edited by John M. Logsdon, 155–177. Washington, DC: NASA, 1998.

Mackrakis, Kristie. "Technophilic Hubris and Espionage Styles During the Cold War." *Isis* 101 (2010): 378–385.

Madden, Frank. "The Corona Camera System: Itek's Contribution to World Security." *Journal of the British Interplanetary Society* 52 (1999): 375–380.

Maddox, Bronwen. "America Wants It All—Life, the Universe and Everything."*The Times* October 19, 2006.

Maher, Neil. "Shooting the Moon." *Environmental History* 9:3 (2004): 526–531.

Mahrane, Yannick, Marianna Fenzi, Céline Pessis, and Christophe Bonneuil. "De la nature à la biosphère. L'invention politique de l'environnement global, 1945–1972." *Vingtième Siècle* 1:113 (2012): 127–141.

Maloney, Sean M. *Securing Command of the Sea: NATO Naval Planning 1948–1954* Annapolis: Naval Institute Press, 1995.

Margulis, Lynn, and James Lovelock. "Biological Modulation of the Earth's Atmosphere." *Icarus* 21 (1974), 471–489.

Margulis, Lynn, ed. *Proceedings of the Second Conference on the Origin of Life.* Washington, DC: Interdisciplinary Communication Associates, 1971.

Margulis, Lynn. *The Symbiotic Planet: a New Look at Evolution.* New York: Basic Books, 1998.

Martin, Brian, and Francesco Sella. "Earthwatching on a Macroscale." *Environmental Science and Technology* 10 (1976): 230–233.

Masco, Joseph. "Bad Weather: On Planetary Crisis." *Social Studies of Science* 40:1 (2010): 7–40.

Mastny, Vojtech. "The 1963 Nuclear Test Ban Treaty: A Missed Opportunity for Détente?" *Journal of Cold War Studies* 10:1 (2008): 3–25.

McCannon, John. *Red Arctic: Polar Exploration and the Myth of the North in the Soviet Union, 1932–1939.* New York: Oxford University Press, 1998.

McCormick, John. *Reclaiming Paradise: The Global Environmental Movement.* London: Belhaven Press, 1989.

McCulloch, Christine. "Obituary: Michael Douglas Gwynne." *The Geographical Journal* 178 (2012): 383–384.

McCurdy, Howard E. *Space and the American Imagination.* Washington, DC: Smithsonian Institution Press, 1997.

McDonald, Robert A. *Corona Between the Sun and the Earth: The First NRO Reconnaissance Eye in Space.* Bethesda, MD: ASPRS Publications, 1997.

McDougall, Walter A. *The Heavens and the Earth: A Political History of the Space Age.* New York: Basic Books, 1985.

McElroy, John H., and Ray A. Williamson. "The Evolution of Earth Science Research from Space: NASA's Earth Observing System." In *Exploring the Unknown: Selected Documents in the History of the U.S. Civil Space Program, Vol.VI: Space and Earth Science*, edited by John M. Logsdon et al., 441–473. Washington, DC: NASA, 2004.

Meadows, Donella H., Dennis Meadows, Jorgen Randers, and William W. Behrens III, *The Limits of Growth. Report of The Club of Roma.* New York, Universe Books, 1972.

Medina, Manuel Alía. "El servicio de investigación geológica." *Energía Nuclear* 6 (1958): 4–16.

Merifield, Paul M., and James Rammelkamp. *Terrain in TIROS Pictures* [NAS 5–3390, LR 17848]. Burbank: Lockheed, 1964.

Merton, Robert K., and Elinor Barber, *The Travels and Adventures of Serendipity: A Study in Sociological Semantics and the Sociology of Science.* Princeton: Princeton University Press, 2004.

Mesco, James C. "Defense Meteorological Satellite Program," in Johnson, *Space Exploration and Humanity*, vol. 2, 873–875.

Miller, Clark A. "Resisting Empire: Globalism, Relocalization, and the Politics of Knowledge." In *Earthly Politics: Local and Global in Environmental Governance*, edited by Sheila Jasanoff and Marybeth Long-Martello, 81–102. Cambridge, MA: MIT Press, 2004.

Miller, D. G. M., N. M. Slicer, and Q. Hanich. "Monitoring, Control and Surveillance of Protected Areas and Specially Managed Areas in the Marine Domain." *Marine Policy* 39 (2013): 64–71.

Miller, Jay. *Lockheed's Skunk Works*. Arlington TX: Aerofax, Inc., 1993.

Misa, Thomas, and Johan Schot. "Inventing Europe: Technology and the Hidden Integration of Europe," *History and Technology* 21 (2005): 1–20.

Mitchell, Timothy. *Carbon Democracy: Political Power in the Age of Oil*. London: Verso, 2011.

Mittman, Greg. *The State of Nature: Ecology, Community and American Social Thought, 1900–1950*. Chicago: Chicago University Press, 1992.

Monahan, Torin, ed. *Surveillance and Security: Technological Politics and Power in Everyday Life*. Abingdon: Routledge, 2006.

Morange, A., Perrodon, A., and F. Héritier, *Les grandes heures de l'exploration pétrolière du groupe ELF Aquitaine*. Boussens: Elf Aquitaine Éditions, 1992.

Morowitz, Harold J. *Energy Flow in Biology*. Woodbridge, CT: Ox Bow Press, 1979.

Morrison, Alastair, and M. Christine Chown, *Photography of the Western Sahara Desert from the Mercury MA-4 Spacecraft*. Montreal: NASA Contractor report CR-126, 1964.

Morrow Jr., John H. *The Great War in the Air: Military Aviation from 1909–1921*. Washington, DC: Smithsonian University Press, 1993.

Mowthorp, Malcolm. "US Military Space Policy, 1945–92." *Space Policy* 18 (2002): 25–36.

NASA Space Applications Program Office, *A Survey of Space Applications* [NASA SP-142]. Washington, DC: NASA, 1967.

NASA, *Astronautical and Aeronautical Events of 1962: Report of the National Aeronautics and Space Administration to the Committee on Science and Astronautics, U.S. House of Representatives, Eighty-eighth Congress, First Session*. Washington, DC: Government Printing Office, 1963.

National Academy of Sciences—International Environmental Programs Committee. *Institutional Arrangements of International Environmental Cooperation*. Washington, DC: NAS, 1972.

National Academy of Sciences—International Environmental Programs Committee. *Early Action on the Global Environmental Monitoring Systems*. Washington, DC: NAS, 1976.

National Academy of Sciences, *Arctic Sea Ice: Proceedings of the Conference Conducted by the Division of Earth Sciences and Supported by the Office of Naval Research*. Washington, DC: National Academy of Sciences-National Research Council publication 598, 1958.

National Research Council, *Human Biomonitoring for Environmental Chemicals*. Washington, DC: NRC, 2006.

Navascués, Otero. "Desarrollo histórico de la minería del uranio desde 1945." *Energía Nuclear* 13 (1960): 92–99.

Navascués, Otero. "Necesidades españolas de elementos combustibles." *Energía nuclear* 5 (1958): 18–34.

Nechtman, Tillman. " '…for it was founded upon a Rock': Gibraltar and the Purposes of Empire in the Mid-Nineteenth Century." *The Journal of Imperial and Commonwealth History* 39:5 (2011): 749–770.

Needell, Alan. *Science, Cold War and the American State: Lloyd V. Berkner and the Balance of Professional Ideals*. Washington, DC: Smithsonian/Harwood, 2000.

Neel, James V., and William J. Schull. *The Effect of Exposure to the Atomic Bomb on Pregnancy Termination in Hiroshima and Nagasaki*. Washington, DC: National Academy of Sciences National Research Council, Publ. No. 461, 1956.

Newell, Homer E. *High Altitude Rocket Research*. New York: Academy Press, 1953.

Newton, Verne W. *The Cambridge Spies: The Untold Story of Maclean, Philby, and Burgess in America*. Lanham: Madison Books, 1999.

Nininger, Robert. *Minerals for Atomic Energy*. New York: D. Van Nostrand, 1954.

Noel, Daniel C. "Re-Entry: Earth Images in Post-Apollo Culture." *Michigan Quarterly Review* 18:2 (1979): 155–176

Noreng, Øystein. *Oil Industry and Government Strategy*. Boulder, CO: ICEED, 1980.

Nouschi, Andrè, "Un tournant de la politique pétrolière française: les *Heads of Agreement* de novembre 1948." *Relations Internationales* 44 (1985): 379–389.

Nouschi, Andrè. *La France et le pétrole: de 1924 à nos jours*. Paris: Picard, 2001.

Nuclear Energy Agency, ed. *The Red Book Retrospective. Fourty Years of Uranium Resources, Production and Demand in Perspective*. Paris: OECD, 2006 [1965].

Nuti, Lepoldo. *La Sfida Nucleare. La Politica Estera Italiana e le Armi Atomiche*. Bologna: Il Mulino, 2007.

Nye, Joseph, and R. Keohane. *Power and Interdependence: World Politics in Transition*. Boston: Little, Brown, 1977.

Nye, Joseph. *Transnational Relations and World Politics*. Cambridge, MA: Harvard University Press, 1973.

Oberdorfer, Don. *From the Cold War to a New Era: The United States and the Soviet Union, 1983–1991*. Baltimore, MD: Johns Hopkins University Press, 1998.

Odum, Eugene P. *Ecology and our Endangered Life-Support Systems*. Sunderland, MA: Sinauer Associates, 1989.

Odum, Eugene P. *Fundamentals of Ecology*. Philadelphia: W. B. Sounders, 1971.

Oldfield, Jonathan D., and Denis Shaw. "V. I. Vernadsky and the Noosphere Concept: Russian Understanding of Society-Nature interaction." *Geoforum* 37 (2006): 145–154.

Oliver, J., and L. Murphy. "WWNSS: Seismology's Global Network of Observing Stations." *Science* 174 (1971): 254–261.

Onkst, David H. "Check and Counter-Check: The CIA's and NRO's Response to Soviet Anti-Satellite Systems, 1962–1971." *Journal of the British Interplanetary Society* 51 (1998): 301–308.

Oppenheimer, Robert J. "Atomic Weapons and American Policy," *Foreign Affairs* July 1953: 529.

Ordóñez, Javier, and José M. Sánchez-Ron. "Nuclear Energy in Spain: from Hiroshima to the Sixties." In *National Military Establishments and the Advancement of Science and Technology*, edited by Paul Forman and José M. Sánchez-Ron, 185–213. Dordrecht: Kluwer Academic, 1996.

Oreskes, Naomi, and Erik M. Conway. "Challenging Knowledge: How Climate Science Became a Victim of the Cold War." In *Agnotology: The Making and Unmaking of Ignorance*, edited by Robert N. Proctor and Londa Schiebinger, 55–89. Stanford: Stanford University Press, 2008.

Oreskes, Naomi, and Erik M. Conway. *Merchants of Doubt*. New York: Bloomsbury Press, 2010.

Oreskes, Naomi, and Homer Le Grand, eds., *Plate Tectonics: An Insider's History of the Modern Theory of the Earth*. Boulder: Westview, 2002.

Oreskes, Naomi. "A Context of Motivation: US Navy Oceanographic Research and the Discovery of Sea-Floor Hydrothermal Vents." *Social Studies of Science* 33:5 (2003): 697–742.

Osborne, Michael Osborne, "Science and the French Empire." *Isis* 96 (2005): 80–87.

Palewski, Gaston. *Mémoirse d'action, 1924–1974*. Paris: Plon, 1988.

Parry-Giles, Shawn J. *The Rhetorical Presidency, Propaganda, and the Cold War: 1945–1955*. Westport, CT: Praeger, 2002.

Patterson, James T. *The Dread Disease: Cancer and Modern American Culture*. Cambridge, MA: Harvard University Press, 1989.

Paucard, Antoine. *La mine et les mineurs de l'uranium français.* Paris: Editions Thierry Parquet, 1994.

Paul D. Lowman, *A Review of Photography of the Earth from Sounding Rockets and Satellites* [NASA TN D-1868]. Washington, DC: NASA, 1964.

Pauling, Linus. "Genetic and Somatic Effects of Carbon-14." *Science* 3333 (1958): 1183–1186.

Péan, P., and Jean-Pierre Séréni. *Les émirs de la République.* Paris: Seuil, 1982.

Pedlow, Gregory W., and Donald E. Welzenbach. *The CIA and the U-2 Program.* Langley, VA: CIA Center for the Study of Intelligence, 1998.

Peebles, Curtis. *Battle for Space.* New York: Beaufort Books, 1983.

Peebles, Curtis. *High Frontier: The U.S. Air Force and the Military Space Program.* Washington, DC: Air Force History and Museums Program, 1997.

Peebles, Curtis. *The Corona Project: America's First Spy Satellites.* Annapolis, MD: Naval Institute Press, 1997.

Peña, Charles V., and Edward L. Hudgins. "Should the United States 'Weaponize' Space? Military and Commercial Implications." *Policy Analysis* 427 (2002): 1–24.

Perrodon, Alain. "Historique des recherches pétrolières en Algérie." In *La recherche pétrolière française*, edited by J. Prouvost, 323–340. Paris: Éditions du CTHS, 1994.

Petersen, J. K. *Understanding Surveillance Technologies: Spy Devices, Privacy, History & Applications.* New York: Auerbach Publications, 2007.

Petersen, Nikolaj. "SAC at Thule: Greenland in the US Polar Strategy," *Journal of Cold War Studies* 13:2 (2011): 90–115.

Pinkus, Benyamin. "Atomic Power to Israel's Rescue: French-Israeli Nuclear Cooperation, 1949–1957." *Israel Studies* 7:1 (2002): 104–138.

Pocock, Chris. *The U-2 Spyplane.* Atglen PA: Schiffer Publishing Ltd., 2000.

Pollock, Ethan B. *Stalin and the Soviet Science Wars.* Princeton: Princeton University Press, 2006.

Poole, Robert. *Earthrise: How Man First Saw the Earth.* New Haven, CT: Yale University Press, 2008.

Porter, Gareth, and Janet Welsh Brown. *Global Environmental Politics.* Boulder, CO: Westview Press, 1991.

Powell, James Lawrence. *Mysteries of Terra Firma: the Age and Evolution of the Earth.* New York: Simon and Schuster, 2001.

Prados, John. *The Soviet Estimate: U.S. Intelligence Analysis & Russian Military Strength.* New York: Dial Press, 1982.

Presas, Albert. "Science on the Periphery. The Spanish Reception of Nuclear Energy: An Attempt at Modernity?" *Minerva* 43 (2005): 197–218.

Quigg, Philip W. *A Pole Apart. The Emerging Issue of Antarctica.* New York: McGraw-Hill, 1983.

Radio Corporation of America, *TIROS: A Story of Achievement* [AED P-5167A]. Princeton: RCA, February 1964.

Radkau, Joachim. *Nature and Power: a Global History of the Environment.* Cambridge: Cambridge University Press, 2008 [2002].

Rainger, Ronald. "Constructing a Landscape for Postwar Science: Roger Revelle, the Scripps Institution and the University of California, San Diego." *Minerva* 39:3 (2001): 327–352.

Rao, P. Krishna. *Evolution of the Weather Satellite Program in the U.S. Department of Commerce—A Brief Outline* [NOAA technical report NESDIS 101]. Washington, DC: Department of Commerce, NOAA, 2001.

Rasmussen, James R. "Historical Development of the World Weather Watch." *WMO Bulletin* 52 (2003): 16–25.

Reed, W. Craig. *Red November: Inside the Secret US-Soviet Submarine War*. London: Harper Collins, 2010.

Rees, Wyn. "The 1957 Sandys white paper: New Priorities in British Defence Policy?" *Journal of Strategic Studies* 12:2 (1989): 215–229.

Reyn, Sebastian. *Atlantis Lost: The American Experience with Charles de Gaulle, 1958–1969*. Amsterdam: Amsterdam University Press, 2010.

Richelson, Jeffrey T. "Undercover in Outer Space: The Creation and Evolution of the NRO, 1960–1963." *International Journal of Intelligence and Counterintelligence* 13:3 (2000): 301–44.

Richelson, Jeffrey T. *America's Secret Eyes in Space: The U.S. Keyhole Spy Satellite Program*. New York: Harper and Row, 1990.

Richelson, Jeffrey T. *America's Space Sentinels: DSP Satellites and National Security*. Lawrence: University of Kansas Press. 1999.

Richelson, Jeffrey T. *Spying on the Bomb: American Nuclear Intelligence from Nazi Germany to Iran and North Korea*. New York: Norton, 2006.

Richelson, Jeffrey T. *The Wizards of Langley: Inside CIA's Directorate of Science and Technology*. Cambridge, MA: Westview, 2001.

Richelson, Jeffrey T., and Desmond Ball. *The Ties That Bind: Intelligence Cooperation between the UKUSA Countries*. London: Allen and Unwin, 1985.

Rioux, Jean-Pierre. *The Fourth Republic, 1944–1958*. Cambridge: Cambridge University Press, 1987.

Rispoli, Giulia. "Between Biosphere and Gaia: Earth as a Living Organism in Soviet Geo-Ecology," unpublished paper delivered at International Congress for the History of Science, Technology and Medicine, University of Manchester, July 2013.

Roberts, Peder. "Intelligence and Internationalism: The Cold War Career of Anton Bruun." *Centaurus* 55:3 (2013): 243–263.

Roberts, Peder. *The European Antarctic: Science and Strategy in Scandinavia and the British Empire*. New York: Palgrave Macmillan, 2011.

Roberts, Royston. *Serendipity: Accidental Discoveries in Science*. New York: Wiley, 1989.

Robertson, Robbie. *The Three Waves of Globalization: A History of a Developing Global Consciousness*. London, Zed Books, 2003.

Robertson, Thomas. "This is the American Earth: American empire, the Cold War, and American environmentalism." *Diplomatic History* 32 (2008): 561–584.

Romero de Pablos, Ana, and José M. Sánchez-Ron. *Energía nuclear en España: de la JEN al CIEMAT*. Madrid : CIEMAT, 2001.

Romero, Ana. "Energía nuclear e industria en la España de mediados del siglo XX. Zorita, Santa María de Garoña y Vandellós I." In *La física en la dictadura. Físicos, cultura y poder en España, 1939–1975*, edited by Néstor Herran and Xavier Roqué (eds.), 45–64. Bellaterra: UAB, 2012.

Romney, Carl. *Detecting the Bomb. The Role of Seismology in the Cold War*. Washington, DC: New Academia, 2008.

Rondot, Jean. *La Compagnie Française des Pétroles. Du franc-or au pétrole-franc*. Paris: Plon, 1962.

Rostow, Walt W. *Open Skies: Eisenhower's Proposal of July 21, 1955*. Austin: University of Texas Press, 1982.

Rowland, Wade. *The Plot To Save The World, the Life and Times of the Stockholm Conference on the Human Environment*. Toronto: Clarke, Irwin & Co., 1973.

Rozwadowski, Helen M. *The Sea Knows No Boundaries: A Century of Marine Science under ICES*. Seattle: University of Washington Press, 2004.

Rudwick, Martin. "The Emergence of a Visual Language for Geological Science 1760–1840." In *The New Science of Geology: Studies in the Earth Sciences in the Age of Revolution*, edited by Martin Rudwick, 149–195. Aldershot, Burlington: Ashgate, 2004 [1976].

Ruffner, Kevin C., ed. *Corona: America's First Satellite Program.* Washington, DC: CIA Center for the Study of Intelligence, 1995.

Rumsfeld, Donald H., et al. *Report of the Commission to Assess United States National Security Space Management and Organization.* Washington, DC: Government Printing Office, 2001.

Ryggvik, Helge. *The Norwegian Oil Experience: A Toolbox for Managing Resources?* Oslo: University of Oslo, 2010.

Sagan, Carl, and I.S. Shklovskii. *Intelligent Life in the Universe.* San Francisco: Holden-Day, 1966.

Sakharov, Andrei D. "Radioactive Carbon from Nuclear Explosions and Non-Threshold Biological Effects." *Soviet Journal of Atomic Energy* 4:6 (1958) [reprinted in *Science and Global Security* 1 (1990): 175–187].

Sánchez Sánchez, Esther M. *Rumbo al Sur: Francia y la España del desarrollo, 1958–1969.* Madrid: CSIC, 2006: 375–394.

Sánchez-Ron, José M. "International Relations in Spanish Physics from 1900 to the Cold War." *Historical Studies in the Physical and Biological Sciences* 33:1 (2002): 3–31.

Sangmuah, Egya N. "Eisenhower and Containment in North Africa, 1956–60." *Middle East Journal* 44:1 (1990): 76–91.

Sarcia, J. A., et al. "Geology of Uranium Vein Deposits in France." In *Proceedings of the Second International Conference on the Peaceful Uses of Atomic Energy*, 592–611. Geneva: 1958.

SCEP, *Man's Impact on the Global Environment Assessment and Recommendations for Action.* Cambridge, MA: MIT Press, 1970.

Schaffer, Simon, Roberts, Lissa, Raj, Kapil and James Delbourgo, eds. *The Brokered World: Go-Betweens and Global Intelligence, 1770–1820.* Sagamore Beach, MA: Science History, 2009.

Schnapf, Abraham. "The TIROS Global System." *Annals of the New York Academy of Sciences* 134 (1965): 149–166.

SCOPE, *Global Environmental Monitoring.* Stockholm: SCOPE/ICSU, 1971.

Scott, William B. "Two-Stage-to-Orbit 'Blackstar' System Shelved at Groom Lake?" *Aviation Week & Space Technology*, March 5, 2006. Available at: http://www.aviationweek.com /aw/generic/story_generic.jsp?channel=awst&id=news/030606p1.xml [accessed February 19, 2010].

Seaborg, Glenn T. *Kennedy, Khrushchev and the Test Ban.* Berkeley: University of California Press, 1981.

Sebesta, Lorenza. "The Good, the Bad and the Ugly: U.S.-European Relations and the Decision to Build a European Launch Vehicle." In *Beyond the Ionosphere: The Development of Satellite Communications*, edited by Andrew Butrica, 137–151. Washington, DC: NASA, 1997.

Segovia, R., and María A. Vigón. "Sobre la construcción de contadores Geiger y circuitos electronicos asociados." *Anales de la Real Sociedad Española de Física y Química* 44A (1948): 686–689.

Silvestri, Mario. *Il Costo della Menzogna.* Turin: Einaudi, 1968.

Simpson, George Gaylord. *The Meaning of Evolution.* New Haven, CT: Yale University Press, 1949.

Singer, Fred S., and Robert W. Popham. "On-meteorological Observations from Weather Satellites," *Aeronautics and Aerospace Engineering*, 1 (1963): 89–92

Snyder, W. P., *The Politics of British Defence Policy* (Athens, OH: Ohio University Press, 1964.

Sohn, Louis B. "The Stockholm Declaration on the Human Environment." *Harvard International Law Journal* 14 (1973): 423–515.

Sokolova, B. T. ed. *High Latitude Air Expeditions "North" [Sever].* St Petersburg: Arctic and Antarctic Research Institute, 2000. Available online at http://elib.rshu.ru/files /img-213115508.pdf.

Sontag, Sherry, and Christopher Drew. *Blind Man's Bluff: The Untold Story of American Submarine Espionage.* London: Harper Collins, 1998.

Sörlin, Sverker, and Julia Lajus. "An Ice-Free Arctic Sea? The Science of Sea Ice and its Interests." In *Media and the Politics of Arctic Climate Change When the Ice Breaks* edited by Miyase Christensen, Annika E. Nilsson, and Nina Wormbs, 70–92. New York: Palgrave Macmillan, 2013.

Sörlin, Sverker. "Narratives and Counter-Narratives of Climate Change: North Atlantic Glaciology and Meteorology, ca 1930–1955." *Journal of Historical Geography,* 35:2 (2009): 237–255.

Soutou, George-Henri, and Alain Beltran, eds. *Pierre Gullaumat, la passion des grands projets industriels.* Paris: Institut d'Histoire de l'Industrie et Editions Rive Droite, 1995.

Sovacool, Benjamin K., and Marylin A. Brown. "Competing Dimensions of Energy Security: An International Perspective." *Annual Review of Environment and Resources* 35 (2010): 77–108.

Sowby, David. "ICRP and UNSCEAR: Some Distant Memories." *Journal of Radiological Protection* 21 (2001): 57–62.

Sowby, David. "Some recollections of UNSCEAR." *Journal of Radiological Protection* 28 (2008): 271–276.

Speth, James Gustave, and Peter M. Haas. *Global Environmental Governance.* Washington, DC: Island Press, 2006.

Spier, Fred W. *Big History and the Future of Humanity.* London: Wiley-Blackwell, 2010.

Stafford, John. *Scientist of Empire: Sir Roderick Murchison, Scientific Exploration and Victorian Imperialism.* Cambridge: Cambridge University Press, 1989.

Stares, Paul B. *The Militarization of Space: U.S. Policy, 1945–1984.* Ithaca, NY: Cornell University Press, 1985.

Stoff, Michael B. *Oil, War, and American Security. The Search for a National Policy on Foreign Oil, 1941–1947.* New Haven and London: Yale University Press, 1980.

Sullivan, Walter. *Assault on the Unknown: The International Geophysical Year.* New York: McGraw Hill, 1961.

Swire, Peter P. "The System of Foreign Intelligence Surveillance Law." *George Washington Law Review* 72 (2004). Available on-line at: http://papers.ssrn.com/sol3/papers.cfm?abstract_id=586616 [accessed 6/15/2012]

Sytinskiy, A., "Seismic observations at Mirny [in 1956]." In *Pervaya Kontinentalnaya Expeditsia 1955–1957. Nauchnye rezultaty,* edited by M. Somov, 153–156. Leningrad, 1960.

Tait, James Brian, ed. *The Iceland-Faroe Ridge International "Overflow" Expedition, May-June, 1960: An Investigation of Cold, Deep Water Overspill into the North-Eastern Atlantic Ocean.* ICES Report 157. Available at: http://ocean.ices.dk/Project/OV60/RPV157.pdf (last accessed October 9, 2013).

Taubman, Philip. *Secret Empire: Eisenhower, the CIA, and the Hidden Story of America's Space Espionage.* New York: Alfred A. Knopf, 2003.

Taylor, Peter J. "Technocratic Optimism, H. T. Odum and the Partial Transformation of Ecological Metaphor after WWII." *Journal of the History of Biology* 21 (1988): 213–244.

Telò, Mario. *International Relations: A European Perspective.* Farnham: Ashgate, 2009.

The NRO at the Crossroads: Report of the National Commission for the Review of the National Reconnaissance Office. Washington, DC: National Reconnaissance Office, 2000.

Thirlaway, H. I. S. "Earthquake or Explosions?" *New Scientist* 18 (1963).

Thomas, Campbell. "Moira Dunbar: Woman Scientist the Navy Refused to Take Aboard," *The Guardian* January 12, 2000. Available online at http://www.theguardian.com/news/2000/jan/12/guardianobituaries1

Thompson, David W. J., and Susan Solomon. "Interpretation of Recent Southern Hemisphere Climate Change." *Science* 269 (2002): 895–899.

Till, Geoff. "Holding the Bridge in Troubled Times: The Cold War and the Navies of Europe." *The Journal of Strategic Studies* 28:2 (2005): 309–337.

Tolba, Mostafa Kamel, and Iwona Rummel-Bulska. *Global Environmental Diplomacy: Negotiating Environmental Agreements for the World, 1973–1992.* Cambridge, MA: MIT Press, 1998.

Tongiorgi, Ezio, et al., "Deep Drilling at Base Roi Baudouin, Dronning Maud Land, Antarctica." *Journal of Glaciology* 4:31 (1962): 101–110

Tongiorgi, Ezio, ed. *Stable Isotopes in Oceanographic Studies and Paleotemperatures. Proceedings of the Spoleto School, 1965.* Pisa: CNR, 1965.

Traynor, Ian, Philip Oltermann, and Paul Lewis. "Angela Merkel's Call to Obama: Are You Bugging My Mobile Phone?" *The Guardian* October 24, 2013.

Trischler, Helmut, and Hans Weinberger. "Engineering Europe: Big Technologies and Military Systems in the Making of Twentieth-Century Europe." *History and Technology* 21 (2005): 49–84.

Turchetti, Simone, Katrina Dean,, Simon Naylor, and Martin Siegert. "Accidents and Opportunities: A History of the Radio Echo-Sounding of Antarctica, 1958–79." *British Journal of the History of Science* 41 (2008): 417–444.

Turchetti, Simone, Néstor Herran, and Soraya Boudia. "Have We Ever Been Transnational? Towards a History of Science across and beyond Borders." *British Journal for the History of Science* 45:3 (2012): 319–336.

Turchetti, Simone, Simon Naylor, Katrina Dean, and Martin Siegert. "On Thick Ice: Scientific Internationalism in Antarctic Affairs, 1957–1980." *History and Technology* 24:4 (2008): 351–376.

Turchetti, Simone. "Sword, Shield and Buoys: A History of the NATO Sub-Committee on Oceanographic Research 1959–1973." *Centaurus* 54:3 (2012): 205–231.

Turchetti, Simone. *The Pontecorvo Affair.* Chicago: University of Chicago Press, 2012.

Turner, Jonathan. "Politics and Defence Research in the Cold War." *Scientia Canadensis* 35:1–2 (2012): 39–63

Turner, Jonathan. "The Defence Research Board of Canada, 1947–1974." Ph.D. diss., Toronto: University of Toronto, 2013.

Turney, Jon. *Lovelock and Gaia.* New York: Columbia University Press, 2004.

Twigge, S., E. Hampshire, and G. Macklin. *British Intelligence Secrets, Spies and Sources.* London: TNA, 2009.

Tyrrell, Ian. "American Exceptionalism in an Age of International History." *The American Historical Review* 96 (1991): 1031–1055.

US Department of Commerce Weather Bureau, *Meteorological Satellites—Global Weather Observers.* Washington, DC: Department of Commerce, 1959.

US Department of Defense, *Soviet Military Power 1990.* Washington, DC: Government Printing Office, September 1990.

US Department of Energy. *United States Nuclear Tests, July 1945 through September 1992.* Nevada: Nevada Operations Office, December 2000.

US Disarmament Administration. *Geneva Conference on the Discontinuance of Nuclear Weapon Tests. History and Analysis of Negotiations.* US State Department, 1961.

US Navy Hydrographic Office, *Ice Atlas of the Northern Hemisphere.* Washington, DC: United States Navy Hydrographic Office, 1946.

US Navy Oceanographic Office and American Meteorological Society. *Arctic Ice.* Washington, DC: United States Navy Hydrographic Office, 1963.

US Navy. *Weather Analysis from Satellite Observations.* Norfolk, VA: Navy Weather Research Facility, 1960.

Vaïsse, Maurice, ed. *La France et l'atome.* Bruxelles: Bruylant, 1994.

Vaïsse, Maurice. "Une filière Sans Issue." *Relations Internationales* 59 (1989): 331–345.

van der Vleuten, Erik, and Arne Kaijser. "Networking Europe." *History and Technology* 21 (2005): 21–48.

van der Vleuten, Erik. "Toward a Transnational History of Technology: Meanings, Promises, Pitfalls." *Technology and Culture* 49 (2008): 974–994.

Van Keuren, David. "Cold War Science in Black and White: US Intelligence Gathering and Its Scientific Cover at the Naval Research Laboratory, 1948–1962." *Social Studies of Science*, 31 (2001): 207–229.

Vasiliev, A. P., et al. *The Service Born in the Atomic Century: The History of the Soviet Special Monitoring Service*. Moscow: SSK, 1998.

Vernadsky, Vladimir I. *The Biosphere*. New York: Copernicus, 1998 [1926].

Vinogradov, B. V. "Kosmicheskaya fotografiya dlya geograficheskogo izucheniya Zemli." *Izvestiya Vsesoyuznogo Geograficheskogo Obshchestva* 98 (1966): 101–111.

Vinogradov, B. V. *Space Photography for the Geographical Study of the Earth* [NASA TT-F-10246]. Washington, DC: NASA, 1966.

Walker, John. *British Nuclear Weapons and the Test Ban, 1954–1973*. London: Ashgate, 2010.

Walker, Mark. *Nazi Science: Myth, Truth, and the German Atomic Bomb*. Cambridge MA: Perseus, 1995.

Wall, Irwin M. *The United States and the Making of Postwar France, 1945–1954*. Cambridge: Cambridge University Press, 1991.

Wallace, Henry A. (with Andrew J. Steiger). *Soviet Asia Mission: 12,000 Air Miles Through the New Siberia and China*. New York: Reynal & Hitchcock, 1946.

Wang, Jessica. *American Science in an Age of Anxiety: Scientists, Anticommunism and the Cold War*. Chapel Hill: University of North Carolina Press, 1999.

Ward, Barbara, and René Dubos. *Only One Earth. The Care and Maintenance of a Small Planet*. New York: W. W. Norton & Co., 1972.

Ward, Barbara. *Spaceship Earth*. New York: Columbia University Press, 1966.

Wark, David Q., and Robert W. Popham. "The Development of Satellite Ice Surveillance Techniques." In *Proceedings of the First International Symposium on Rocket and Satellite Meteorology*, edited by Harry Wexler and James E. Caskey, 415–418. Amsterdam: North-Holland, 1963.

Wark, David Q., and Robert W. Popham. "TIROS I Observations of Ice in the Gulf of St. Laurence." *Monthly Weather Review* 88 (1960): 182–186.

Waterman, Alan T. "New Horizons for the Atmospheric Sciences." *Annals of the New York Academy of Sciences*, 95 (1961): 688–696.

Watt, D. C. "Britain and North Sea Oil: Policies Past and Present." *The Political Quarterly* 47:4 (1976): 377–397.

Watt, Donald C. *Succeeding John Bull: America in Britain's Place 1900–1975*. Cambridge: Cambridge University Press, 1984.

Weart, Spencer. *The Discovery of Global Warming*. Cambridge, MA: Harvard University Press, 2003.

Weeks, W. F. (with W.D. Hibler) *On Sea Ice*. Fairbanks: University of Alaska Press, 2010.

Weeks, W. F., and O.S. Lee. "Observations on the Physical Properties of Sea-Ice at Hopedale, Labrador." *Arctic* 11:3 (1958): 134–55.

Weir, Gary E. *An Ocean in Common. American Naval Officers, Scientists, and the Ocean Environment*. College Station: Texas A&M University Press, 2001

Welch, Mary. "AFTAC [ex AFOAT-1] Celebrates 50 Years of Long Range Detection." *The Monitor*, October 1997, 8–32.

Weston, Scott A. "Examining Space Warfare: Scenarios, Risks, and US Policy Implications." *Air & Space Power* 23:1 (2009): 73–82.

Wexler, Harry. "Observing the Weather from a Satellite Vehicle." *Journal of the British Interplanetary Society* 7 (1954): 269–276.

Wexler, Harry. "The Satellite and Meteorology." *Technical Session Preprints of the American Astronautical Society*, Preprint no. 104254 (1956): 1–15.

White, Frank. *The Overview Effect*. Reston, VA: American Institute of Aeronautics and Astronautics, 1998.

White, G. F. "SCOPE. The First Sixteen Years." *Environmental Conservation* 14 (1987): 7–13. See also Greenaway, *Science International*, 176–177.

Widger, William K. "Satellite Meteorology—Fancy and Fact." *Weather* 16 (1961): 47–55.

Widger, William K., and Chan N. Touart. "Utilization of Satellite Observations in Weather Analysis and Forecasting." *Bulletin of the American Meteorological Society* 38 (1957): 521–533.

Wiener, Norbert. *Cybernetics, or Control and Communication in the Animal and the Machine*. New York: Wiley, 1948.

Williams, Glyn. *Voyages of Delusion: The North West Passage in the Age of Reason*. New Haven, CT: Yale University Press, 2003.

Wilson, Harold E. *Down to Earth: One Hundred and Fifty Years of the British Geological Survey*. Edinburgh: Scottish Academic Press, 1985.

Winkler, David F. *Cold War at Sea: High-seas Confrontation Between the United States and the Soviet Union*. Annapolis, MD: Naval Institute Press, 2000.

Wittner, Lawrence S. *The Struggle against the Bomb*. Stanford: Stanford University Press, 1997, 2 Vols.

Woodlief, Ann. "Lewis Thomas." In *Dictionary of Literary Biography* vol. 275. Available at: http://www.vcu.edu/engweb/LewisThomas.htm.

World Meteorological Organization, *International Cloud Atlas: Manual on the Observation of Clouds and Other Meteors* [WMO-no. 407]. Geneva: WMO, 1987 [1975]).

Worster, Donald, "The Vulnerable Earth: Towards a Planetary History." In *The Ends of the Earth: Perspectives on Modern Environmental History*, edited by D. Worster. Cambridge: CUP, 1988.

Worster, Donald. *Nature's Economy: A History of Ecological Ideas* (Cambridge: Cambridge University Press, 1994.

Wright, Edmund A. *CRREL's First 25 Years: 1961–1986*. Hanover, NH: United States Army Corps of Engineers, 1986.

Wright, Thomas. "Aircraft Carriers and Submarines: Naval R&D in Britain in the Mid-Cold War." In *Cold War Hot Science: Applied Research in Britain's Defence Laboratories 1945–1990*, edited by Robert Bud and Philip Gummett, 147–183. London: Harwood/Science Museum, 1999.

Ziegler, Charles A., and David Jacobson. *Spying without Spies. Origin of America's Secret Nuclear Intelligence Surveillance System*. Westport, CO: Praeger, 1995.

Zimmermann, Peter D. Zimmerman, "Remote-Sensing Satellites, Superpower Relations, and Public Diplomacy." In *Commercial Observation Satellites and International Security*, edited by Michael Krepon et al. London: Macmillan, 1990.

Zimmermann, Robert. *Genesis: the Story of Apollo 8*. New York: Random House, 1998.

Zuckerman, Solly. *Men, Monkeys and Missiles: An Autobiography, 1946–88*. London: Collins, 1988.

Index

Page numbers in italics denote information in figures.